WORKING *in the* WOODS

WORKING *in the* WOODS

A HISTORY OF LOGGING ON THE WEST COAST

Ken Drushka

Harbour Publishing

First Print-On-Demand edition 2017

Harbour Publishing Co. Ltd.
P.O. Box 219, Madeira Park, BC, V0N 2H0
www.harbourpublishing.com

Edited by Daniel Francis
Designed by Fiona MacGregor
Cover painting by Bus Griffiths, photographed by Bob Cain
Cover design by Roger Handling
Printed in the USA

Graphics on pages 305–308 courtesy of Opsal Steel, from their 1942 catalogue.

Photograph and oral history source credits: BCARS—British Columbia Archives & Records Service; BCFM—British Columbia Forest Museum; CDM—Courtenay & District Museum & Archives; CRM—Campbell River Museum & Archives; CVA—City of Vancouver Archives; FERIC—Forest Engineering Research Institute of Canada; NCMA—Nanaimo Centennial Museum & Archives; PRM—Powell River Museum; UBC—University of British Columbia Library, Special Collections; VPL—Vancouver Public Library; WOTW—Woodsmen of the West, M.A. Grainger (first ed.), London: Musson, 1908.

Canada Council for the Arts Conseil des arts du Canada

Harbour Publishing acknowledges the support of the Canada Council for the Arts, which last year invested $153 million to bring the arts to Canadians throughout the country. We also gratefully acknowledge financial support from the Government of Canada through the Canada Book Fund and from the Province of British Columbia through the BC Arts Council and the Book Publishing Tax Credit.

Cataloguing data available from Library and Archives Canada
ISBN 978-1-55017-072-6 (cloth)
ISBN 978-1-55017-763-3 (POD)

Ken Drushka worked as a journalist before spending sixteen years in the forest industry as a logger, a silviculture contractor and the operator of a custom sawmill. He is the author of *Stumped: The Forest Industry in Transition* and *Against Wind and Weather: A History of the BC Towboat Industry.* He has written many articles and essays, and is co-author with Ian Mahood of *Three Men and a Forester.*

For Laura, who insists on a good job

Framed between two big Douglas fir trees growing at the intersection of Robson and Burrard streets is the first Hotel Vancouver, 1897. (VPL 13009)

Contents

CT·686

Introduction

Previous page: Modern truck logging begins. Frank White with one of Bill Schnare's trucks, Vedder Mountain, 1935.

The question arises immediately, why publish a history of logging now? The simple answer is that it has never been done. There are several books that deal with certain aspects of logging or logging in particular areas, or books that come at the subject from various topical points of view. But no one has attempted to describe the evolution of logging methods used in the coastal forests of British Columbia, along with the people who created the industry.

Anyone who knows anything about logging realizes this is a monumental task. Over the years it is likely that more than half a million loggers have toiled in BC's coastal forests. As well, there are no two acres of this forest, no two trees, logs or logging sites that are the same. Every log hauled out by every one of those hundreds of thousands of loggers was different. It is a risky business to undertake a history of the industry because so much has to be left out.

To simplify matters, I have deliberately avoided several aspects of the forest industry often included in books about logging. There is nothing here about the sawmills, pulp mills and other conversion plants that for 130 years have consumed the river of logs that flows down the coast. Nor is this book about the social life of loggers, replete with nostalgic descriptions of bunkhouse life and the inevitable glowing tribute to cookhouse cuisine. There are enough of these accounts in print already, and besides, having eaten in many camps, the best thing I can find to say about the food in most of them is that there was lots of it.

This book is not about the Glory Days of Logging. I don't think it ever was that way; at least I've never met anyone with more than a little logging experience who describes it as glorious. Logging is hard work. On the BC coast it also tends to be wet and muddy much of the time. Until recently most loggers lived in far-from-ideal camps, isolated from family and friends, engaged in a physically demanding task that was also extremely dangerous. The kind of people who talk or write about the glory of logging are the same kind who speak of the glory of war—they've never been close to the front lines. In fact, logging has often been compared to war. Peter Trower, who is both a poet and a logger, published a poem titled "Like a War." It begins:

> No bombs explode, no khaki regiments
> tramp
> To battle in a coastal logging-camp.
> Yet blood can spill upon the forest floor
> and logging can be very like a war.

I don't think Trower is suggesting, and neither am I, that logging is an awful job. In my limited experience, I found it to be the most exciting, challenging and, in the end, satisfying work I have ever done. But glorious? No. It's a job one does to feed the family, pay the rent and keep the wolf from the door. At its very worst, it is still better than the purgatory of working in an office or, God forbid, the mill.

When you look at old logging pictures it is easy to see how the romantic notion of logging arose. A bunch of guys stand smiling at the camera, with huge logs scattered here and there and gigantic machines poised for action. And it is always sunny, or at least never raining. But when the rain came and the photographers left, the loggers kept on working. This is one of the problems with any coastal logging history that contains photographs, including this one. The pictures tell only half the story; the other half of the time it was pouring rain.

For many people, the history of logging is primarily a history of railway logging. This is understandable because of the enormous interest in railways of any kind. This book does not dwell long on railway logging, partly because it has been covered well in Bob Turner's recent book, *Logging by Rail*, but also because there is a great deal more to the history of the industry than what occurred during the railway phase.

Instead, I have elected to concentrate much

of my effort on documenting the development of logging systems that replaced rail and steam, as well as spending some time on a relatively unexplored facet of the industry, the manufacturing of the tools and machines used in the woods. The emergence, and relatively recent demise, of this supplier sector is a perfect illustration of a disturbing contemporary phenomenon—what could be called the de-industrialization of British Columbia. While the thousands of jobs that have disappeared in the logging business in recent years have been well noted, very few people are aware that perhaps an equal number have disappeared among those businesses that provide loggers with the tools of their trade.

This book was made possible by a loan from the Truck Loggers' Association. Since its formation fifty years ago, this organization, more than any other, has understood that the interests of loggers and the interests of the public in the province's forests are the same. The TLA, from its first convention, has provided the only forum where issues of concern to both the forest industry and the public are openly debated. Characteristically, in funding this book no one in the TLA even suggested the organization retain any control over its contents. Its directors and members have been unstinting in their support and generous with their time in helping bring the book to publication.

Production of this book would have been impossible without the active support and assistance of many people. First and foremost I would like to acknowledge the unflagging support of Don Williams, President of the TLA, without whose efforts it would never have happened. There is a long list of people in the industry, or retired from it, who have helped immensely, including Monty Mosher, Murial Knowles, Viv Williams, Vern Wellburn, Ted Arkell, Bus Griffiths, John Casanave, John McRae, Bill Manson, Frank Rustad, Harry Foster, Mike Acres, Robin Clark, Art Allison, Ian Mahood, Pete Genberg, Harper Baikie, Marvin Lee, Tom Everett, Frank Berto, Bill Hanna, Diane Ellison, Owen Henniger, Bob Woods, Tom and Ted Dunn, Bob Heffring, Doug Ruxton and Charlie Parsons. In addition, I want to express my appreciation to those listed in the Sources section for their time and patience in granting interviews and for lending their photographs.

I would also like to give special thanks to those who also write about logging and have been generous with their information and knowledge, beyond all reasonable expectations—Curley Chittenden, Joe Garner, George McKnight, Bill Thompson and Richard Rajala.

And, finally, I express my gratitude to the dedicated librarians, curators and archivists whose efforts have preserved most of the documentation of the industry's history that survives. In particular I am, once again, indebted to Jeanette Taylor of the Campbell River Museum, whose passion for BC coastal history is inspiring. I also owe a great deal to George Brandak at UBC Special Collections, Kris Anderson at the BC Forest Museum, Bob Turner, Bob Griffin and Robin Patterson at the Royal BC Museum, Brian Young and Alan Specht at the BC Archives & Records Service, Susan Smith and Glen Duncan at the Link & Pin Museum, Barbara Simkins at the Kaatza Station Museum, Lola Westell at the Elphinstone Pioneer Museum in Gibsons, Catherine Siba at the Courtenay & District Museum & Archives, Don Olson at the Camp Six Logging Museum, Teedie Kagume at the Powell River Museum, Lillian Weedmark at the Alberni Valley Museum, and Sue Baptie at the City of Vancouver Archives.

Finally, my thanks to Howard White, publisher and friend, for his active encouragement and assistance; Fiona MacGregor for a great design; and Dan Francis for a painstaking job of editing.

I The Coastal Forest

Previous page: Mixed stand of Douglas fir, western red cedar and western hemlock owned by Clark and Lyford on south side of Seabird Lake near Blenkinsop Bay. One of a series of photographs taken on a timber cruising expedition by BC's first forest engineering firm, established in 1910 by Judson Clark and Peter Lyford. June 1917. (John Cress photo; author's collection)

When the first Europeans began arriving on the Pacific Northwest coast in the eighteenth century, they found a forest which had been growing for almost fifteen thousand years. It was a forest, or series of forests, that had reclaimed a landscape scoured by glaciers for five thousand years. One hundred and fifty centuries ago, as temperatures warmed and the ice retreated, the various species that make up this forest began repopulating the barren land. Some moved in from the south, where they had retreated in front of the advancing glaciers. Others returned from refuges on the Queen Charlotte Islands and the Brooks Peninsula on Vancouver Island, places the glaciers had not covered.

The unique geography of the region determined the character of this new forest. By European standards, the land itself was forbiddingly rugged. Mountains thrust up out of the sea creating an intricate maze of islands, inlets, steep valleys, lakes and raging rivers—all in a narrow band, running from Alaska to the deserts of central California, squeezed between the ocean and mountain ranges less than 100 miles inland.

In British Columbia all but thirty-odd miles of the coast are protected from the influence of the open Pacific by islands. The largest of them, Vancouver Island, creates a huge inland sea composed of the Gulf of Georgia, Puget Sound and the Strait of Juan de Fuca. The prevailing winds along this coast carry vast amounts of moisture sucked from the Pacific Ocean. As this soggy air sweeps inland and rises to cross the mountains, it releases large amounts of rain. Even by the standards of the British explorers, this was a wet country. Most of the rain falls between October and May, leaving a relatively dry summer, particularly in the southern part of the region.

This unique geography creates a spectacular forest—the temperate rain forest—the most obvious feature of which is gigantic coniferous trees. Most of the dominant tree species of this forest age well; some live for 500 years, with individual trees living to 3,500 years. These trees maintain high rates of growth over long periods of time.

The most immediate consequence of this longevity is the growth of enormous trees. Trunks 6 and 8 feet in diameter and 250 feet tall are common. Large individuals of several species exceed 300 feet in height and may be 20 feet in diameter. Some scientists theorize that the existence of these gigantic trees is due to the north-south orientation of the coastal mountain ranges which has allowed the trees to retreat south during successive cycles of glaciation, thus maintaining unbroken millennia of natural genetic selection. Other northern tree species were unable to evolve in the same way, the argument goes, because their genetic development was interrupted when their retreat was blocked by the east-west mountain ranges of eastern North America, Europe and Asia. Further support for the theory that gigantism among Pacific Northwest trees is genetic and not environmental is found in the fact that exotic species grow no larger here than in their place of origin. And plantations of some Pacific Northwest species in other countries indicate they will grow as well there as they do here.

Significantly, all of these big trees, and the dominant species in every region of the coastal rain forest, are conifers. The widely accepted explanation for this is found in the relatively low summer rainfall. Most forests growing in wet climates are dominated by deciduous species, as is the case in the United Kingdom or the tropical rain forests. But in order to thrive, deciduous forests require a high moisture level during their peak growing season—the summer—and they do not get that on the BC coast. Further, winters here are too cold for the deciduous species to keep their leaves and maintain their growth rates, as they do in the tropics. As a result, conifers have dominated the region.

Opposite: Timber cruising camp near Port Renfrew in 1919, showing five species of trees. (Leonard Frank photo; BCFM 1–5)

DOUGLAS FIR
HEMLOCK
WHITE PINE
HEMLOCK
HEMLOCK
HEMLOCK
RED CEDAR
BALSA

Major changes occur in the composition of the coastal forest with changes in latitude and altitude. As one travels north, or climbs higher up the mountain slopes, the dominant species change, sometimes dramatically. The redwoods of northern California do not grow outside that state, giving way to the Douglas fir which occupies a zone reaching almost to the northern end of Vancouver Island and, on the mainland, north to the Gardner Canal near Kitimat. Much of the rest of the British Columbia coast, all the way to Alaska, is dominated by western hemlock, at times in close partnership with western red cedar. Parts of this zone contain large amounts of balsam, a catch-all label for three species of true fir—amabilis fir, grand fir and sub-alpine fir. The outer coast of Vancouver Island and much of the Queen Charlotte Islands are covered by a Sitka spruce forest, along with cedar and hemlock.

These are very general classifications. Every species, and many other minor tree types, are found on many parts of the coast, surviving and even thriving in locations where the climate and soil are favourable.

The timber was of the most magnificent description. Within the area comprehended by our eyes lay an easy fortune for any man of moderate means. Spars of Douglas fir and hemlock 100 to 200 feet in height and from two to six feet in diameter, without a twig for 100 feet, stood in every direction.

Robert Brown, a biologist and explorer, writing about the Cowichan Valley in his diary during an exploration of the interior of Vancouver Island in 1864.

Opposite: A typical stand of western red cedar growing at Orford Bay in Bute Inlet, 1926. (H.W. Roozeboom photo; VPL 1401)

The first Europeans to visit the BC coast, British and Spaniards, understood little of the biological makeup of the forests. Most of the trees were unknown to them and they never ventured far from the shore into the dense, dark coastal rain forests. No doubt they were quick to note the many ways in which the native inhabitants of the country utilized one species, western red cedar. Very quickly, though, they

It was not until 1880 that I began to see the possibilities of holding timber for an advance. In twenty-five years I have seen timber increase from $1 an acre to $100 an acre. I have seen vast fortunes made by men who bought my timber for a song. They are reputed millionaires, and I am still cruising timber.

I went to Major Downie and told him my scheme. He saw through it, and told me to get away and stake some timber. I knew of a large lot of timber at Powell Lake, and I staked it and we got 20,000 acres at $1 an acre, payable in three years and Crown granted. Immediately after this the act was passed raising the price to $2.50. I surveyed it and found 7,000 acres of good timber and 10,000 of fair timber, the rest scrub. I sold my interest for $500 and the Major sold out for $1,000.

I then started out to hunt up more, and I came across the finest bunch of timber man ever saw at Malaspina Inlet. I went to the late W.P. Sayward, as fine a man as ever lived and a man of his word. I said, "I have a particularly fine piece of timber, but I want my price." He said, "What is your price?" I said, "Two thousand dollars." He said, "Well, Mose, I'll take it, but of course I want to send my cruiser in to cruise it."

This was in 1882, and he told James Miller, a man working for him, to go up and cruise it. In the meantime, J.J. Hunt, who knew me in 'Frisco, told Mr. Merrill, a young fellow, 21 years of age, out this way from Michigan, to go over to Victoria and hunt up Mose Ireland. Young Merrill hunted me up and asked me about timber. I said I had got a fine piece, but Sayward was going to take it. Miller failed to go up, and I took Merrill up to see the timber. Well, he just ran through the woods and looked around and waved his arms and said, "It's great; it's fine! I want to buy it. I'll give you $2,500."

I told him that it was sold to Sayward for $2,000, but if Sayward did not want it he could have it for the same price. Sayward's cruiser reported unfavourably, said there was only a quarter of a section there, and Merrill bought it. He paid me $2,000 for 5,000 acres and he paid the government $2.50 an acre, about $17,000 including survey expenses. He has been offered this month $1.75 stumpage, and he has been offered more than once a half million dollars for what I sold him for $2,000. He has refused to sell at any price. I want to add that when young Merrill went back to Michigan he sent me $500 as a bonus, which stamped him as a generous man. Merrill was pleased with what I did and he engaged me to cruise 15,000 acres at $1.25 an acre. He is holding all his timber still, and has made a great fortune out of that venture alone, and is now only forty-four years old—still young enough to enjoy it. He is now a member of Merrill & Ring on Puget Sound, and all he has touched has helped to enhance his fortune, and he and his brother are multimillionaires.

Mose Ireland, an original partner with Moody and Van Bramer in the first sizeable sawmill on the BC mainland, built in 1861. Born in 1830 in Maine, he was a legendary BC timber cruiser and recorded his memoirs in 1905.

Douglas fir stand of McDonald Murphy Logging Company at Cowichan Lake, 1929. (BCARS 73747)

would have realized that cedar was of limited use in the global forest economy of the day.

During the last half of the eighteenth century, the strategic centre of the timber industry was shipbuilding. A nation's economic and political power in the world was directly proportional to the number and quality of sailing ships it possessed. The building of ships consumed enormous volumes of the best-quality timber. Prior to the use of steel and steam machinery, the critical importance of timber in ship construction was a major factor in the early establishment of forest conservation and management laws in what today are the established forest nations of the world.

Of critical importance to shipbuilders were the trees that made the best spars, or masts. The biggest and fastest ships needed strong, straight spars, 100 feet and more in length. The deciduous forests of Britain and western Europe did not provide these spars, so throughout much of the colonial era the great shipyards obtained them from the coniferous forests surrounding the Baltic Sea—in Sweden, Norway, Finland and the politically unstable states of Latvia, Estonia and Lithuania. But a diminishing supply, and the ease with which the Baltic could be sealed off from the eastern Atlantic ports, created a need for other sources of high-quality spars.

Early visitors to the West Coast were well aware of the value of good spar timber. The first British sea captain to arrive on the scene, James Cook in 1778, replaced the rotting spars on his ships, quite likely with Douglas fir. Ten years later a maverick trader, John Meares, came to Nootka Sound with a crew of Chinese woodworkers who hewed a deckload of Douglas fir spars he intended to sell in China. Unfortunately his ships encountered a storm in mid-Pacific and the crews had to pitch the spars overboard.

Over the next fifty years a trickle of spars left the Pacific Northwest, including BC. By 1847, the first sawmills in Washington and Oregon

Opposite: A mixed stand in the Capilano Valley, 1920s. (VPL 3734)

were providing lumber to the growing California market. In that year the commander of British forces in the Pacific dispatched a shipment of Douglas fir to the naval yards in Portsmouth for comparison with Baltic timber. Tests found that BC fir was superior even to timber produced in Riga, Latvia—then considered the best available. This focussed world interest on the northwest forests, and on Douglas fir in particular.

Most of the coastal Douglas fir forest is found in the United States, between the Canadian border and the central coast of California. In BC the most prolific sites surround the Gulf of Georgia. The species likes a mild climate with relatively dry summers. It favours well-drained soils and, in BC, seldom grows above 5,000 feet elevation. It grows in mixed stands, particularly with cedar and hemlock, and under the right conditions in almost pure stands. Douglas fir is strong, straight grained, highly workable and decay resistant. It makes ideal lumber, veneer and pulp stock. For the past century it has probably been the most valuable commercial timber species in the world.

When the BC forest industry was struggling into existence in the latter part of the nineteenth century, a number of factors had to be taken into account by prospective mill owners and loggers. The first was the condition of the markets for forest products. Another was the availability of a timber supply. And, finally, there was the ease with which the timber could be harvested, using the relatively primitive logging equipment of the time. What launched the industry in BC were the magnificent stands of Douglas fir surrounding the Gulf of Georgia.

There were and still are several distinct, major Douglas fir forests around the Gulf, each of which attracted the province's first loggers and, to some extent, its own distinctive approach to logging.

The Lower Mainland, where the first major milling complexes were located, was primarily a Douglas fir and cedar forest. Hemlock was also mixed in, but in smaller amounts. The mix of cedar and fir was significant for the establishment of the early export sawmills on Burrard Inlet and the Fraser River, because in addition to the developing markets for fir, there was a ready market for cedar lumber in California, and for shakes and shingles. This area takes in the Fraser Valley below Lytton, including the heavily timbered shores of five large lakes along the north side of the Fraser—Harrison, Stave, Alouette, Pitt and Coquitlam—and the timber lands of the Skagit and Chilliwack valleys. Before they were cleared, stands of fir covered most of the Fraser Valley, except for low-lying prairie lands, all the way to the salt chuck at

A huge cedar tree at a Comox Logging & Railway Company camp, Ladysmith, 1935. (Wilmer Gold photo; IWA Local 1-80)

White Rock. The Lower Mainland area also includes the heavily forested slopes of Burrard Inlet, Howe Sound and the Squamish valley.

To the north, up the mainland shore, was a spectacular stand of fir and cedar covering the area between Jervis Inlet and Desolation Sound. The existence of several lakes enhanced the commercial value of this forest immeasurably. They provided easy access into the furthest reaches of the forest and, where they dropped to the sea, substantial hydroelectric power potential that was eventually harnessed to drive the Powell River pulp mill.

Immediately to the north of the Powell River peninsula is an area long known to loggers and towboaters as the Jungles. It is comprised of the mainland coast and the scores of islands between Vancouver Island and the mainland, up as far as Drury Inlet on the north shore of Queen Charlotte Sound. This is a complex maze of islands, channels, inlets, lakes, rivers and just about every other formation known to geographers. Much of the shoreline is steep enough that if a tree was properly felled, it stood a good chance of sliding right into the water. There it could be bucked into logs, boomed up and towed away, with practically no investment required in logging equipment. There were no huge stands here, but smaller pockets of fir, cedar and hemlock were enough to keep an individual or a small enterprise going. The complex network of land and water meant that most of the timber did not have to be hauled long distances to the salt chuck for towing to market. The Jungles, more than any other area, was the birthplace of independent loggers for the first five or six decades of the industry's development.

Along the east coast of Vancouver Island, south from the Salmon River valley almost to Victoria, is some of the most productive forest land in the world. For the most part it is a flat, accessible, heavily forested coastal plain, divided into two distinct areas by the Beaufort range that squeezes close to the shoreline between Cumberland and Qualicum.

The northern portion includes the forests behind Sayward, Rock Bay, Campbell River and Courtenay—almost 2,000 square miles of one of the most valuable fir-cedar forests in the Northwest. If ever a land was made for large-scale railway logging, this was it.

To the south, below Parksville, behind Nanaimo, Ladysmith and Duncan, lie a number of lakes from which rise heavily forested slopes. The largest of these, Lake Cowichan, cuts through the mountainous heart of the island and

Opposite: Mixed cedar, fir and hemlock stand at Camp O, East Thurlow Island. Timber cruiser Peter Lyford, left, and Charles Rooth. Fir on right, at 400-foot elevation, is five feet in diameter. June 1917. (John Cress photo; author's collection)

About 300 years ago, that portion of Vancouver Island lying between Salmon River and Victoria, bounded on the West by the mountain range, was by all indications heavily burnt over with very few seed trees left. The stand that has been cut lately from this area averages about 300 years old, with only an odd tree of the old stand occurring in some of the River bottoms. It would be interesting to know what caused this great fire. The large pieces of charcoal turned up by the roots of fallen trees, or exposed by grading for the logging roads, indicate that it was almost a complete burn, and that natural re-seeding produced a new forest of an even growth of healthy timber. It may be that the old stand was over-mature, insect killed and wind fallen which, with a particularly dry season and a strong west wind, produced the necessary conditions for a devastating fire.

Eustace Smith, the coast's most renowned timber cruiser. He began logging at age fourteen, in 1890, near Courtenay. He turned to cruising in 1904 and worked at this task from a Vancouver office until 1949.

Cruiser Charles Rooth leaning against a fifty-four-inch cypress at Booker Lagoon, Broughton Island, June 1917. (John Cress photo; author's collection)

empties through a broad, timbered valley into the protected waters of Cowichan Bay. The mild climate and rich soils of the Cowichan and Chemainus valleys attracted some of the Gulf basin's first settlers, farmers who learned how to log when confronted with the task of clearing their land. From the onset of industrialization this area developed an integrated forest products economy, replete with large-scale logging operations and some of the province's major export mills. It has always attracted the attention of big-money operators. Today, with most of the prime old-growth fir and cedar gone, the big companies are closing down their mills and departing for opportunities elsewhere.

On the other side of the island, in a huge lake-strewn basin, lies another distinct Douglas fir forest—the Alberni Valley. Two major lakes, Great Central and Sproat, provide easy access to large stands. The long reach of Alberni Inlet provides deep-sea access through Barkley Sound and into the heart of an integrated wood-based industrial complex. In some respects this area can be seen as extending southeast along Juan

Opposite: Mixed stand with several five-foot diameter firs, near Seabird Lake. June 1917. (John Cress photo; author's collection)

de Fuca Strait, all the way to Victoria. The Nitinat–Port Renfrew–Sooke region consists of another predominantly fir-cedar forest. Unlike the rest of the West Coast, it is protected from the worst weather of the open Pacific and was open to logging at a relatively early stage of development.

In addition to these major Douglas fir zones, there are a number of other discrete forest regions on the BC coast. The west coast of Vancouver Island, north of Barkley Sound, is one. It slowly changes from a fir forest in the south to a Sitka spruce forest in the Kyuquot and Quatsino district. One hundred and fifty miles to the northwest, and 50 miles from the mainland, lie the Queen Charlotte Islands. This area is completely out of the fir zone. Its forests consist of heavy stands of Sitka spruce, hemlock and cedar. The spruce here is some of the best in the world, huge trees, 8 and 10 feet in diameter and 250 feet tall, clear and straight. For early loggers the difficulty lay in getting timber off the Charlottes, across the dangerous waters of Hecate Strait. This obstacle would have prevented development much longer than it did had not World War One created a critical demand for spruce to use in airplane construction, and no effort was spared to gain access to the Queen Charlotte stands.

Along the adjacent mainland, from Seymour Inlet northwest to Prince Rupert, is another immense area of mountainous country, shielded from the Pacific along most of its length by scores of outer islands and penetrated by a series of long inlets. The forests of the outer islands and continental headlands in this region are sparsely covered with stunted forests, mostly cedar and hemlock. But at the heads of most of the inlets are rain-shadowed valleys, heavily forested with fir, hemlock, cedar, spruce and balsam. With some notable exceptions, the forests here are pulp forests in a commercial sense. Until the development of log barges it was difficult and expensive to move logs south across Queen Charlotte Sound, and most of the logging was done for the Ocean Falls pulp mill until it closed in the 1970s. The quality of timber on the North Coast, combined with geographical conditions, caused distinctly different types of logging systems to evolve.

Together, all these forests constituted an almost unimaginable commercial opportunity. Just how much usable timber exists at any given time is a difficult thing to determine. Quite apart from the staggering task of going out and measuring the forest, ways of assessing a commercial timber volume are in a constant state of flux. For a timber stand to have commercial value, someone has to be willing to pay as much as and preferably more than it costs to fall the trees and haul them to market. Timber is bought and sold in a global market, so its price rises and falls with changing economic circumstances. When prices rise, larger volumes of timber in harder-to-reach areas become commercially valuable. Or, if a logger figures out a way to log more cheaply, that too makes more timber available. Thus, over time, there is a tendency for commercial timber volumes to increase, even though large amounts are logged during the intervening period.

A good case in point is the Alberni Valley. The first coastal sawmill of any significance commenced operations there in 1861. Three years later it closed because of a shortage of logs. Clearly, as more than a century of heavy logging in the Alberni region indicates, there was lots of timber in the area. But the mill's owners did not find an economical way to get that timber to their mill, so from their point of view they were out of timber.

The first serious attempt to calculate the amount of timber available in BC was under-

I first came to BC in 1907. I was sent out by a syndicate of people in Ontario that had conceived the idea of taking up special timber licences. The leader of the party was the late Roland D. Craig, who was in charge of the fire-fighting department of the Dominion Forest Service. He resigned from that department to head this party that came out. He hired as his assistant in the party Mr. H.R. MacMillan, who was then a student at Guelph, or had just finished his course, and who had spent a couple of summers during his holidays as a fire ranger in the Prairie Forest Reserves, for the Dominion government. The syndicate sent me out to represent them. Craig bought a boat, hired a cook and procured a batch of Government timber maps of the Coast. I was eighteen when I came out and had my nineteenth birthday on June 15th sitting in Menzies Bay on the boat, waiting for the Seymour Narrows tide. MacMillan was about three years older.

We used the Admiralty charts and Coast Pilotage to tell us where we were going up the coast, and the government blueprint maps to show the timber staked. We had a subscription to the *Provincial Gazette*, so every week when the *Gazette* came out we had to bring our maps up to date. Sometimes we had to wait until someone went back to Vancouver and picked up back copies of the *Gazette*, and then we had to enter on the map all the new stakings in the coast district in order to avoid duplicate staking. The woods were full of other cruising parties, and this was the only way we could avoid staking someone else's timber.

Aird Flavelle, who returned to BC in 1912 and, with R.J. Thurston, established a series of sawmills in the Lower Mainland.

Examples of the legendary Sitka spruce trees in the Queen Charlotte Islands, 1918. (VPL 3837)

taken by the federal government in 1918. Using utilization and recovery standards of the day, it estimated there was, in 1913, about 230 billion board feet of commercially useful timber on the coast. This volume constituted about two-thirds of the standing timber in BC, even though the Interior forest covers an area almost twenty percent larger than that on the coast. One measure of the significance of the coastal BC forest is that at this time it held almost one-quarter of the entire Pacific Northwest's standing volume of timber. Given the nature and scale of this forest resource, by the last half of the nineteenth century logging was an industry whose time had arrived on the coast of British Columbia.

They talk about this big spruce that's down there in Carmanah. There was all kinds of those spruce down in the lower Nitinat, Sarita River. Lots and lots of spruce down there, all fifteen feet through. We thought nothing of it in those days except it was a damn nuisance. They were so big you had to cut them to twenty-four-foot lengths for the machine to move them. You'd have what you called three-log loads. You'd put the two little logs on the outside so it wouldn't roll and then plunk this great big monster in.

Jack Bell, who started logging in 1928 for Nanaimo Lumber Company. He worked for Bloedel, Stewart & Welch at Franklin River from 1935 to 1966, where he was logging superintendent, and retired at MacMillan Bloedel's Northwest Bay Division.

II Beginnings

Previous page: BC Mills Timber & Trading Company oxen teams at Camp A in Shoal Bay, East Thurlow Island, 1890. Logs were skidded to this spot, then rolled into the water with peaveys or jacks. (BCARS 14286)

One day in 1860, at the head of Alberni Inlet, the first real loggers on the BC coast headed into the woods. The head faller, axe in hand, sized up a likely looking tree, probably a Douglas fir, 4 or 5 feet in diameter and maybe 200 feet tall. Gazing up into its branches as he walked around it, he calculated which way it would fall. Then, without any ceremony to mark the event, he set to work. That's when it started.

Prior to this, at several other locations, trees were felled and logs were bucked and sent off to one of the small sawmills built to meet local needs for lumber in the colonial era. But this early logging was not undertaken by people who thought of themselves as loggers or spent most of their time working at that occupation.

The Alberni sawmill, a substantial steam-powered operation, was financed by a group of British investors brought together by a pugnacious English sea captain, Edward Stamp, who came to the Pacific Northwest in the 1850s to obtain lumber and spars in Puget Sound. One of these investors, James Anderson, eventually took over the mill and it became known as the Anderson mill.

After delivering the mill equipment, Stamp made additional trips to the inlet and brought in a logging crew, complete with oxen and the standard logging equipment of the day. These men set to work in the late fall of 1860, making spars and building an inventory of logs for the mill, which began cutting lumber the following May.

It is likely this crew came from somewhere in Puget Sound where logging had been under way for a decade or more. The head logger, who may have been working for Stamp as an independent contractor, was Jeremiah Rogers. Not a great deal is known about BC's first professional logger. Rogers was from New Brunswick, where he was born in 1818 and where he worked in the woods. Sometime during the 1850s he moved to Puget Sound to log for one of the dozen or so mills that had opened there. He had a family when he went to Vancouver Island, and a daughter born there was named Alberni. A son, Lincoln, does not appear to have followed in his father's occupational footsteps. The existence of a nephew, Willie Rogers, indicates a brother may have been here as well. Over the next twenty years Rogers became the best-known and most innovative logger in BC.

Rogers's logging operation began around the Anderson mill site, at the foot of what is now Argyle Street, and extended through the 15,000-acre timber limit Stamp had negotiated in Victoria. As fir and cedar in the area were depleted, crews built skid roads along the Somass and Stamp River flats and eventually along the south shore of Sproat Lake. At first they skidded logs directly to the mill, but as that distance increased to a mile or so, they hauled them to the river or lake and floated them to the mill.

The timber accessible to this ox-powered enterprise lasted only four years, during which about 35 million feet of lumber was milled. Increasing logging costs, combined with growing competition from U.S. mills after the conclusion of the American Civil War, led to the closure of the mill in 1865. Gilbert Malcolm Sproat, who succeeded Stamp as mill manager, explained the situation to the Colonial authorities, and suggested that a railway might be the solution. He continued, in a lament familiar to coastal loggers for the next 130 years: "the greatest difficulty is experienced in getting supplies for man and beast to these places, and in removing trees from the rough hillsides and benches. . . Our arrangements are liable at any time to be upset by the weather."

The company lacked the funds to build a railway and Sproat predicted this would be the last industrial enterprise of its kind ever built on Vancouver Island. Of course, a half century later a logging railway was built, one that operated at great profit for forty-five years before giving way to truck logging.

The major factor confronting Sproat, and

Oxen logging at Alberni for the Anderson mill in 1864. The first commercial logging operation on the BC coast lasted from 1860 to December 1864. Under the direction of Jeremiah Rogers, it supplied logs to the mill established by Captain Edward Stamp. (BCARS 22915)

others thinking about getting into the timber business on the BC coast, was the state of the market. The only word to describe this market during the 1860s is "unsettled." The Civil War from 1861 to 1865 disrupted the flow of timber from the US to European and Asian markets. Demand was growing dramatically around the Pacific—in Australia, Hawaii, Chile, China and San Francisco—and reports of shipments from Vancouver Island earning several hundred per-cent profits were common. The BC industry's chief competition came from the more well-established American mills in Washington and Oregon, which had a ten-year head start, sold the same products and were closer to the booming California market. This natural advantage was bolstered by a one-dollar-per-thousand-board-feet tariff on lumber imported into the US that was in place until 1894. Because the government did not yet see the province's timber reserves as a source of public revenue, BC lumbermen enjoyed substantially lower stumpage rates than those prevailing in the more competitive US log market. The BC lumber industry was born under a US tariff, and ever since has lived under the threat of a tariff. The major bone of contention has always been stumpage—the price at which standing timber is sold.

In spite of the difficulties, and the failure of the Anderson mill, BC was ideally situated to respond to the global timber boom of the 1860s. Even as the Alberni mill was closing, Stamp and others like him were in the process of launching a lumber export industry on the mainland. As was the case in Alberni, the financing for these new mills came mostly from Britain.

The well-connected Captain Stamp had asked for additional timber rights from the authorities in Victoria, and had lined up the money in London to build another mill, under the convoluted name of the British Columbia and Vancouver Island Spar, Lumber and Sawmill Company. Initially he intended to set up shop well up the mainland coast at Port Neville, and Rogers and his logging operation relocated there in 1864. Stamp, however, changed his mind about the mill site, deciding instead to build it in Burrard Inlet. He asked for 15,000 acres of timber in several locations on the south shore of the inlet, along the Fraser and up Howe Sound, as well as permission to log on the government reserves at what later became Stanley Park and Jericho.

Originally, the only means of obtaining timber was to buy or otherwise negotiate the acquisition of the land it was on from the Colonial authorities. Then, in 1865, a land ordinance provided for leases that allowed the logging of

Jeremiah Rogers logged for Edward Stamp at Alberni and Port Neville before establishing a logging camp at Jericho on English Bay, to supply the Hastings Sawmill on Burrard Inlet. He died in 1879 at age sixty-one. (BCARS 14553)

timber off Crown lands without acquiring ownership of the land. Leases could only be obtained by those actually in the logging business, subject "to such rent, terms and provisions as shall seem expedient to the Governor." After 1888 these leases were limited in term and carried annual ground rentals of between five and ten cents an acre. In 1897 a royalty on timber volumes was brought in.

Not one to be intimidated by authority, Stamp did not wait for his concessions to be granted before getting under way. After a false start on a site at Brockton Point in 1865, the mill was relocated to a small peninsula in downtown Vancouver, between what are now Dunlevy and Heatley avenues, and Stamp started cutting in April 1867. The site eventually became known as the Hastings Mill, and even after its name was changed in later years, it was referred to up and down the coast as the Hastings Outfit.

This was not the first mill on the inlet. Pioneer Sawmill Company had started up on the north shore in 1863, cutting timber logged from the land around the mill. It was taken over two years later by a group that included Sewell Moody, a butcher from New Westminster, and timber cruiser Mose Ireland. They bought the mill and 480 acres of timber for $6,900, then increased its capacity and began cutting in 1865.

The almost immediate success of these two mills created a big demand for logs and, as a result, the development of a logging industry. It was the feeding of these mills, and the logging of the lands around Burrard Inlet and south to the Fraser River, that established logging in BC as a clearly definable occupation or business.

Pre-eminent among the many loggers drawn to this focus of industrial development was Jeremiah Rogers. In April 1865, he moved from Port Neville and set up a logging camp on the naval reserve on the south shore of English Bay, naming it Jericho. Rogers logged most of what is now called Kitsilano from the Jericho camp and another set up at Kitsilano beach. He also logged Douglas Park (at West 20th Avenue and Heather Street) from a camp on False Creek, and several lots of his own along the North Arm of the Fraser, east of Boundary Road.

Rogers's operation, while probably the most sophisticated one around, was typical of several others working in the area over the next two decades. The two big export mills obtained all their logs from contractors like Rogers, hired to cut the timber granted these companies. Some contractors, including Rogers, soon obtained leases of their own, and it appears that many of them worked for or sold logs to both mills.

The big-time logger on the North Shore was Austin Cottrell, a contractor for the Moody company, who had an ox camp at the mouth of Lynn Creek, just up the inlet from the mill. He logged the Lynn Valley and probably other Moody land. Angus Fraser was another Hastings contractor, taking over Rogers's operation and equipment when the latter died in 1879. Fraser, as well as the partnership of Daggett and Furry, began logging Point Grey and the south slope of the Hastings claim, from about West 25th Avenue down to the North Arm of the Fraser. Furry went on to run his own camp for Hastings at Rock Bay until the late 1920s.

Austin Cottrell's bull teams were of special interest to me. There was the large hovel that housed the cattle and hay—the sling where the oxen were shod was a frame affair. On each side was a roller, up about four feet, to which was strapped a strong canvas. After the ox was in and secured by the head, the sling passed under his tummy, a bar was used to turn the rollers and hoist him off his feet so he was helpless. Then, by a rope, a foot would be fastened to a timber and all was ready for nailing on the shoes to the cloven hoof. Most of them would bawl and try to tear loose. I never knew an ox that would allow itself to be shod like a horse.

Jack Warren Bell, early Vancouver pioneer and logger.

Oxen skidding hewn spars for Gray Brothers. Date and location unknown. (NCMA P1–35)

I remember how I was impressed by the mill at Hastings, and by Granville, one summer day in 1876, when I was first taken to the Inlet by my father, then a practising physician at New Westminster. Nine miles through the cool, untouched forest we had driven in a little buggy till we came to the sea at the "End of the Road," rounding a little ravine over which was built Maxime Michaud's hotel.

At first there was only a trail leading from Hastings to Granville, or Gastown, as some called it after Gassy Jack Deighton, who ran a hotel next to the Skookum House, then in charge of Constable Jonathan Miller. In addition to Deighton's Hotel, there was another conducted by Joe Mannion, and there was the Sunnyside.

At all these hotels there was usually a drift of handloggers coming and going. The shanties and log cabins of these men were to be found from remote distances up the Gulf all the way down to the Inlet. Generally two or three men would be in partnership. They would cut the timber with their double-bitted axes on their own account, and build skidways and chutes on the hillsides to slip the big logs down to the water. They were all big logs, for they had the pick of the timber. The trees cut would be only those which ran 50 or 60 feet clear without a branch. Flooring was in demand without a knot showing in it; and there was a very profitable market in China for square timber of exceptionally large size. The handloggers would make up their own booms and wait for a tug to come along up the coast after it. The logs would be scaled after they reached the mills on the Inlet, and it was the quality of these logs which made the name of British Columbia famous in the lumber markets of the world.

At the immemorial winter feast of the Sun, observed in some fashion by nigh all men far north of the Equator, and held sacred by us as Christians, there would be a general reunion on the Inlet of all handloggers scattered along the coast. They would begin to arrive a week or so before Christmas Day, and they would remain till after New Year's Day, which was more insistently celebrated than Christmas by the Scotch and French and Dutch among them. But on both days the hosts of the various hotels and logging camps—Joe Mannion, Ben Wilson, Angus Fraser, Old Dad Daggett and the like—would take the head of their respective tables, close lined with guests for all the long length of them, and would order the feast gorgeously, with never a thought for profit.

Tom MacInnes, early Vancouver resident and journalist.

Handfallers making falling cut with a crosscut saw on a large spruce. The undercut has been made with axes. This method with these tools was used throughout the coastal area from about 1890 until the 1940s. (Link & Pin Museum)

The essence of logging involves two basic tasks: knocking 'em down, and dragging 'em out. Everything since the day Jerry Rogers first went to work at the head of Alberni Inlet has been a variation on those themes.

The skills and techniques that Rogers and others like him brought from eastern Canada and the US were not always successful in the coastal rain forests of the Pacific Northwest. In the East a certain approach to logging had evolved. The primary system used for a century or more entailed logging during the winter, using animal-powered sleds to haul logs to the nearest suitable river, and waiting for spring breakup to drive the logs downstream to the mill. That system did not work very well in the Northwest. Logs were a lot bigger. It rarely froze or snowed enough to provide a reliably hard surface over which they could be transported. And the coastal rivers were not subject to predictable freshets—they were liable to flood at almost any time of the year. As well, the waters were too rough in most of them and they were too rocky, cutting through jagged canyons and over too many rapids and falls for good river driving.

Even the primary job of the logger, falling, had to be modified on the West Coast. When Rogers began, the basic falling tool consisted of a single-bitted or pole axe. It was not until the late 1870s that the longer-headed, double-bitted axe was developed by a Seattle company for use in the Northwest. Its finely tapered head bit deeper into the western softwoods, and the more bluntly filed second bit was used for rougher work, saving the razor-sharp falling bit for the task of cutting through six or eight feet of solid timber.

Although crosscut saws had been in use since the middle of the fifteenth century, they were not used for falling until about 1880. Even then, they were one-man saws from three to six feet long—too short to be of much use on the big fir, cedar and redwoods of the Northwest. It was not until sometime in the 1880s that a longer,

The falling cut completed, the fallers are using steel wedges and sledgehammers to lift the tree, which has a slight backwards lean, so it will fall where they want it. (Link & Pin Museum)

two-man saw was developed for use in these forests. The earliest recorded use of saws in BC was at Lake Cowichan in 1883. These saws were made in various lengths, the standard models being ten feet long, with a variety of shapes, thicknesses, widths and tooth designs. By 1904, for example, the Atkins saw company offered fourteen models of two-man crosscuts, each in thirteen different lengths.

The teeth in the first saws, including the early two-man saws, were known as plain or scratch teeth. They both cut the fibres and cleared out the shavings. About 1900, soon enough to be a part of the development of Pacific coast two-man saws, raker teeth were invented which sliced out the fibres severed by the cutting teeth.

Fallers in the 1860s had no option but to tackle the five- or six-foot firs with only a single-bitted axe. The first problem they encountered was the flared, pitch-filled butts on the trees. To get above the flared butts they used springboards, two-by-eight planks four to six feet long, inserted into a notch cut into the tree about four feet above the ground. Sometimes fallers went up two boards, or eight feet. On the end of the springboard was attached a horseshoe or, later, a special metal device to hold the board firmly in the notch, while allowing the faller to pivot as the job progressed. Because of the size of the trees, axemen usually worked in pairs, each with his own springboard.

They began by cutting a deep notch, or undercut, facing the direction they wanted the tree to fall. The shaping of this undercut determined where the tree would fall. This was of more than passing interest. If fallers were going to spend three or four hours bringing down a big fir, the last thing they wanted was for it to hang up in another tree. And if the tree hit a stump or a rock and broke, the buckers had to make an extra cut to get rid of the broken end.

Once the undercut was completed, the fallers cut new springboard notches on the backside of the tree and made the falling cut. Because it was not possible to fall against the lean with an axe, early fallers had to select their trees carefully. In

Fallers Art Bellerby (right) and Joe Casey fall a big fir for BC Mills Timber & Trading at Heydon Lake, 1919. (Henry Twidle photo; CRM 4065)

most stands this meant that a lot, if not most, of the trees were left. As a result, fallers became adept at dropping selected trees through standing timber. Here, as elsewhere, the use of primitive tools required a higher level of skill on the part of the worker.

By the 1890s, as stumps in Stanley Park, the UBC Endowment Lands and various locations on Vancouver Island indicate, saws were used for making the falling cuts. Falling cuts were positioned on the back of the tree and a few inches above the the undercut to prevent the tree from kicking backward off the stump when it began to fall. Fallers carried a whisky bottle full of kerosene, with a hook attached, to lubricate the saw when falling fir or spruce because of the sticky pitch in these trees.

Fallers soon learned that once the saw was well into the tree a steel wedge inserted in the kerf behind it prevented a backward leaning tree from tilting back and pinching the saw. Driving the wedge in with a sledge could even raise a tree

When that tree was laid down, the bucker went in there and he measured that log, that tree. Suppose he was going to make three logs out of it; he would measure it and he would mark it off—we'll say they wanted a log forty feet long. He would mark it off forty feet, thirty feet or whatever length he wanted, but mostly around about forty; from twenty-eight to forty was the majority of logs. He would cut them that length.

Then following him would be the knotter and the sniper. He was the same man, the knotter and the sniper. They went in there and they cut any knots off from that tree. The sniper would cut a kind of ring around that tree so it would climb the skids—make a sleigh-runner type on the end, sort of half pointed.

Then the next man to come along was the rider. He had to have a good eye because he'd take that log and he would sometimes have to have it rolled over until he got the straightest and the best side of the log for that log to ride on going down the skidway. He would take the bark off that right along, and if there was a hump on it he would cut the hump off. When they came to haul it out of the woods they would put the hitch on it, and then when they pulled it, flipped the log on that ride, and then it didn't have to drag over the bark. Then they hauled them out.

Albert Drinkwater, who logged in the Surrey area in 1910.

quickly to cover. Once everything settled down, they gathered their tools and moved on to the next tree.

After the fallers completed an area, the buckers or spar makers moved in. For the first few decades, loggers like Rogers spent much of their time making ship spars. From the late 1860s well into the 1880s, large numbers of spars were made at several locations on the beaches at English Bay. Rogers operated a spar camp in the 1870s near Marine Drive and Boundary Road, above the river flats. After falling the straightest Douglas fir trees, the men bucked them to the desired length and skidded them with oxen to the spar yard, where they were hewn into octagonal spars up to 100 feet long and floated down against the lean, making it possible to fall a lot more timber.

A new system of specialization quickly developed with the use of saws. Undercutters using axes cut all the notches, then fallers came along later with saws and dropped the trees. In the larger camps which appeared early in the twentieth century, separate saw-filing shops were built, and the fallers and buckers took their saws in each night to be sharpened.

When a tree began to fall, all hell was liable to break loose. The first thing that might happen was for a branch to tear loose, high up in the tree, and drop onto the fallers below. This became one of the common causes of death and injury. When the tree hit the ground, the butt end might bounce and hit a poorly positioned faller. The tree could smash into another tree or snag, breaking off branches or its top, which could fly back at the stump and hit the faller.

Smart fallers—the only kind to stay alive for long—learned to clear escape routes and locate a safe spot to hide when the tree came down. As soon as it began to move they let out a mighty yell to warn others in the area, dropped their saw, leaped from the springboards and moved

About '90, the saws started to come in. When Comox Logging and Railway came here, they brought in the first saws with rakers because the saw without the raker wouldn't clear. We didn't know enough in those days to use wedges and hammers to free the saw. Later, that came in. Later on you would find a man travelling through the woods with a seven-foot saw on his back, fifty pounds of wedges in a sack, a hammer and an axe and springboard.

They bucked the trees with the axe, cutting into 32 feet and 40s as a rule. I'm speaking now of our little humble operation—I don't know what they were doing further down, in Washington.

A lot of falling was still done with axes. I could take you out where we operated—we cut out millions of feet in that Trent River valley. There are stumps there today that are three boards high, to get away from the pitch when the saws first came in because they couldn't pull the saw through the pitch. An old fellow, Bill Reed, who was a very old man, worked for Grant & Mounce as long as I can remember, till he died. He wouldn't put a saw in a tree.

It took him a long time to get a double-bitted axe. He wanted a pole axe. He eventually became roadmaster of Grant & Mounce, building skid roads. He had a twenty-four-foot straight edge level, a peavey, a saw, mattock and pole axe. That was his whole equipment.

I can take you to those roads to this day. There are miles of them up near Cumberland. He would set a couple of poles up with a bit of rag on them, and sight them. How he got his levels I don't know—didn't do it with level poles as we do nowadays. He used to put this straight edge over three skids. The skids were eight feet apart. He would dig a ditch and fall a tree across this road close to the ditch so that he would get three pieces out of it for three skids, chop them up, then throw them with the peavey to their place, bark them and cut a little gouge in them so that the log would lead.

Charles H. Grant, born in Nanaimo in 1881. He lived all his life in Royston. His father began logging with oxen at Union Bay about 1879–80 and established a steam-powered sawmill at Union Mines for the Dunsmuirs in 1889.

Two handfallers at Port Neville Logging, 1947, drop a big fir across the hillside to reduce the possibility of it breaking. A springboard, on the downhill side of the stump, was needed by one of the fallers. (CRM 10744)

a ditch to the Fraser. They were then made into rafts and towed to Burrard Inlet for loading onto ships.

The barking crew peeled the spar log, maybe 100 feet or more long, on the spot. The longer the spars, the higher the price paid per foot. In the spring or summer, when the sap was running, this job was much easier and done with spuds; the rest of the year the men used axes. The spar makers, using a plumb bob and chalk line, marked off the dimensions and set about hewing the spars with big, sixteen-inch-faced broad axes. They shaped some into tapered timber squares, others into octagonal timbers.

If sawlogs were wanted, buckers cut them to the desired lengths, at least twenty-four feet and usually no more than sixty. Bucking was a one-man occupation. In most cases there was not a safe downhill spot for a second bucker to stand. In bucking the thick trees it was usually necessary to make part of the cut from the bottom up. One means of accomplishing this with the long bucking saws was to support the back of the saw in a groove carved on the handle of an axe strategically stuck into the tree. Later, a special undercutting device served the same purpose.

Once bucked, logs were prepared for skidding. Any remaining limbs were removed, and the ends sniped by cutting a bevel around the front end so they could be skidded more easily. Then the bark was peeled off along the ride, that side of the log that tended to face down when it was skidded. Now they were ready for the dragging-'em-out phase.

In the beginning of a logging operation like the ones around the Anderson, Hastings or Moody mills, skid roads radiated from the mills out into the surrounding forest. Usually these were no more than a mile long, because it was easier to skid the logs to the water farther down the beach or up a nearby river. The roads were built by a skid-road crew before the fallers began working. With mattocks, picks and shovels, they made a right-of-way on a level or slightly uphill grade. About every eight feet, they dug a trench across the road. Then they dropped a convenient tree, maybe twelve or fourteen inches in diameter, and bucked it into skids which were rolled into the trenches with peaveys. Dirt and rocks were tamped in around them, and using a twenty-five-foot straight-edge split out of cedar, the men levelled the skids and adzed them out in the middle to create a trough that kept the logs skidding down the centre of the road. In some cases, or when the skids wore down, hardwood—usually maple—gluts were hewn and mortised into the skids for the logs to ride on. On corners the skids were sloped down,

toward the outside of the curve, so the logs would stay on the road when they were being pulled.

Trees were felled parallel to the road when possible, so that logs could be rolled onto the skids. Otherwise, they were felled away from the road so that when a turn of four or five logs was made up, the bigger butt log was at the head of the turn.

Until the railway arrived at Burrard Inlet in 1886, the major source of skidding power was cattle, including oxen or neutered males, and bulls. The great virtue of oxen was that they were available, docile, very powerful and inexpensive—a third the price of horses, or less. And they could, finally, be put to good use in the cookhouse. On the other hand, they were not nearly as smart or fast as horses.

The camps run by Rogers, Cottrell and other contractors of the day probably had up to forty or fifty oxen each. Teams of four to six yoke, or pair, were used to yard the logs from the stump to the skid road, where teams of eight to ten yoke took the larger turns down to the dump at the beach. The animals were housed in a barn, called a hovel, and tended by a roustabout who fed, watered and brushed them. They were shod in a special frame, with two crescent-shaped shoes on each foot. While oxen got along without shoes on farms, they needed them in the woods, where they pulled heavy loads over rough ground littered with sticks and stones.

Harnessed together in wooden yokes attached by chains, each team was worked by a teamster or bull puncher. He was assisted by an apprentice or greaser. They carried their gear on a sled, called a pig, that was dragged up the hill by the team and attached at the rear of the turn coming down. The greaser carried a five-gallon

Unlike horses, oxen did not willingly accept iron shoes. The delicate task of affixing two shoes to each fragile hoof was facilitated by strapping the animal into a frame. (BCFM 3–4)

Long timber sticks to be hewn for the British market were drawn to a yard where the broad-axe men would chalk line them to the desired size. All logs were barked in the woods by "barkers." First they were score-hacked by cutting a deep notch every few feet, to the chalk line. The wood between notches chopped, split or wedged off but leaving enough timber so the broad-axe man could finish a clean smooth surface with his sixteen-inch-faced axe. After the two sides were hewn the timber was turned on its flat side with screw jacks so as to have the remaining two sides hewn. Even the sawn ends had to be hewn to show the axe marks. Many a bundle of the hewn chips I carried home for the kindling in the stoves.

Jack Warren Bell

Oxen skidding logs for Royal City Planing Mills, possibly along the Fraser River in South Vancouver or New Westminster, 1889. (Link & Pin Museum)

The first work I done when I got out of school in 1908 was in a logging camp, greasing skids for a horse on Pender Island. Then I went to Victoria and got a job on the boats, a freight boat on the west coast of Vancouver Island. But I got seasick and went back to logging at Pender. Then I went with the outfit—they were a bunch from the States, from the San Juan islands—to Galiano Island. I went falling.

There was very little steam equipment around then. The first time I ever saw logging done was on Pender Island with oxen. They only took the best logs, nothing with knots in it at all. Only the clear logs and left the others. Horses and oxen were at about the same time.

The skid roads were the same for horses as for oxen. A log put in the ground and a notch put in it, in the middle of the log, so they would stay in the trough.

I had a bucket and a stick with a rag on it. I would dip it in and out onto the skids. Keep ahead of the horses. It was a thick black oil. It wasn't fish oil because it was black. You didn't do much greasing if it was downhill.

There was a tongue. We had four horses. Some of them used six, but we had four. They were harnessed to this tongue. It was just a pole with a hook on it that you drove into the log. You attached the tongue to the log. We skidded two or three logs at a time, depending on how heavy they were.

We had a thing dogged onto the last log called a pig. It was two saplings with boards on it and two-by-fours around it to hold the tools we used. That went onto the last log and went down, then the horses pulled it back. You had a peavey, a jack, a hammer. There was just two of us, the driver and me.

Cliff Brackett, born on Pender Island in 1892. He logged for almost forty years.

kerosene can of oil or tallow and a stick with a rag on it to grease the skids. On steep grades or in wet weather he often traded in his oil can for a can of sand to throw on the skids to slow the turn. For a two-year period in the early 1880s, a fish plant in Burrard Inlet made skid oil from herring caught at spawning grounds around the present site of the Vancouver Yacht Club.

At the top end of the road, the hook tender and his helper, the rigging slinger, lined up the logs on the skids. They used chains with dogs on each end that were driven into the logs. When the turns were made up, the logs were set end to end, no more than a foot apart, with two or three feet of slack in the chains connecting them. Whenever possible the lead log was the heaviest butt log. When it was set in motion, its momentum started the other logs in the turn moving.

With much effort the teamster, having brought his team up the hill, turned it around and hooked it up to the logs. Under the chain, in front of the first log, he placed a sampson—a piece of cedar a foot or two longer than the diameter of the first log—and tilted its top back toward the log. Then he readied his team so they would all pull at once. On his command they surged ahead. As the top of the sampson moved forward it lifted the front end of the lead log, helping the team to start it moving. When the chain to the second log tightened, it was set in motion, and so on down the line. Urged on by the teamster and aided by the application of grease or sand, the team headed for the beach, maintaining a slow, steady pace. If the pace

Mike Doyle loved his oxen, he was not cruel like many are. When he stuck, he would quietly go to the leaders, lift the chain, have them step up a bit if it was slack, then on to the next on a tour of inspection, talking in a low voice—perhaps prod the off-ox a poke and tell him to "wake up." When he reached the bull team, they would be in perfect line and all the slack taken up. Then he would place the samson in position with the aid of the skid greaser, climb up on a stump, log or mound, where the team could see him. Mike would seem to be restless—making quick steps back and forth, take his hat off and slap it down, pick up a stick and break it to small bits. All the time he was watching the team. They too were uneasy, some would moan, some stamp back and forth. At the psychological moment, Mike yelled, "Yip, Yip, come on you so-and-so," grab his hat and goad stick and rush toward them, prodding any laggard as he ran from leaders to turn. As a unit they bend their necks, the samson swings the first log, it comes back and the load is started.

Jack Warren Bell

slowed too much, the turn might stop and the logs would have to be levered into position by hand, the sampson re-set and the turn started again. If the pace was too fast, the logs might overrun the animals, breaking legs and turning the team into the loggers' dinner.

The end of the skid road ran parallel to the beach. When the turn reached this point the logs were disconnected and rolled into the water with a peavey. The pig was hooked on and the team headed back up the hill for another turn.

For twenty-five years or more this was one of the two major methods used to log in BC. In its first stage, oxen logging was big-time logging. It took a lot of capital to put together and operate a logging camp such as the one established by Rogers at Alberni and later at Jericho. If Rogers did not have the capital, which he probably didn't, he would have had the financial resources of the Hastings outfit behind him.

It was a system that worked well under certain conditions, the major one being the existence of long, rather gentle slopes down to the

Oxen were a little slower than the horses to get started. They were always slower to get into the harness. But they didn't pull with a collar like a horse did, of course. The yoke was bent, made out of vine maple usually. It was a piece of wood bent right over and put down through a big piece of timber, about five by eight or something like that, and it was shaped, you know. It went right over the ox's neck and he pulled on the breastbone. The weight came on his breastbone and on the top of his neck.

A horse and an ox being the same weight, I believe the ox would outpull the horse. He was a brutal thing for power. He had terrific power. And they would have a string of them, six or eight teams, whatever it was they were hauling. You might have as many as sixteen oxen on the string, possibly more at times. But I've seen ten and thirteen hauling on these roads. Another peculiar thing about them: they very, very seldom stopped to give them a rest. They nearly always kept on walking until they get done, because he don't walk fast enough to hurt his wind.

If the puncher, as they called him, sees one lagging and getting behind, he'll jab him with this here prod—all it does is make him take a little longer step, and he catches up. They don't plunge like a horse will. They don't get excited like a horse.

Albert Drinkwater

The horses themselves became a part of the man that drove them. He was more interested in his team of horses than he was in the meals he ate. That's not stretching a point at all, because a horse was, to him—well to him it was part of his business; his life was the horse, and he would do anything for his horse—stay up at night, and doctor it, fix it up, and he'd wash it every Sunday. It was a regular thing to take the horses and give them a nice, warm bath with warm water and wash them all over, rub them dry, clean them up, and they'd come out on Monday morning shining like velvet, you know. It was a sort of a—what would you call it? It's something between the man and the horse that you couldn't separate, and these people would do anything for their horse.

It's something I couldn't tell you because I can't explain it. It is how a man could take a team of twelve horses—that's six teams, you see, in line—and make every horse step into harness right on the second. Every horse. And every horse bent down to pull just like one horse. That's something that—it takes a teamster to do that.

A man that takes a stick and prods the horse, he won't pull half the load as the other man would that knows how to handle horses. He wouldn't pull half the load because the horses would not pull together. You take a man that knew his horses and handled them, he could pull a load, an enormous load, and every horse pulled his weight. They were all big horses, weighing from about 1500 to 1800 pounds, some more. The big heavy type—Clydes and that type of horse. They were horses that got used to the woods, and once they got used to it they knew more about that skid road than you did.

Of course, maybe I'm soft with horses. I like horses.

Albert Drinkwater

There was another form of logging they call "short logging." They would, oh, be hauling logs from here, we'll say, and they wanted to get them down to New Westminster or somewhere, or to some mill or wherever it happened to be. You'd load them logs onto a wagon, a logging truck you would call it, with the team—you did it with cables—and you hauled it down to the mills. That was a form of logging where two horses handled the wagon. And if the mill was close enough and the horses didn't have to walk too far, it wasn't so bad, because big horses don't walk very fast as a rule. That was what they called short logging. We did a little bit of that at one time.

Albert Drinkwater

water. Probably one of the big reasons the forest industry developed so rapidly along Burrard Inlet and the lower Fraser River was the existence of extensive stands of fir ideally situated for oxen logging and other kinds of ground skidding on what is now Vancouver and the lower slopes of the North Shore.

The other method of logging that developed coincidentally with ground skidding by oxen involved much more rudimentary methods. An individual, often with one or more partners, selected a stand of high-value timber in a location where trees would fall or slide directly into the water. Failing that, human muscles were required to coax the trees, weighing as much as fifty or sixty tons, off the hill and into the salt chuck. This was and still is known as handlogging. It provided the means for individuals, partners or families with little capital to establish themselves as loggers. It was handlogging, in the years to come, that enabled the independent logging sector to establish itself in coastal BC.

The teamsters would be out about 4 o'clock in the morning working with the horses. The teamster, the grease monkey and the roustabout—the fellow that looked after the stables—got the horses ready in the morning. Every one of the horses had to be brushed and cleaned before it went out, and harnessed. They would do that and then come in and get their breakfast. The teams were fed and everything else was done. Well, when they come out after breakfast, they'd swing their team in the line and away they'd go.

Albert Drinkwater

Handlogging developed on the coast out of the opportunities provided by the environment. Beginning on the north shore of Burrard Inlet and continuing all the way north to the sixtieth parallel, the Coast Mountains rise right from the saltwater shore. By comparison, the relatively gentle terrain of coastal Washington and Oregon did not provide the conditions that allowed an individual, equipped with only a few hand tools and a lot of determination and ingenuity, to earn a living as a logger.

In the upper reaches of Burrard Inlet the land rises steeply from the shore, and in the 1860s it was covered with stands of high-grade cedar and fir. At about the same time Rogers and the other big logging contractors began work, a stream of immigrants began to trickle in from the logged-off areas of eastern Canada, the northeastern states, Washington and Oregon. There was a ready market for logs at the Hastings and Moody mills, where orders quickly outstripped the ability of the oxen contractors to maintain the supply.

One of the first to arrive, in 1867, was a family of New Brunswick loggers—George DeBeck, his four sons and three daughters, along with some of their husbands who also knew how to log. Beginning with practically no capital, the DeBecks established themselves in the forest industry, where they thrived. Because they were a pioneering family unit, with daughters and sisters who married young men doing the same work, they established connections throughout the logging and milling industry. One of George DeBeck's daughters, Olive, married James Bell, also a New Brunswick logger, as was his son Jack. Jack's cousin was Norman MacDougall, whose sister was married to Andrew Haslam, who owned a sawmill at Nanaimo and was one of the first to log with oxen up-island, at Elk Bay. A DeBeck ran that camp.

In 1877 George's sons cobbled together a sawmill, the Brunette Mill, that eventually grew into one of the largest mills on the coast, cutting 100,000 feet a shift. They established a pattern for the independent logger—the individual, family or partnership that starts, if necessary, with nothing but a willingness to work, an entrepreneurial spirit and access to some timber. The DeBecks were followed by many others like them—the Moshers, Andersons, Bracketts, Bendicksons. The descendants of these people have formed the backbone of the coastal industry ever since.

But when they began handlogging in the late 1860s, they were faced not with a predictably prosperous future, but with hard, dirty, dangerous work. On some handlogging shows it was easy to get started. In a little bay or cove, where there was water and some protection from the wind for the logs, the handloggers bucked a windfall cedar into six- or eight-foot lengths and split out boards with which they built a cabin. A dozen or so of the first logs obtained were chained end to end in a bag or bull pen, where succeeding logs were stored. Then the handlogger, alone or with his partners, felled suitable trees into the water. If the butts hung up on the beach, they were nudged into the water with a heavy-duty jack—Bokers, at first, from Germany, and later Gilchrists and Ellingsons made in Vancouver. In the water, the tops were cut off at the first branches and the tree-length logs stowed in the bull pen.

As they moved up the hillside the task became harder. The jacks—two or three of them—weighed sixty or seventy pounds each and had to be wrestled up slopes that often turned into bluffs and cliffs. On these hillsides the task of falling a six-foot Douglas fir was intimidating. Before tackling it, loggers dropped smaller trees across its path to prevent it from breaking and to help it slide downhill. Successive springboards got the axemen into a suitable position to notch the tree—which could be fifty or sixty feet straight down to the ground below.

The only independent logger that existed before trucks was the handlogger. He didn't need any equipment except a Gilchrist jack, or an Ellingson jack. The Ellingson jack was far superior to a Gilchrist jack as far as handlogging was concerned. You had a hell of a lot more leverage on the ratchet, and it had a cloven hoof on it, it would stick on rock or anything else. The Gilchrist had a metal cup on the bottom of it, and had a tongue which was a damn nuisance in the bloody woods. It wasn't good for anything.

I knew the Ellingson boys well, they were up on Pitt Lake when I was up there. Otto Ellingson was building boats up in Pitt Lake. On those jacks you'd use your ass on the handle. His ass slipped off, it hit his jaw and he busted his jaw in four places, the handle coming up.

Al Hendrickson, who started logging in 1923 at the head of Pitt Lake. He began H&W Logging in 1943 with Earl Watt, and he retired in 1969.

We were getting more with the horse. The horse would make two extra trips a day as compared to oxen. Oxen would mosey along. An ox would take a notion to swing off and eat a salmonberry bush or something, and hold up the whole works. Then you would have to poke him in the ribs with a goad stick to get him to go. The horses didn't do that; they went right along and they were well taken care of and well harnessed.

A good bull puncher didn't make much use of that goad. He talked to them. In fact, he would get up against their shoulders and shove them over. Duncan McAllister was one of the best we ever had. He was a humane old Scotsman. He would talk and blether away to them at their shoulder, just back at the ribs, and pretty soon they would go—lean into the yoke.

The ox, you know, was a peculiar animal—he had no speed in him. He would just plod along. He had his plodding gait and that was all. All the cursing and swearing and goading with the goad stick made no difference. You could draw blood on his ribs and all the rest, but it made no difference. He would let a bellow out of him and still he rolled one shoulder to the other.

Often, to get the load started, they used a samson—backed the oxen to get enough slack in the draft chain and set a samson three or four feet from the log. When they tightened up it had a tendency to lift the front of the log up to reduce friction.

It was used more with the horses than it ever was with the oxen in our experience, because when it got to horse teams, we would have four and five logs in a string—cut back the whole tree as it fell. They were dogged together with dog chains and dog hooks, a hook driven into that log and four or five feet of chain driven into this other log. Between the logs there was a play of a foot and a half, or maybe two feet. When they stopped to give the horses a breather on a long haul, these chains were all tight and it was pretty hard to start the whole load. So, to get away from that, they had allotted places on the road where they were to stop and at that place they had a long samson laying there, and a long timber. This timber is put under the connecting chain and the slack pulled up, jacking each log until you got eighteen to twenty inches of slack between logs. Then you put your samson down and the first log was the only one you had to start; it was a snap and a jerk on the second one. If the dog didn't pull out, the log came. Then you got them all moved.

As the samson went over, it fell down between the skids and the log pushed it to one side. Usually it was a piece of cedar about four or five feet long with a nick cut in it. Depending on your log. If your log was four feet, the samson had to be longer than that.

Charles H. Grant

The banks of the Fraser were not particularly adaptable to handlogging owing to the flat delta lands which bordered the waterways. It was soon perceived that the big source would be Burrard Inlet which had the properly sloping banks to the water. The beginning of the City of Vancouver, Gastown, could well be said to have originated as a settlement available to handloggers or, even before Gastown, there was Maxie's situated at what was later known as Hastings townsite, and it was from Maxie's that the trail and later a road was built to New Westminster.

The felling, at first, was done entirely with axes, even for the large six-, seven- or eight-foot trees. The use of saws for felling came in the late seventies. This was probably due to the fact that the use of small angled wedges enforcing the felling of a tree in a given direction was appreciated. Great skill was acquired by the axeman. The felling axe was developed, a double-bitted axe, with a long narrow blade, sharpened to a razor edge with a whetstone. Every axeman carried his own stone, which was a natural stone to be used with saliva to touch up the edge every few minutes. The edge was carefully ground, very slightly rounded so that the axe would not bind in the cut, but would come away freely. Proper use of the axe was somewhat like that of a golf club or tennis racket where the whip given to the stroke gives it its greatest force at the point of impact.

The trees were felled almost invariably from the stump six or eight feet up, or higher, for two reasons: one was to get away from the gnarled or twisted grain near the ground in order to get a chip and split readily; the other reason, of course, was that from the greater elevation the better direction and the better jump or throw of the tree when it left the stump could be utilized. If the tree made a clean run to the salt water and floated, then all that was necessary would be to trim off the branches and buck off the top, and buck the tree into log lengths.

If, however, the tree lodged then there was the problem of freeing it and getting it to run again. The tree would probably have to be barked. This was simple enough in the spring or early summer when the sap was running and the bark could be removed with a slice bar, which was something in the nature of a giant metal chisel, about one-half the length of a crowbar. If, however, the bark was tight then it was entirely an axe job. Great skill was required here. A log barked in this manner would give the same regular appearance as the scales on a fish. Then the front end of the log would have to be sniped or rounded so that it would ride toboggan-like over small obstructions. Then the log had to be started, and that was where the artificial power was used. This was almost solely the jack screw, by means of which the log or front end of the log could be raised to the proper height and started. Then it had to be blocked up and started with the application of force elsewhere. This was very often a dangerous procedure, particularly if a man was working under a log at the bow end, and when the bow was raised sufficiently to clear the obstructions it would often start to run and the logger would have to have a clear way to get out of its path.

Very often a log would plunge head first into the gravel or mud and lodge with the upper end still resting against the bank or bluff, as the case may be. For many years afterwards it was a common sight around Burrard Inlet, particularly the north (Indian) arm, to see these logs still lodged and leaning up against the cliff or bank. It was also common practice to try to free those logs. If they could not be jolted loose in some way, or by the use of the lifting effect of other logs attached at low tide to use the power of their buoyancy in lifting the log free, the only recourse they had was to cut it off. This, however, was a tricky and hazardous job.

My father tells of one time when he was offered a job by a chap he always described as "the white Frenchman," a handlogger of some renown. When they were about to start out to work the first morning, they were going to try to free some logs which were stuck in the mud, my father was asked if he had a rope for his axe. He was told that he needed a rope to tie to the axe because when the log was cut off and started to run he would probably fall into the water and they could not afford to lose the axe. Needless to say, my father did not take the job.

E.K. (Ned) DeBeck, born in Vancouver in 1883. He worked as a timber cruiser early this century and later was appointed clerk of the BC Legislature. His grandfather, George DeBeck, a New Brunswick logger and sawmiller, came here with three sons in 1867 and engaged in some of the first handlogging of Burrard Inlet. The sons later established the Brunette Sawmill, and succeeding generations of DeBecks have worked in the forest industry, until the present day.

Denman Island logger Joe Fitzgerald, using a Gilchrist jack to roll a log into the water at Henry Bay, 1906. (CDM D500c)

When the falling cut was put in, the tree, all fifty tons of it, leaped forward off the stump and crashed down through the standing timber onto the cross-logs. With luck, it continued down the hill into the water. As often as not, it failed to get that far. The top plunged into the ground, or it came up against a stump, or the hill was not steep enough to maintain the momentum. This was where the hard, dangerous part of the job began.

The usual procedure was to cut off the top, snipe the end, then roll the log or use the jack to push the downhill end to one side. Various means were then used to set the log in motion. Merely rolling it a bit or jacking up the front end might work on a steep slope. If it did, the handlogger had to scramble out of the way to avoid being swept away with the log. On easier slopes, and even on flat ground, another method was used. The logger employed one or two jacks to lift the front end of the log, a precarious task on a downhill slope. Peeled saplings were placed under the log. Then he placed another jack at the back end of the log to nudge it ahead. With luck, on a slope, the log ran to the water. If not, the procedure was repeated. On flat ground this technique moved a log six or eight feet forward each time.

It could take a few hours, or a few days, to get a log into the water. The financial reward for a day's work varied, and the system favoured the ingenious and the skilled. The fact that a great deal of this work was performed in a steady rainfall did not make it any easier.

Once the logs were in the water they were boomed into orderly rafts and moved to the mills. In the early handlogging days on Burrard Inlet, one of the early steamboats sometimes did the towing. The *Beaver*, an old sidewheeler originally owned by the Hudson's Bay Company, was occasionally employed at this task. Hastings Mill owned a tug, the *Isobel*, as did Sewell Moody, the *Etta White*. If a tug was not available, the booms were often kedged to the mills—a laborious process in which a special anchor was taken ahead by boat and dropped, then the boom was pulled toward it with a hand-powered windlass or winch. Rogers is known to have kedged booms from Kitsilano through First Narrows to the Hastings Mill site.

Until 1888 there were no licences or leases specifically allowing handlogging. In Burrard Inlet some of it was done on leases held by the two mills, and some without any legal sanction. For the first twenty years or so, no one really cared who owned the timber the handloggers were taking. There was lots more out there and if anyone objected, the itinerant logger moved on to another stand.

By the early 1880s, however, the industry was gaining momentum. With the growing export market for lumber, the timber supply in the

This type of horse-powered windlass was used in the early part of the century to pull stumps and, occasionally, to yard logs. The horse was trained to step over the cable as it walked in a circle around the machine, pulling the wooden arm attached to a drum. (Link & Pin Museum)

immediate vicinity of the Burrard Inlet and Fraser River mills was beginning to run dry. Partly to control a growing logging industry, but primarily to raise revenue, the provincial government introduced a system of handlogging licences costing ten dollars a year plus a royalty on timber cut. There were no restrictions on where a handlogger could operate; they were free to cut any ungranted Crown timber. There were no restrictions on the equipment used until 1906, when loggers were prohibited from using steam-powered machinery to yard logs.

Before the turn of the century there were a few other techniques used to move logs, apart from the oxen and handlogging systems. After the railway arrived on Burrard Inlet in 1886, horses became more readily available and began to replace oxen. They were much more costly to buy and required more expensive harness, but they used the same sort of skid roads as oxen, and were much faster and smarter. By the time horses became easily available, skid roads had been built well back from the mills and the log dumps on the beaches, and the switch from oxen made it possible to extend these roads back even farther.

Only rarely were animal-powered, wheeled machines used to haul logs on the BC coast. One device, called big wheels, was used in the relatively dry pine forests of California and the Interior. These were two-wheeled log haulers which raised one end of the log when pulled by a team of oxen or horses, making it easier to skid. The only recorded use of a big wheels on the BC coast was for a brief period near the turn of the century, at Genoa Bay near Duncan. The ground was too soft to use the wheels effectively, a condition that also discouraged the use of animal-drawn wagons to haul logs. Rogers used a mule team to haul a rubber-tired wagon down a hewn log road from Douglas Park to False Creek. Ben Brown used horses and a wagon at Coombs about the turn of the century when the Esquimalt & Nanaimo Railway Company would not let him skid logs over its tracks. These and similar endeavours were exceptions.

The log chute was introduced at about the same time as horses. It was a simple trough made out of four or five logs, and it ran from a high point—usually a bench or plateau—down to the water. Skid roads were built back from the head of the chute. Often a chute was used only in wet weather or if a stream could be diverted to run down it, both to aid in log movement and to prevent friction fires from starting.

Two of the earliest chutes on the coast were used at Wreck Beach and the tip of Point Grey by Daggett and Furry, when they logged the government reserve before 1890. Chutes were also used extensively along the Sunshine Coast between Langdale and Sechelt. Much of the land in this area rises sharply from the beach, then levels out onto the easy slopes of Mount Elphinstone—ideal oxen and horse-logging country, covered with dense stands of fir and cedar. Charles Dupuis, an early Gibsons Landing oxen logger, had a chute running down the hill in what is now downtown Gibsons. J. McMyn, another oxen logger, built a chute at Grantham's Landing. He was killed when hit by a log that bounced out of the chute. Other early loggers in this area built wide rollways down

Logging with horses, 1895, at the present site of Macdonald Street and Point Grey Road in Kitsilano. By this time much of the area had already been logged using oxen. (CVA Log P7 Neg 6)

which the logs rolled into the water. An advantage of this method was that logs could be decked at the top of the rollway and put into the water together when the exposed beach was calm.

In 1886, when Angus Fraser started logging the Hastings timber on the south slope, he resorted to another expedient to move logs across the boggy flats along the Fraser River. He dug ditches and used water from a stream to float the logs to the Fraser. A similar method was used by Daggett and Furry when they logged the Dunbar–Kerrisdale area, using a ditch near the Musqueam Reserve.

Jeremiah Rogers, when he was still logging at Jericho in 1874, introduced what is arguably the first mechanized logging equipment in the Pacific Northwest. Four years earlier the BC General Transportation Company, owned by a member of the first provincial legislature, Francis J. Barnard, had purchased four Thompson Road Steamers from Scotland to haul freight on the Cariboo Wagon Road between Yale and Barkerville. At this time steam-powered tractors were used in many parts of the world, mostly on farms to pull agricultural implements, but some were designed for use on roads and, for a period, they threatened to provide serious competition to the steam railways. Steam tractors were highly regarded by small entrepreneurs wanting to establish themselves in the transportation business.

The road steamers depended on the existence of well-built, preferably paved, roads. In the British Isles they were barred from many roads, probably at the behest of the railway companies. Their weakness lay in their weight and the lack of suitable tires. R.W. Thompson, the builder of the steamer bearing his name, first invented the pneumatic tire which would one day allow trucks successfully to challenge the supremacy of the railways. But he was unable to use this invention to advantage at the time he was producing road steamers. Instead, they came equipped with hard rubber tires, which were their downfall on the Cariboo Wagon Road. When they got wet, the steel wheels tended to slip inside the tires, and the Confederation-era roads were neither smooth nor firm enough to provide the machines with good footing. Two of the steamers were shipped back to Scotland, and the other two ended up in Rogers's hands.

He enjoyed limited success running the tractors on flat-hewn logs at Douglas Park. A parallel set of these logs were probably laid end-to-end on top of the cross skids on the old

Chutes like this one used by Hastings Sawmill Company oxen loggers on East Thurlow Island—possibly in Bickley Bay about 1891—provided a means of skidding logs down slopes too steep for oxen or horses to work safely. (BCARS 14285)

Opposite: This model of the Thompson Road steamer is very similiar to those imported for use on the Cariboo wagon road and used by Jeremiah Rogers to skid logs in Kitsilano about 1874. (From A Century of Traction Engines, London: Percival Marshall & Co., 1959, p. 103)

Ox team at Hastings Sawmill camp in 1890, near the present site of Arbutus Street and West 33rd Avenue in Vancouver. Oxen skidded the logs to the landing where they were rolled onto rail cars pulled by a small locomotive to a log dump on English Bay. (CVA Log P35 Neg 22)

ox roads, and the logs skidded between them, behind the steamer. But they clearly did not work very well, as no other loggers picked up on the idea and the last that was heard of the steamers was a report that one of them was being used to power a pile driver at Hastings Mill. Rogers later used the wheels on his mule-drawn wagon.

Another logging procedure, tried with limited success, was the log drive. The first operation of this kind was probably the one on the Somass River in the 1860s, bringing logs out of Sproat Lake to the Anderson mill. In 1884, Lewis Casey and the King brothers started driving logs down the Tsolum River at Courtenay after skidding them to the river with oxen. This was one of the more enduring river drives on the coast, lasting for several years and aided by a string of piles at the mouth of the river to catch the logs.

The first drive on the Cowichan River, by Hewitt and McIntyre, took place in the winter

of 1890–91. The Cowichan River provides a perfect illustration of why river driving never became a favoured logging method on the coast. The timber coming off the hillsides above Cowichan Lake was big fir and cedar, six and eight feet in diameter, in logs usually sixty feet long. They had to go down a river that drops 550 feet over its thirty-five-mile run to Cowichan Bay. It is a twisted, winding river, laced with rocks, narrow canyons and gravel bars, with 130 sets of rapids and about 30 falls, some fifteen feet high.

In addition to their own logs, Hewitt and McIntyre contracted to move other loggers' timber down the river. They had "improved" the waterway by blasting out rocks and other potential hangups, and intended to charge a toll for others using the river. A sluice dam was built at the outlet of the lake, through which the logs could be fed and flushed down the river with the dammed-up waters of the lake. A large crew was employed, including a contingent of Natives with their dugout canoes used to carry winch lines to stranded or jammed logs.

The first drive was a disaster. Heavy winter rains scattered logs along the banks and bars down the length of the river. A road bridge at Duncan was swept away and the Esquimalt & Nanaimo Railway bridge was threatened. Subsequent attempts were more successful, and several loggers, including Joe Vipond and George Lewis, made many drives. In 1899 Lewis took out 8 million board feet of knot-free timber, all in sixty-foot logs. The last drive on the Cowichan was in 1908.

The Victoria Lumber & Manufacturing Company made one unsuccessful attempt to drive the Cowichan and, in 1906, another failed attempt on the Puntledge River at Courtenay. This last effort created a massive logjam, and an experienced young river driver from Wisconsin named Matt Hemmingsen was brought in to clear it out.

Sometime around the turn of the century, before it was dammed at the outlet of Coquitlam Lake, the Coquitlam River was used for driving logs. And about 1910, an attempt was made to drive Lillooet River, at the head of Harrison Lake, probably by Brooks-Scanlon, the first big company on the Harrison. For almost twenty years, starting in 1934, Trethewey Logging drove the Lillooet, using Cats in the shallows and donkeys in the deeper water to keep the logs moving. This operation, first under Edgar Trethewey and later under his son, Alan, was

At the time that I was working at Port Neville (1895)—and for several years following—most of the logs used by the few sawmills that were operating on the coast were bought from small operators who were logging by skid roads with oxen and horse teams, while a large quantity of logs was produced by handloggers who could be found dotting the foreshores of nearly every big bay and channel where sufficient shelter could be obtained for a small boom pond. In those days a handlogger obtained a licence for $10, permitting him to remove timber from unalienated lands, while a team logger paid $50 annually for a similar privilege.

It was the time of the small operator. They were for the most part financed by mills, which supplied provisions, tools, boom chains and other essential equipment, and also provided for the towage of the booms to the sawmills. The only mills with which I had dealings were the Hastings Mill at Vancouver and the Brunette Mills at Sapperton. The standard price for a long time was $3.25 per thousand to the logger, plus the 50 cents royalty to the government and the towage. There was little or no sale for cedar and hemlock, and only occasionally for spruce.

Protestant Bill MacDonald was logging at the head of Port Neville, just below the Indian Reserve and above the Narrows. He did not have a partner. Bill was very proud of his ability as a logger, and had handlogged by himself for a number of years. On one occasion, I remember, he located a big fir—about 10,000 fbm—which was growing on very flat ground near the beach. But between the place where he must fell the tree and the beach was a large Indian midden—or ridge of clam shells and earth—forming an adverse grade against running the tree to the water.

Nothing daunted, Protestant Bill set to work and dug a wide cut through the midden, giving a flat grade to the sea. He laid skids and waited until the sap was running; then he felled the tree with precision, just where he wanted it—headed for the opening to the beach. After he had greased the skids and barked the log, he set to work to get it out. He set two big jack screws, with a slight tilt toward the opening. Then, setting up his big landing crank jack at the butt, he gave the tree a slight lift and push forward. The leaning jacks would topple, and the tree would be thrown anything from six to ten feet toward the water. Then Bill would start all over again. After three weeks of patient and prodigious labour he got the big log afloat and into his boom. No log was going to beat him.

Eustace Smith

probably the longest lived river-driving operation in coastal logging history.

In the late 1950s and early 1960s, two Trethewey-related companies, Trethewey-Wells and Cattermole-Trethewey, drove logs down the Fraser River from the Chilcotin area. During the winter, tree-length bundles were dumped in the Fraser and during the spring freshet, they floated down through the Fraser Canyon to Hope, where they were caught and boomed. One winter the river froze, creating an enormous logjam of about 10 million board feet. When the water rose in the spring, the jam broke loose and swept down the river, through the holding booms at Hope and beyond. Every beachcomber and tug on the river was busy for days collecting loose logs from the burst bundles.

The success of river driving in other parts of North America was due to a combination of deeper, slower rivers and logs small enough to manhandle with a peavey. In coastal BC, these conditions did not exist. The nature of the rivers, combined with unpredictable weather and erratic water levels, spelled the abandonment of the system as soon as better methods came along.

Between 1860 and 1870, lumber exports from coastal mills increased from 250,000 board feet a year to about 20 million, and 5–6,000 spars were shipped out. Most of this production came from the two pioneer mills on Burrard Inlet. Over the next twenty years, several more mills were built in the Lower Mainland, spreading from Burrard Inlet to False Creek and along the Fraser. More small mills appeared up the valley and at various points on Vancouver Island. This activity produced a steadily growing demand for logs and, in response, loggers began to move into the outlying areas of the coast. Once the timber supply began to decline in the general vicinity of the Lower Mainland mills, and logs had to be boomed so they could be towed, it did not cost much more to bring them from the most distant corners of the inside waters. Instead of choosing the closest timber, the mills went in search of the best stands within the area where logs could be safely towed in flat booms.

This upcoast migration took two general forms, establishing a configuration for the industry that prevailed for more than a century. One component of the move was undertaken by the mills, which obtained timber rights at various locations around the Gulf of Georgia and up Johnstone Strait. For the most part the mills engaged contractors to do the logging, at least in the early stages before it became necessary to invest large amounts of money in railway logging systems.

Joe Vipond was the only one who ever logged the Cowichan Lake right. He run the river. People logged that for a long, long time and logs were scattered all along the bank and, by golly, Joe took off about 14 to 15 million feet and he got every one of the other fellows' logs. He hired a bunch of real Frenchmen to get on the boom that knew something about river driving. He got everything out of the river. Vipond cleaned up about $50,000 on the timber he took off the other fellows' lots. The next fellow that went in there was Harry McArthur and he lost everything, just because he didn't have the right men to handle that river. That's another job. The ordinary fellow here doesn't do that job.

Richard Fiddick, born in Nanaimo in 1879. He began logging at age sixteen, and he worked in the woods on Vancouver Island most of his life.

I worked on the river drive in 1906 driving tote wagon, and Mr. Joe Vipond, a tall red-headed man, was contractor and boss of the drive. Now it was very important to start sluicing the logs at the proper time. They had to be let go when the water was not too high.

It soon became time for Joe Vipond to watch his marker on the river at McCallum Landing for the right water. Then the tug *Bute*, operated by Joe Jordan of Sathlam, towed the logs into the mouth of the river and the sluicing started. In sluicing it is important not to let too many logs go at once or you are liable to have a logjam. The Indians are in readiness with their canoes and as the last logs are let go they start down the river, dislodging the logs from the banks.

They had only got started when word came through that the logs were not coming by Duncan, so Joe Vipond and I started down the river to find the source of the trouble. When we got to the canyon below Scutts Falls what did we see but the canyon full of logs for almost its entire length, and what a sight—with the water bubbling through.

What had happened was that when the men were sluicing they had let about four boomsticks get away intact and they had got crosswise in the canyon. We got some men and powder and lowered a man down with a rope, who placed a charge of dynamite, and lo, behold, when the charge went off all the logs started to move and they very nearly took the white bridge out at Duncan going through.

Bazil Kier, born at Somenos, near Duncan, in 1888. He worked logging most of his life in the Cowichan area and left a detailed description of life in the Cowichan Valley in the early years of this century.

After 1888, to encourage manufacturing in the province, leases were made available to sawmill companies at lower rates than to non-mill owners. By the time these leases were discontinued in 1905, they covered 688,000 acres. The various Lower Mainland mills, along with others like Haslam's in Nanaimo, acquired timber in the biggest and best stands around the Gulf and up through the Jungles.

The most active of these companies was the Hastings Mill. Stamp had already acquired timber for the company in Port Neville and he obtained additional leases at Granite Bay, East Thurlow Island, Salmon River and Rock Bay. There was also an isolated Hastings camp at Lang Bay, south of Powell River. The first camps at each of these locations were oxen shows, under the direction of contractors or semi-autonomous foremen. Each foreman was assigned a letter for his camp, and when the camp moved, its name went with it. The letter O, for example, designated the camp run by Saul Reamy. The first Camp O was set up at some point during the mid-1880s in a small bay on the south shore of East Thurlow Island, where Nodales Channel meets Johnstone Strait. Skid roads were pushed north and oxen hauled the logs out. During the course of the next few years, before and after the company was sold and became the BC Mills Timber and Trading Company in 1889, several other camps were set up on leases in the same area. By 1890, Camp A was operating on the other end of East Thurlow at the mining town in Shoal Bay. A third camp on the island opened at Hemming Bay at about the same time.

In 1900 a camp opened at Granite Bay on Quadra Island, and eventually employed between 400 and 500 loggers. A boom camp was located at the beach, and the main camp was set up about three miles up the railway. At one point the camp was serviced by a floating bordello staffed in part by women from the Cape Mudge reserve. When chaplains from the Columbia Coast Mission boat and Indian Agent Ward DeBeck tried to curb the social activities of the loggers, their attempts were not appreciated. On Vancouver Island camps were started in Palmer Bay at the mouth of the Bear River in the early 1890s, and south at Rock Bay in 1900. The latter became the Hastings outfit's main logging camp on the entire BC coast—and, in later years, it was legendary.

Once the Hastings camps had logged the timber accessible to oxen, rail lines were laid up the skid roads and the oxen skidded logs to the railway. Although there is much confusion and conflict in the historical records, some evidence indicates that the first locomotive used in these camps was "Curly," the first steam rail engine brought to BC in 1881 to work on the CPR. It was moved to Palmer Bay about 1893, around the same time that Saul Reamy moved Camp O to Vancouver Island from East Thurlow. Although records are sparse, it is possible to piece together an account of Reamy's career that illuminates the early history of the coast's first logging camps.

There is no record of Reamy working for Hastings Mill on Burrard Inlet, or any indication of where he learned to log. In 1893 or 1894, when the first Camp O timber was logged off, Reamy moved his camp and crew to Palmer Bay and up the railway to a series of camps, moving farther and farther back into the woods. Later he reopened the East Thurlow Camp O site and used "Curly" on a three-mile railway line to Simmons Lake. Oxen skidded the logs to the lake, and they were towed to the railhead with a small steam tug and loaded onto cars.

Reamy was renowned as a traditional, conservative logger. He refused to fire someone because of old age or injury, and he maintained that if a man was able to sit up to eat, he had a job at Camp O. Over the years Reamy and his aging crew worked their way to the remote site of the last Camp O, near Pye Lake. As long as they continued to load out logs, the company continued to push the rail line in behind them and to haul the logs down to the beach camp at Rock Bay.

At some point Reamy accepted a single-drum steam winch, probably to yard logs out to the

In 1888 my father was logging for Andrew Haslam of Nanaimo, about twelve miles above Seymour Narrows, on Vancouver Island. It was sort of a standard size camp of those days, consisting of somewhere between ten and twenty span of oxen. We went up on the tug *Rustler* and had two scows in tow. On the scows were the cattle, the hay, and all equipment and supplies necessary for the camp. Prior to our arrival there had been a small group to put up the cookhouse, bunkhouse and a cabin for my father, mother and us three children. We arrived late at night—at any rate it was dark—and everything had to be lightered to shore in rowboats. I do recollect, however, that in landing the cattle, four or five of the men with a big 2-by-12 plank would push each one off into the water and it would swim ashore. There was a man in a boat to see that they did not head across Johnstone Strait. Apparently, the operation was carried out without casualty.

E.K. (Ned) DeBeck

skid road where the oxen skidded them to rail. On one occasion the company shipped him a two-drum yarder, the second drum for a haulback line that would dispense with the horse—or, in Reamy's camp, maybe even an ox—used to pull the line back into the woods. As legend has it, Reamy and his crew hauled the yarder up to their camp, removed the haulback drum and went back to work logging the way God meant men to log—with a single-drum yarder.

The Moody mill went into a decline following Moody's death at sea in 1875 and the arrival of the transcontinental railway at the competition's mill across the inlet in 1886. But the big North Shore mill also moved out to obtain timber and established a camp at Gibsons Landing. George Gibson, who logged his preemption for many years, had an association with the Moody mill, and his family and possessions were moved to the homestead on the company tug, the *Etta White*, in 1886.

The early Hastings camps, particularly at Rock Bay, and some of the other mill camps were the forerunners of the big four- and five-hundred-man railway camps that flourished in the 1920s and existed in different forms at various locations for fifty more years. They were the antithesis of the other type of logging operation that began to spread over the lower coast in the mid-1880s.

The extension of the mills' logging operations up the coast between the mid-1880s and the turn of the century led to the building of a fleet of tugboats to service the camps and tow the log booms to the mills. These boats made it possible for small, independent handlogging operations to spread up the coast because they now had a means of getting their logs to market. By 1900, handloggers were at work on every suitable bay, cove, inlet, channel and hillside between Vancouver and Drury Inlet. And while this was a widely dispersed phenomenon, the centre of handlogging activity was Minstrel Island, deep in the heart of the Jungles.

One of the first settlements in the Jungles, Minstrel had a store, post office, saloon, whore-

---

The Union Steamship Company maintained a most useful passenger and cargo service to the logging and mining camps and canneries of the coast. The little *Comox*, especially, pioneered the trade between Vancouver and the logging camps. She was licensed to carry sixty-two passengers, but usually managed to reach Vancouver with more than that number, the majority of whom were well primed for their visit to the city.

The *Comox* had a bar, and it was inevitable that loggers with money in their pockets should take full advantage of it. The result: fights, songs and hilarity galore. Rarely was there anything mean or cantankerous behind the fights; they usually started from an excess of animal spirits, or a desire to show off. "I can lick any man on this boat," some red-blooded, broad-shouldered fellow would yell, and instantly the challenge would be accepted. Seldom was there much damage done; the roll of the vessel and the bubbling of the booze usually brought both gladiators to the deck, from where they would be helped up to go back to the bar to cement a new-found friendship.

Those men—and they were real men—were proud of their strength and of their woodscraft. They boasted of their accomplishments with axe and saw, and of their ingenuity in yarding logs from difficult situations, and they were always willing to back their pretensions with their money.

Billy Dineen, of Nanaimo, was one such champion. He was a mighty man with the axe, and what he could do with it was almost unbelievable. On one occasion he accepted a challenge to cut 40,000 feet in a single day, using only an axe. He selected four big trees, each scaling a minimum of 10,000 feet, and in eight hours he had felled them all!

The courage and endurance of the loggers of fifty years ago was extraordinary. I have been told of two brothers who were handlogging on Thurlow Island. They were both caught under a tree, one of them by the foot. In order to free himself to assist his brother, he chopped off his own foot. There was another happening that I recall. It was a handlogger on Toba Inlet. He was using a jack on a log, when it started to slide unexpectedly, and he was badly crushed by the butt. He crawled down the hill to his boat and rowed six miles down the inlet to another camp. He died as they were lifting him from the boat.

Loggers were good spenders, and open-handed with their money. No charitable appeal met with disappointment in a logging camp. In the woods the logger dressed in the rough clothing of his calling, but in the city the younger men donned expensive dress. There was one article that every logger coveted—a good pair of boots for his work. For several years, Cook brothers of Vancouver operated a store boat called the *New Era*. In addition to being fitted as a small department store, she carried a shoemaker, who measured the feet of his customers on one trip and delivered the boots the following visit. While the *New Era* was in operation, a fad for alligator hide boots costing the then outrageous price of from $20 to $30 ran through the camps.

Eustace Smith

house and all the other modern conveniences of the day. In nearby bays the small booms of individual handloggers were made into larger booms for the long tow down the protected routes of the Inside Passage to Burrard Inlet and the Fraser River.

The character and evolution of handlogging in this area was vividly described in Martin Allerdale Grainger's book *Woodsmen of the West*. Grainger worked around Minstrel Island during the late stages of the handlogging era. He went on to write the first Royal Commission report on logging and became BC's second chief forester, after H.R. MacMillan. Like MacMillan, Grainger believed that the energy and ingenuity of the handlogging community was a key element in the future development of the coastal logging and milling industries.

A fierce independence motivated the handloggers, a desire to be their own boss, balanced by a dislike of working in company camps, the only alternative for a logger. It was here, among these people, that the independent logging sector, the market logger with no interest in mills, was born. Here, techniques, attitudes, skills, and many of the coast's next generation of loggers were conceived, and matured.

There was a downside to handlogging that in later years led to attempts to abolish or restrict it. From silvicultural and utilization points of view, it was often destructive and wasteful. In a lot of cases it was a form of high-grading. The 1918 Commission on Conservation report estimated that forty percent of the timber felled by handloggers never made it to the water. The report went on:

> During the last 28 years, hand-loggers have destroyed the timber on over 1,000 miles of shoreline extending back from one to twenty chains, averaging perhaps five chains, and covering an area of 50,000 acres. Though no figures are available as to the amount of timber cut by hand-loggers, it is estimated, from personal observation, that they have marketed perhaps 500 million feet, cut and allowed to go to waste, 300 million, and

Top: Handlogging crew at work in the slash, rolling a fir log to bark it so it will slide easily when pushed with a jack. (WOTW)

Centre: A situation handloggers sought to avoid. The log ran down the hill and into the water, and embedded itself in the bottom. (WOTW)

Bottom: "Where handloggers once worked." An illustration of the wasteful logging practices which occurred on some handlogging shows. (WOTW)

A typical handloggers' camp in the Jungles, just after the turn of the century. The buildings are constructed from logs and hand-split cedar. (WOTW)

> directly caused the destruction of an additional 800 million feet, through fire and windfall resulting from their operations . . . It would appear that the usefulness of the hand-logging system has passed, and that it should be discontinued, as inimical to the object of forest conservation.

While the logging industry spread throughout the south coast, the sawmilling industry also began to expand, moving to new locations. The logging of Surrey began in the 1870s to feed the New Westminster mills and small local mills. The best and most accessible timber was along the Serpentine, Nicomekl and Campbell rivers. Ditches were used extensively in the wetter areas of this region.

Much of the lower Fraser Valley was logged by farmers clearing land. The first major logging operation in the area was around White Rock, when Gilley Brothers opened their camp near Elgin in 1885. As the demand for logs grew, so did the need for agricultural land, and logging slowly moved east up the valley on both sides of the Fraser. Completion of the CPR line on the

By the early 1890s, when this photo was taken at Port Kells on the Fraser River adjacent to Barnston Island, horses were beginning to replace oxen because they were faster and more intelligent. (BCARS 45572)

The crew of this Port Kells camp, probably owned by BC Mills Timber & Trading, gathered in front of their bunkhouse in 1891. (BCARS 45571)

north shore of the river in 1885 and the Great Northern line on the south side of the valley during the 1890s spurred settlement, logging and the building of sawmills as far away as Chilliwack and Harrison by 1900.

The Harrison area, including the lake, had seen a small but steady amount of industrial activity from the time a small water-powered mill opened at Port Douglas on the Lillooet River in 1858. By the end of the century, with a growing demand for logs down the valley and a market for lumber via the CPR on the prairies, this activity accelerated. In 1899 the Trethewey brothers took over a small sawmill at Harrison Mills. They sold it five years later to the Rat Portage Lumber Company in Manitoba, which went on to build one of BC's larger mills on the site, but the Tretheweys carried on as a major force in the logging business in the area until the 1970s.

> It was about this time (1902) that J.S. Emerson, who was a considerable figure in lumbering circles for a few years, opened a large camp at Greenway Sound. It became a sort of rendezvous for loggers, for in addition to his own operations he purchased handloggers' booms. It was quite a tough place, with a poker game going most of the time. Men who had sold booms after working for months would risk the whole proceeds of their toil on the luck of the cards. It was a busy place, too, for it became a sort of shipping and supply centre to which loggers could come to meet the steamers and exchange news and views.
>
> *Eustace Smith*

On the east coast of Vancouver Island, small mills and logging camps opened at a steady pace. King and Casey moved their operation north from Courtenay to Duncan Bay in the 1880s and started logging the Sayward Forest, one of the largest continuous Douglas fir forests on the coast. In 1886 all of the unalienated east coast land, from Campbell River south to Victoria, was granted to the Esquimalt & Nanaimo (E&N) Railway. This set the stage for some enormous developments in the future, when the Dunsmuir family sold parcels of the E&N lands to well-heeled American timber companies, launching the second major phase of the coastal forest industry.

While previous developments had had a predominantly British and eastern Canadian character, the new players were mostly Americans. The first to arrive was John Humbird, representing a timber syndicate that included Frederick Weyerhaeuser. Like others to come, these industrialists had established themselves in the northeastern forests and were now moving into the Pacific Northwest, including BC. Humbird's group bought 100,000 acres of E&N fir-covered lands in the Chemainus and Nanaimo valleys in 1889. The company, Victoria Lumber & Manufacturing, opened one of the largest, most modern mills on the coast at Chemainus, replacing a local mill established almost thirty years earlier by George Askew. The company set up its own logging operations and quickly imported some of the first steam-powered logging machinery on the coast.

A new era had begun.

Benjamin Brown (left) was born at Nanaimo in 1874. The father of Alvin and Gunny Brown, he was one of the last oxen loggers on Vancouver Island. He is shown here in 1907, a mile in from Departure Bay at the Savage and Arbuthnot camp, logging for the Red Fir Lumber Company, which had a sawmill at the present site of the Nanaimo arena. In the background is an early two-drum steam yarder of the type that was then replacing oxen and horses. (Brown brothers' collection)

In the nineties I got tired of working for someone else and decided to log for myself. I had a few dollars that I made handlogging in Gray's Creek when I handlogged there in 1887, and I decided to try horse logging. I never liked logging with bulls as they are stupid and very slow. My people were great hands for horses, as we raised them. There were few teams in BC, so I went to Washington.

There I met Steve Tingly, a former coach driver I knew in Ashcroft. He told me of a fine team he had driven that had no more timber to log and were for sale. I saw this team, and one could not wish for a better outfit. The wheel team weighed over 1900 pounds apiece, and none of the rest were under 1600 pounds apiece. I paid $1000 for the wheel team and $600 for the other teams; because I paid cash I got the equipment thrown in with the horses.

I took the outfit to Vancouver and then went out to my friend, Captain Harry Trim on Westham Island, as he had the best hay and oats in the valley. I bought ten tons of hay and five tons of oats. We loaded everything on a scow and went up to Naiatt Bay.

We landed, and first thing we did, we put up a big tent that I had rented for the horses and feed. We cut some cedar and made shakes, and we built a barn and feed shed. We slept in with the horses till we got a bunkhouse built. I had a full crew and, as was the custom of the times, no one was on the payroll till the camp got into production. This custom went out when the rails came in.

We laid out the roads. These were sixteen feet wide, with twelve-foot skids. The fallers felled the timber parallel with the road, but not on the road. The logs could easily be parbuckled into the road, and there was little guarding. Where a tree was on the road, hand skids were laid. A hand skid was any skid two men could carry. The undercutter went through first and put in the undercut to guide the tree's fall.

If there was pitch in the butt he used his Methodist Axe—double-bitted axe, two-faced like a Methodist—to fall this tree as the two-toothed saws the fallers had were too much work with pitch.

The barkers came after the fallers and cut off the bark on the ride, the side on which the log ran on the skid road. If the log was pointed and had no ride, they hewed one on this log.

On the grades we used split-edge cedar skids as they would take grease in dry weather, or sand in wet weather. One Sunday we took two rowboats and all went to Phillips Arm where we cut maple for gluts to be inserted in the skids on the flats. The gluts were dovetailed into the skids, against the pull. My brother Alec was a good axeman and did this job. We used mutton tallow and dogfish oil to oil the skids and make the logs run easily. The oil companies put out a skid oil, but it did not work as well as it did not have the body.

I used a little tongue for the wheel team. This had a dog in the end and held the turn from running over the rigging. On the grades the teamster and greaser rode the wheel team. Steve was very good. He would ease the turn down the grade and the head logs would be close to the flat before the team had to trot. When at the landing, the greaser would jack the logs into the water and walk back. He had a little fire up in the works where the logs were. Here he kept his grease on a little fire to keep it warm and melted.

I worked on the boom and scaled the logs. Some of the butts had a quarter ring shake, and the scaler took off four inches to make up for the loss and difficulty of manufacture of these logs. Seventy logs or 140,000 was an average day's work, and there was no stumpage or royalty. I had a handlogger's licence that entitled me to work on any Crown timber if there was no one else working in the area. Horses or bulls were allowed on these licences.

Hiram McCormack, pioneer Quadra Island logger.

III Early Steam Power

Previous page: An early photo of "Old Curly" at the Port Kells log dump, Parson's Channel, 1891. (CVA Log P52 Neg 36)

The arrival on the scene of the Victoria Lumber and Manufacturing Company (VL&MC) marked the beginning of great changes in coastal BC, particularly in the forest industry, which had come into significant existence only thirty years earlier.

> The industries of the Coast are not yet beyond the initial stage. Logging, requiring the least capital and presenting a maximum of return for a minimum of skill, has made the greatest progress. As yet not much timber is being cut which cannot be put into salt water with a donkey engine. The general procedure has been to stake a piece of timber, or in other words pay the government an annual rental for a licence to cut the timber from a certain area, then secure a small force of men, a donkey engine and start a logging camp. The donkey is placed on the shore, preferably in a small bay where the logs may be boomed without being exposed to the full force of the waves. A skid road is constructed with as easy a grade and as few curves as possible, leading back to the timber to be cut. The line, an inch or inch and a quarter steel cable, leads from the drum of the donkey to the timber. In order to facilitate the handling of the line, a smaller cable, the haulback, is attached to the second drum of the donkey, and together with the hauling line forms a loop, one end of which is reeled in as the other is paid out. Thus the heavy line may be pulled back to the woods for the next haul. This represents the simplest outfit. Usually two donkeys are used, one situated half a mile or so from shore, which yards the logs to a central point, the other the "road" donkey at the shore, which hauls the logs in "turns" or towes of eight or ten logs each over the skid road to the shore. By means of this massive machinery, huge logs are jerked through the small trees and underbrush, and often develop a lurching sprightliness which one seldom sees—save in a runaway buggy. When boomed in booms 100 to 125 yards long, containing each a quarter of a million feet, they are towed to Vancouver to the mills. There are as yet few mills north of Vancouver. At present railroads are being built to reach the less accessible timber, and with the completion of these, logging will receive a great impetus.
>
> *H.R. MacMillan, reporting on a trip up the coast while he was Chief Forester, in 1912 or 1913.*

First, the company represented a shift in the primary source of investment capital in the industry, from Great Britain to the United States. Stamp and his contemporaries were tied in to the capital markets of London; VL&MC was built on the first flood of US money into the province. There were several reasons for the American influx. Most of the prime timber lands in the US northwest had already been taken up. Conservationists had managed to get much of the remainder placed in federal reserves that would eventually become the US National Forests. In part, the move into BC was a continuation of the cut-and-move-on philosophy that had permeated the North American industry for a century or more.

Today, Canadian nationalists tend to view this influx of American capital, people and technology as an invasion by a foreign power, an early example of American neocolonialism. In fact, it was merely another stage in the historic ebb and flood of the forest industry across the international boundary. Previously there had been an "invasion" of the US northeast woods by loggers from Ontario, Quebec and the Maritime provinces. The greatest American logger of them all, Paul Bunyan, was a Québécois who fought in the Papineau Rebellion of 1837 before settling down to a marginally more respectable occupation as a logger.

This period, from the 1890s until World War One, was a very turbulent one in BC. Apart from a lot of undeveloped real estate, there was not much happening economically in the early 1880s, and little of the kind of boosterism that accompanies rapid industrial development. Things began to change quickly after the arrival of the railway, but economic development was kept in check by the difficulties of obtaining investment capital from the usual British sources. When events to the south made that capital available in BC, politicians and businessmen were ready to respond.

In 1903, the government of Richard McBride came into power in BC. The premier, ambitious

You might say that the problems you met in logging were nothing like the problems you met when you came to sell logs to sawmills. You could get them out and get them into the chuck, but after we got them to town was to sell them to some mill that wasn't pledged to some bank under Section 88, and I would say the loggers—the open market logger—probably lost more money to sawmills than they ever lost in the operation of their camps.

In the experience of Bloedel, Stewart & Welch, they never lost money in logging. The only money we really lost, which was part of our operating cost, was what we lost to sawmills that couldn't pay for the logs—the banks took them over. But even with those losses, there never was a year that Bloedel, Stewart & Welch didn't make money.

The cedar logger—I think very few of them ever made any money. If they made money they'd be living on top of the world and the next year the market would be washed out completely from under them. Aird Flavelle was an exception to the rule.

Cedar is a specialty wood. You can't log it like you can fir. It's easy to break. There's too much waste and low-grade timber in a tree. I kept away from cedar. I never did like cedar. We were what you call fir loggers.

Sidney G. Smith

and unhindered by an excess of scruples, adopted all the conventional techniques to encourage development and keep his Conservative Party in office. He was an enthusiastic proponent of land grants to railway promoters and looked upon the forests of the province as a heaven-sent source of public revenues. The U.S. corporate invasion had been encouraged in 1901 with laws banning the export of logs cut on Crown land, a move that blocked the growth of Puget Sound mills using BC logs. In 1905, McBride introduced changes in Special Licences for timber, turning them into transferable options to log for a twenty-one-year period on square-mile tracts, leaving open the amount payable to the Crown for these rights. This move precipitated a frenzy of timber staking, and within three years the number of Special Licences increased from 1,500 to 15,000. Public revenues increased proportionally and an unstated policy was established that prevails to this day—the provincial government retains control of the forests in order to sell them off, over and over again, sometimes to the highest bidder, but usually to its corporate friends. Whether these payments have been according to law, or under the table into party or cabinet ministers' pockets, has been a matter of choice, depending on who is in power. The principle has remained the same; the allocation of timber rights is one of the key instruments to getting and hanging on to power in Victoria.

The nature and scale of this new capital investment had also undergone a change. The VL&MC was not just a U.S.-financed company, or even a large U.S.-financed company; it was a syndicate involving several large American interests, including Frederick Weyerhaeuser. There was a notable difference in the character of these companies compared to earlier ones. Weyerhaeuser's own company, one of the largest in the US, still bore the stamp of his personality and the benefits of his attention; he had a direct interest and participation in it. The VL&MC, on the other hand, was owned by a syndicate of which Weyerhaeuser was only a single and distant member. It was a much more remote entity, under the control of a professional manager, E.J. Palmer.

This was an era in which the advantages of

Employing of loggers in those days was done right on the street. The old Hastings Mill had a man walking around the waterfront there, hiring people for the logging camp. They'd just hire you and give you a ticket to get on the boat. We boarded the old *Cassiar* to Rock Bay. It's an overnight trip. It finally arrives at Rock Bay and everybody piles out. It was on the sixth of January 1919 that this took place. They loaded them all onto boxcars and they had a railroad running from Rock Bay right out to Robert's Lake and there they unloaded onto the lake, and transported everything on a float across the lake to the other corner, to where the log dump was. Then they had a mile spur of railroad into an artificial lake they had dammed up for logging into.

There was about a foot and a half of snow that day when we went in there. And of course the men are all piled into the boxcars just like sheep, and fights going on all the time with the brakeman because they wanted the door open and he insisted on closing it. Quite a do.

We arrived at the camp, Camp D. The foreman there was a fellow by the name of John Morgan, he was from Oregon. They all unloaded and you fought for a place to sleep. They had bunks, three tiers high, and no springs, just slats of cedar were laid down, and you carried your own blankets. Sometimes you were lucky if they had a pad there, and they were lousy. It wasn't the warmest of places, lots of fresh air coming in all over—you needed it too. Hell of a way to live, though. Almost like a concentration camp.

Art Bellerby, born in Winnipeg in 1900. He came to BC in 1918 and worked as a logger before becoming an RCMP officer. He later worked at Vivian Diesel and retired to Campbell River, where he died in January 1992.

big corporations were constantly paraded before the public. "Large capital and effective organization are not only necessary, but fundamental," F.H. Parks of the International Timber Company told delegates to the 1911 Pacific Logging Congress. "In addition, conditions would seem to warrant the assumption that this is only a beginning and that future logging operations, as the timber areas become further diminished and distances from tidewater increase, must of necessity become more complex, demanding increasing skill and capital." This was one of the first salvos in the coming struggle between big corporations and smaller, independent logging companies. The central business ethic was: the bigger, the better. The only worthwhile goal of a small company, according to this ethic, was to become a big one.

An equally significant innovation of VL&MC was its early adoption of steam-powered logging equipment. E.J. ("Old Hickory") Palmer, the company manager, brought many new traits to the industry. He was the prototype of the office-bound "bean counter" long despised by practical men who spent their days in the woods. He was also a firm believer in the advantages of new, modern equipment and introduced what some people have maintained was BC's first steam-powered yarder, as well as its first logging railway. Regardless of the historical accuracy of these claims, the company was one of the leaders in the reorganization of logging methods.

With the use of oxen and horses, the task of moving bucked logs from the stump to the mill or the beach was essentially one continuous operation. Some logging shows used separate teams to move logs from the stump to the skid road, and then down the skid road, but the same technology was employed. The introduction of steam yarders and railways broke the logging operation up into a series of operations, or phases.

Falling and bucking operations remained

I regret to say that we have made no material reduction in the cost of putting in logs. It is simply Hades to try to do anything with men in this country. Yesterday we had but twenty-six men to work, and only two engineers—running the other donkeys with boys, or anyone we could pick up—the bookkeeper running one of them. They will simply get up and go, giving you no notice or warning. To make matters worse, the government recently passed a law compelling us to employ licensed engineers. When this is put in force it looks as though we might as well stop trying to do business, for knowing that they had us in the hole, they would certainly take advantage of it.

Hastings have just adopted a new system (at Rock Bay), on the same lines as some of the larger operators on the Sound. That is, they have established a saloon and boarding house, and allowed prostitutes to come in, at their salt water landing, which is six miles from their camp. They allow no whiskey to go to camp, but they sell the men all that they want at the landing. They will give them whiskey and board as long as the proceeds of their time cheques last, but absolutely refuse to give them a single meal, after they have spent all that they had. They say the results are, that by keeping between three and four hundred men around, they are enabled to have 150 men to work all the time. They say the men will come down to the landing, and instead of getting to Vancouver and leaving them with a large plant (four locomotives, ten miles of railroad etc.) idle, as they did formerly, that by the time the steamer comes in, they have no money to pay their fare to Vancouver, and will go back to work. They say they employ every man that applies for work, that this is what the men seem to want, that they have tried faithfully for twenty-five years to deal honourably with them, but that they cannot do it. Mr. Alexander remarked the other day that a $2000 whiskey bill would pay a $20,000 payroll. It seems very hard lines when a business firm has to resort to this. The Simpson Logging Co. have bought up all the shore rights for six miles, at the big operation of Hood's Canal, and will allow no other steamer to land at their dock, except their own. They allow no whiskey in camp, which is ten miles from the landing, but run a saloon at the landing. Mr. Anderson told me that he paid off, on the morning of the 24th, with about $6000, and the steamer left there on the afternoon of the 25th for Seattle, and in that time, he had taken in over the bar $2700, with a profit of about $2000. They also lease land to a house of prostitution.

The Lord knows what the results are going to be, if this state of affairs continues, as the men will go there and work, when they will not come here, where they get their cash at the end of every thirty days, and are treated as men.

E.J. Palmer, Manager of Victoria Lumber & Manufacturing Company at Chemainus, in a memo to J.E. Glover, January 3, 1902.

Reputedly the first steam donkey in BC, this Dolbeer spool-rig steam donkey was purchased by Victoria Lumber & Manufacturing for its logging operation at Chemainus in 1892. (BCFM L–2)

much the same, except that the addition of rakers to the saw teeth meant that axes were no longer used to make the final falling cut. The real change occurred after the logs were bucked. The work of moving the log from the stump to the beach was separated into at least two and, before long, four distinct phases. A yarding machine, or donkey, dragged the log to the road. If this was a railway, and not just a skid road where animals or another steam yarder took over, then the logs had to be loaded onto cars which hauled them to the log dump. In time these phases—yarding, loading, hauling and dumping—were all performed by specialized machinery requiring new skills from their operators. In practice, however, the transition was much more haphazard and subject to the availability of machinery, logging conditions and a variety of other factors.

Almost every region of the southern coast has its claimant to the title of "first steam yarder in BC." The earliest record of a donkey in BC was the one purchased by VL&MC in 1892, almost a year after the company started logging at Chemainus with oxen and horses. If any of the several photographs purporting to be of this first donkey are authentic, it was probably a Dolbeer steam logging donkey.

The first Dolbeer donkey was patented in 1882 by John Dolbeer, a partner in the Dolbeer and Carson Lumber Company of Eureka, California. This machine had an upright boiler and a single-cylinder, horizontal engine that drove a horizontal shaft with a gypsy, or capstan, on each end. Dolbeer's first model had a horizontal drum between the gypsies. A year later he brought out what he called an "Improved Logging Engine" with a single, vertical gypsy head, which was probably the model acquired by VL&MC.

In spite of the documented evidence on behalf of Dolbeer's invention, there is good reason to think that other steam donkeys were at work even earlier in BC and other locations in the American northwest. For many decades before 1882, steamships of various sizes and configurations were active on the coast. They, as well as many of the visiting sailing ships, were

A Dolbeer-type spool donkey with engineer at left and spooltender at right. This was the first steam donkey used in the Surrey area, perhaps as early as 1887. (BCARS 45514)

My dad had sold his mill and got this contract to clear all of that land in Coombs for the CPR, that long straight stretch. They had the first donkey I ever saw, a steam donkey. I can remember very well. It was like a capstan on a boat. It was vertical, and there was no gears. At least no friction. It just turned when the engine went but they had drums with it. You couldn't pull it out of gear. They wrapped the line around the capstan same as they do on a boat. You put another wrap on it to get more power, and hang onto the end. They were using wire cable.

You had a spooltender. He didn't have to pull on it, he just had to hold the slack and coil it up. When you got the log in, they threw the wraps off, and the line horse, old Bill, he'd take it out by himself, they never drove at all. He would go back to where we got the last log. And then he'd follow the next log in. Old Bill, he would just be eatin' blackberry bushes and huckleberry bushes, following the log into the landing. Then we'd unhook the log and hook the line onto him again and he'd walk back out to the woods. He did that all day long.

Well, it would come about a quarter to twelve, and Bill figured it was too damn close to eatin' time. "I ain't goin' back there again." So he wouldn't go and get any more. He'd go over and stand beside a hollow log where we'd put the oats in for him, and had a sack full of hay. He'd go wait there, he wouldn't go out again. And he'd do the same at quitting time. Come quittin' time, about five o'clock, the hell with it. And he would quit. That effectively shut the whole damn outfit down when he quit, nobody could log. My uncle said he was the best logger in camp.

Al West, born in 1904. He grew up at Coombs and logged in the Alberni area from 1919 until he retired in 1965.

equipped with steam-powered engines with vertical gypsies very similar to Dolbeer's, used to raise anchor, hoist sails and handle cargo. These engines were called donkeys long before they were used in the woods. It seems reasonable that some of them found their way to shore and were put to work logging.

For example, in 1883 the steam vessel *Grappler* was wrecked off Duncan Bay, just north of the site of Campbell River. One of the survivors was a Scottish steam engineer, Kenneth Henderson. He remained in the area and, according to his son Jim, was logging in Duncan Bay soon after, using a "spool rig," as these devices became known, with a horse to haul the line back to the next log. There are numerous reports of similar machines at various locations before the turn of the century. Some historians state that the Hastings Mill camps on Johnstone Strait obtained four steam donkeys in 1891, and recently loggers at Tahsis found the remains of a hand-powered capstan off a sailing ship that had apparently been used in the woods to winch out spars.

The type of logging made possible with these machines is known as "ground-lead logging." In a typical ground-lead setup the donkey's boiler was fired up with some choice, clear Douglas fir—easy to split and giving lots of heat. With a head of steam on, the line—a two- to three-inch manila rope, up to three hundred feet long—was pulled up the skid road and tied to a tree or stump. The donkey was started, which set the

An early model single-drum roader at work, possibly on Denman Island. It is an unusual setup, with an ox used to haul the line back for the next turn. (CDM P200–1001)

About 1903 or 1904 I got a job from a very fine fellow named Elijah John Fayder. He backed Frank Gallagher, a red-haired Irishman, to open a little camp, all told five men, and I was one of them. We were to get out hemlock bark for the tannery at Westminster. The lease was what was known as Tan Bark Lease No. 432, at the junction of Toba Inlet and Price Channel behind Double Island.

The cabin for the crew was a log house, the lower logs about four or four and a half feet diameter, and about forty feet long. The next tier would be smaller and the top logs smaller still. It had a shake roof and earth-and-gravel floor. There was one window at one end where somebody had chopped a hole in some of the logs about a foot square and through which no one could see for grime. The window at the other end was just a piece of glass. That was the only light except through the door, which was made of split cedar. The dining room was a little shed outside—four poles and some crosspieces and roof of shiplap. There were two benches, one each side of a shiplap table. When it was fine we'd eat out there. When it was raining we'd sit inside with a potbellied stove which would take a cord of wood. A hurricane lamp was our only light.

The bunks were made of poles cut out of the woods, three or four inches through, with bits of branches left here and there which served as hooks to hang clothes on. Poles were lashed to uprights and then others laid across from one end to the other and lots of hemlock, cedar or spruce brush spread on them for a mattress. There were two tiers. When the fellow in the upper bunk turned over, the dried brush would fall on the fellow in the lower bunk, so the lower bunks were not very popular. We carried our own blankets.

The highest paid man was the head faller. He got $2.50 for a ten-hour day. I think my pay was $2.00, less board. Board was really worth about $1.00 a month, but we paid two bits a day for it. Provisions would come up every week or so. The meat would sometimes nearly walk off the ship but by long boiling could usually be eaten. We had hardtack mostly but sometimes bread and butter. Also Climax jam with nails and bits of sacking mixed up in it, but more or less edible.

The logging equipment consisted of a single spool donkey, a little bit of a thing with a 300-foot line—no haulback—and a horse. The mainline was pulled back by the horse after the donkey had pulled in the log on the mainline, after which the process was repeated. We would strip the hemlock logs of their bark and load it on a sled, then the horse would pull the sled down to a scow. The tug which picked up the scow was usually the *Royal City*, an old torpedo boat. Gallagher sold the logs, boomed and all, for $4.25 per thousand board feet.

All hands and the cook would help pile the bark onto the sled, and the horse would be hitched on. When there was rough ground to go over, the whole thing would almost capsize. There would be wild yells to the horse to stop, which he would do quite readily, and hand peaveys and poles would be put against the sled to stop it tilting and it would finally make its way to the beach if we were lucky. If not the whole thing would capsize and we'd have to pile it up again.

Cecil Harlow Edmond, a logger who arrived in Vancouver in 1903.

A two-drum yarder used as a roader with a six-horse team skidding from the stump at Howard MacFarlane's camp on the north end of Denman Island, 1906. (CDM P200–1001b)

gypsy turning. The spooltender made two or three wraps of the line around the gypsy and pulled the loose end tight. As long as he maintained the pressure, the line held onto the gypsy, and was reeled in. A spool-off man coiled the line as it came in. Thus, the donkey yarded itself up the road to the bucked timber.

The donkey was tied down to convenient stumps or trees. An ox or a horse dragged out the line and the hooktender hooked it onto the log with a grab hook or tong. On signal, the spooltender reeled in the line. When the log came up against an obstacle, the line was slacked off and run through a snatch block tied to a tree or stump. Since only the best-grade fir and cedar trees were felled and bucked, it required a great deal of tricky manoeuvring to snake the log out through the standing trees and high stumps.

If horses or oxen were used on the skid road, the yarded logs would be rolled onto the skids in bunches of five, ten or more, depending on the log size, the number of animals in the team and the lay of the road. These bunches were called turns. Very soon after the first donkeys appeared, they were put to work yarding the logs down the skid roads as well. These machines became known as "roaders." If there were curves in the road, rollers were attached to convenient trees or stumps, around which the skidding line ran. Soon the roads themselves were modified, with troughs or chutes replacing the cross-skids in downhill sections.

About the turn of the century, donkeys with horizontal drums began to appear in BC. This modification most likely coincided with the availability of suitable wire rope. Twisted wire rope had been in use for centuries, but it was not until the 1890s that manufacturers began producing steel cable capable of withstanding the abuse of a logging show. Most machinery using wire rope ran it through large sheaves, off the ground and in relatively protected positions. The need for portability of logging equipment necessitated the use of small sheaves, which quickly wore out early types of wire rope. It was dragged through the dirt and mud, worked in the rain and embedded with bark and wood—all of which also wore it out.

At first loggers used a single-drum donkey, combined with a line horse on the haulback, with the line reeled onto the drum during the yarding-in phase. This improvement dispensed with the need for someone to coil the line. Very soon, a second drum was added. In most cases this drum held a second line that was strung out through blocks around the edges of the setting and tied on to the outer end of the mainline. Known as the haulback line, the second line was used to pull the heavier mainline, with its bull hook—to which the grab-hook lines were attached—back into the setting. In some of the early railway shows, the second drum was used to parbuckle logs onto rail cars. The first double-drum yarder used in BC was bought by Shawnigan Lake Lumber from Washington Iron Works in 1898, and used steadily for the next forty-five years.

A ground-lead crew typically consisted of one or two wood cutters and perhaps a fireman to feed the boiler. The chaser unhooked the logs as they came in to the landing. An engineer, sometimes called a leverman or donkey puncher, ran the donkey in response to signals relayed to him by the whistle punk. The whistle punk ran a light line from a whistle on the boiler, out through the woods to a location near the active logging site. At this end of the line the hooktender was in charge. He instructed a rigging slinger and one or two chokermen (as they were later called) how to hook up the logs and work them out to the landing. He gave orders to the whistle punk, who jerked the whistle wire in the appropriate combination of short and long blasts on

Engineer at work on "Old Faithful," a two-drum roader built at Victoria Machinery Depot and a copy of a seven-by-nine Washington, the first double-drum yarder used on Vancouver Island. The twenty-five-horsepower steam donkey was used for about fifty years and is preserved in the British Columbia Forest Museum. (BCFM 2–1)

Below: An early Comox Logging Company operation with a two-drum yarder ground skidding logs and using the second drum to parbuckle logs onto the rail cars. (BCARS 43051)

Above and opposite: A turn of logs dragged along a skid road with a steam roader at the Greene Point Rapids camp of A.P. Allison Logging. The logs are held by grab hooks attached to cable, which is hauled in by wood-fired engine at the beach. (Art Allison collection)

The old ground yarder system was the most frustrating and irritating business that you could imagine. The hooktender had to have the patience of Job and few men had that kind of patience. The turmoil of getting those logs out was terrible, and remember that stumps were still six, nine or ten feet high. Think of dragging the logs over the ground through all that to where they were picked up at the road or on salt water. The chains wouldn't hold and the chokers came off and the men would get so irritated that their curses would expand and have no relation to the log itself. The yarder would haul a log some 1500 feet if it had room to do it, but the stumps were so thick on the ground that it probably wouldn't haul it more than fifty feet on the first lap, then they had to change the choker and go another fifty feet. There was a lot of jumping backward and forward. It's difficult to convey all the frustrations of that day. The high lead, while far more destructive than the ground yarding system, was God's gift to the logger. If you go out on any of the old areas that were ground yarded, you will find luxurious second growth. It was much better forestry than the high lead that superseded it.

R.V. Stewart, who worked for the BC Forest Branch from 1914 to 1927.

the whistle, indicating to everyone on the crew what was about to occur.

Once a log was attached to the butt hook and the signal to go ahead was given, the mainline was reeled in. It was side-lined around obstacles, up to the butt-chain block. This large block, also known as a Tommy Moore block, hung on a stump about 75 or 100 feet ahead of the donkey and in lead with the drum, so the line spooled onto the drum evenly. When the butt hook passed through the Tommy Moore, the line was slacked and the choker unhooked, passed around the block and rehooked. The log was then dragged to the landing where the chaser unhooked it.

It was difficult to log downhill with such a system. The logs outran their chokers and fell out, or ran behind stumps on the side opposite to that on which the *line lay*. The men would have to drag them back with the haulback line, or disconnect the chokers and run them around the obstacle.

Hooktenders and rigging slingers became very skilled in the art of ground skidding logs through standing timber. They were among the many opponents of later logging systems that replaced ground lead, pointing to their ability to log an area selectively and leave a residual stand

to grow. While there was some validity to this argument, early ground-lead logging was really a form of high-grading, and the trees left did not necessarily constitute a healthy or valuable forest. Today, in various locations around the Gulf of Georgia where ground-lead logging was practised, there are stands of almost pure hemlock, left when the fir and cedar were taken. Many of these stands are badly diseased and are a poor testament to the supposed virtues of the ground-lead system.

The continuing evolution of steam donkeys led to the development of bigger and more sophisticated roaders. Until they were supplanted

The first logging camp there was in the Comox district except for ox and horse outfits was up the Courtenay River about halfway between Courtenay and Comox Lake. It was the first steam donkeys that were in this country. They were river driving to Courtenay, or to the Comox Bay, like.

I went there and I landed there just about dusk this night. Of course I wasn't gally when I was a kid. I was kind of shy. Here was a big bunkhouse and I could hear a lot of ya-yaing in there. I crawled up on a stump about forty feet from the bunkhouse door and I sit on this stump. I didn't have nerve to go in, you know. Finally a guy came out and he says: "Hey, kid, do you know it's time you was in bed?" I says: "I haven't got a bed." "Well, you come in here," he says, "and I'll give you a blanket." So I did, I went in. The bunkhouse, I don't think, was any more than twice as long as this room, and there were sixty men in it. They were three bunks high. "Well," he says, "you can climb up on that." He gave me a blanket and another guy gave me a blanket.

So anyway, the next morning the bell rang and, of course, everybody climbed out of bed. "Come on, kid, you've got to eat." So I went into the cookhouse. In those days, as soon as they eat breakfast, they'd all hit whatever direction they were going to work. Of course I didn't talk to anybody. I sit on a block of wood—I didn't know what to do—on the edge of the trail going to the woods. Pretty soon a big Norwegian by the name of Matt Hemmingsen who was foreman of the camp, he came out and he says: "Hey kid, you run away from home, eh?" I says: "No, I didn't have any home to run away from." "Oh, they give you a licking." "Yes," I says, "they give me a licking." "Ah, pooey," he says. "I got lots of lickings when I was a kid. That's nothing." I says: "You never got a licking like that." "Oh, rats," he says. So I just had a little shirt on and I flipped it off. Of course I had these black marks as wide as your thumb right clean from the back of my neck to my heels. Well, he just about had a fit. He asked me who did it. Of course he didn't know them. He said: "Come on with me. I'll teach you how to be a whistle punk."

So we went to the woods. There was an old fellow, I should imagine about sixty, who was punking whistle. He went up to this old guy and he says: "Oh, you go and help the chokerman. I'm going to stay with this kid for an hour or so and teach him how to punk whistles." So that's how I got started in the woods. I was just past fourteen then.

Ben Ployat, logger. He got his first job as a whistle punk at Victoria Lumber & Manufacturing's camp near Courtenay, about 1900.

The first system I recall was when we used a nine-by-ten steam donkey using a ground lead, that is a bull block or "Tommy Moore" today, hung on a stump for lead, and yarding the logs to a three-pole fore-and-aft chute to send them into the water. There was no strawline, so the haulback was pulled around an entire quarter then dropped in for each road. Later this system was extended until we had up to five steam donkeys in line relaying the log turns in to fore-and-aft skid roads, one to the other, for distances of up to one and a half miles.

The yarder was ground leading, as above, with a swing machine pulling the logs to the road and making up the turns of eight to ten logs, depending on size and conditions. These turns were dogged or grabbed together with grabs and the road donkey pulled them by the first log. A dugout or pig was hooked on behind the turn to take the grabs back to the woods for the next turn. Ever hear of the PF ["pig fucker"] man? He was the man in charge of this part of the operation.

About 1918 we introduced a high-lead yarding system much the same as we use today, but still using the fore-and-aft skid roads which had been much improved, so that the dogging up of turns with grabs was eliminated. Only a choker (open face type) was attached to the last log in the turn, which then pushed or bunted down the roads rather than pulling on the first log.

Another system we used was the drop chute. This was a spar tree rigged at the top of a steep chute with a steep apron built around it. The logs being yarded up to the tree were raised until the other end swung into the chute, then dropped, which would loosen the choker, the log sliding on down to the water.

Art Bendickson, son of Hans, who started punking whistles in 1917 and logged all his life. He was a director of the Truck Loggers' Association for ten years.

by railways, roaders were used to reach farther and farther back from the beach. In some places, when the capacity of the main-line drum was reached, a second donkey was positioned along the skid road to skid logs to the first machine. Third and fourth machines might also be added, with spool rigs, oxen or horses skidding to the main skid road.

Above: An Empire 11 by 14 donkey, owned by Clark and Lyford, swinging logs into Booker Lagoon, 1917. (John Cress photo; author's collection)

An eleven-by-fourteen Empire roader, with 4500 feet of line on its main drum, working at BC Mills Timber & Trading Company Camp H at Rock Bay, June 1917. A special drum, optional on this Empire model, is used to parbuckle load rail cars with logs up to seventy-two feet long. (John Cress photo; author's collection)

Even more difficult to pinpoint than the use of the first steam donkey is the site of the first logging railway. There is little doubt that the first locomotive to arrive was the often rebuilt "Old Curly," brought in by CPR railway construction contractor Andrew Onderdonk in the early 1880s. According to Edward Austin, who rebuilt the locomotive at Yale, it was built by Union Iron Works in San Francisco about 1869. After the railway was completed, the locomotive was sold to Royal City Planing Mills and used to haul logs on logging lines in Surrey. After Royal City and the Hastings Mill were combined into BC Mills Timber and Trading Company by John Hendry in 1889, "Old Curly" became part of the Hastings outfit's operations at Palmer Bay in 1894, and moved to Granite Bay in 1901. From there the locomotive went to various BCMT&T camps, including Camp O on East Thurlow Island, where it was photographed by BC Chief Forester H.R. MacMillan in 1913. Today a restored version of "Old Curly" is on display at the Burnaby Village Museum.

BCMT&T replaced "Old Curly" with a bigger locomotive after connecting its logging rail line to the main line of an early Seattle–Vancouver line, the New Westminster Southern Railway, in 1891. This line ended at Port Kells on the Fraser River, where the company dumped logs for towing to its mill in New Westminster. Photographs of this operation indicate it employed a combination of horses, two-drum roaders and an elegant looking locomotive. During the same period, BCMT&T built a short line in Kitsilano equipped with a small four-wheel locomotive.

Meanwhile, at Chemainus, things were not going well for "Old Hickory" and the crew at VL&MC. A series of setbacks, including poor markets, a smallpox epidemic and teredos in the pilings holding up the sawmill, resulted in a mill closure for much of the 1890s. When things got rolling again, the accessible timber was beyond easy reach of the horses, so in 1899 a railway was built up the Chemainus Valley. The company began by using the locomotive as a skidder, dragging the logs between the rails.

According to records, this Royal City Saw & Planing Mills locomotive is "Old Curly" at an unknown location in 1895. But it may be a much earlier photograph, as Royal City was combined with Hastings Mill in 1889 to form BC Mills Timber & Trading, and other photos of the locomotive portray a much altered version. (CVA Log P40 Neg 25)

Below: "Old Curly" at a somewhat later date, possibly at Camp O or Granite Bay, loading logs skidded by the large roader at left. (CVA Log P51 Neg 35)

BC Mills Timber & Trading camp at Bear River, 1896. The locomotive at rear, to the right of the pile driver, is "Old Curly." The camp was opened in 1894 at a site also known as Palmer Bay and Bear Bay, for a brief period the company's centre of operations in the area before it was moved a few miles south to Rock Bay. (BCARS 59747)

Below: Another early BC Mills Timber & Trading locomotive, built by the Canadian Locomotive Company. Photographed at Camp O on East Thurlow Island by H.R. MacMillan when he was BC's chief forester, 1912–16. (H.R. MacMillan photo; BCFM L-3)

When "Old Curly" was sent to Palmer Bay, it was replaced with this locie, a mainline engine bought from Great Northern, shown near Port Kells in the early 1890s. (BCARS 45531)

Near Camp 5, west of Ladysmith, Victoria Lumber & Manufacturing's locie #3 skids a turn of logs between the rails in 1901. They have been skidded to the railway by the ten-by-twelve roader at left. Left to right: Bob Laird, Joe McDonald and Dutchy Reid. (BCARS 40332)

Alan & South built this wooden-rail logging railway at Little River, east of Comox, in 1901. (CDM P200–1001a)

Planks were laid on the ties, with six-by-six guides along the inside of each rail, and turns of ten to fifteen logs trailed behind the locie. Even later, when log cars were used, this method was employed on steeper spurs up the side of Mount Brenton.

It is largely a question of definition whether this was the first railway logging in the area. Much earlier, in 1887, the E&N Railway loaded logs along its right-of-way for transport to the Chemainus mill. Some consider this to be the earliest railway logging on south Vancouver Island; others insist that since the E&N was not a logging railway, it does not qualify.

The fourth instance of early railway logging was at the BCMT&T camp at Wolfssohn Bay (later Lang Bay) south of Powell River some time in the early 1890s. This line eventually went back eight miles along Whittall Creek and burned coal from a small open-pit mine. A few miles to the north, at Sliammon Creek, another, shorter line, 2.5 miles long, hauled logs out of the Wildwood area around the turn of the century. The locomotive may have been the same one used by the company in Kitsilano.

E.J. Palmer came directly to Chemainus from a job as a conductor on a mixed train running in Weyerhaeuser territory. The "old gentleman," as we called Weyerhaeuser senior, saw quality in him. Although E.J. had no feeling of interest in timber, logging or sawmill (he went to the woods about one day yearly and walked around the yard two or three hours weekly and very seldom went into the mill), he was interested in commercial opportunities and in new methods. He brought the first donkey to Vancouver Island—a "spool" donkey. He also built the first logging railroad in British Columbia. This railroad did not have cars to carry the logs. Planks were laid longitudinally between the rails, the logs were "sniped," then were "dogged" together at the ends so that the locomotive could pull a string of them, (about six to twelve according to size) mostly downgrade, sliding on the planks, to the dump at salt water in the upper end of Ladysmith harbour.

H.R. MacMillan, BC's first chief forester. He worked for E.J. Palmer at Victoria Lumber & Manufacturing Company in 1917.

This pole railway was used at Steelhead Camp south of Stave Lake, about 1920. (Ryukiche Miyake photo; Japanese Canadian Centennial Project)

There is evidence that in 1894 the Moodyville Mill Company put in a short line from Powell Lake, using a locomotive purchased in Pittsburgh and cars built at the BC Iron Works in Vancouver. If so, this was a short-lived operation, in spite of the fact the company reputedly held 8,000 acres containing an estimated 300 million feet of high-grade Douglas fir.

Around the turn of the century, a variety of unusual railway machinery was tried out at several locations. The Shawnigan Lake Lumber Company built a wooden pole road west of the lake in 1900. A team of horses was used to pull two four-wheeled wagons over this road. Eventually a primitive steam-powered locomotive constructed in the company shop replaced the horses. Cast-iron wheels for these pole roads were available from several foundries in the U.S., where they were used extensively before conventional railways came into use in 1879. The Shawnigan Lake wheels were cast by the A.H. Steeples Foundry in Seattle.

American loggers Alan and South built a second pole road at Little River near Comox in 1900. A steam locomotive with cast, concave wheels pulled log cars similarly equipped. In 1904 this operation was taken over by Lyman Hart and Tom Bambrick, who replaced the poles with steel rails. This was the first logging railway north of Nanaimo. A pole road was used during the latter part of this era between Powell Lake and Michigan Landing, although it may not have been used for logging but to haul clay for plugging leaks in the dam at the lake. The cast-iron wheels from this operation are now on display at the Powell River Museum.

In 1908, on the north shore of Burrard Inlet, the McNair–Fraser Lumber Company installed a cable locomotive—probably the only one in BC—as part of a conventional railway show. These machines were used in several locations in the U.S. and at least two models, the Willamette Cable Locomotive and the Climber Locomotive, were in production for a few years. McNair–Fraser used a Willamette, which dragged the fifteen-to-twenty-log turns over the crossties. The timber on this claim was on Hollyburn Ridge, two and a half miles from the beach, at elevations of 1,200 to 1,600 feet. It was far too steep for a conventional railway, although the company had tried it and was reluctant to write off the road it had built. The locomotive consisted of a large vertical boiler, engines equivalent to those on a fourteen-by-fourteen road engine, and a large differential gear attached to two bull wheels five feet in diameter. It ran on a standard-gauge track and hauled itself along on two wire ropes, one inch thick, strung on the outside of the rails and held in place by hardwood pegs set in the ties. The cables were tied to a stump at the top of road, and at the bottom were wrapped around tightening wheels at the end of the log dump. Four wraps of cable were wound around each bull wheel. When power was applied, friction prevented the bull wheels from slipping on the cable and the locomotive moved up or down the track at about five miles an hour. The climbing locomotive, called the "Walking Dudley" by the public, was used by McNair–Fraser until about 1912 when it broke loose and ran down the track, off the log dump and into the water.

To lay out the railway we just used a staff compass and an Abney level. We did a pretty good job. Before that we had worked on different main line railways. The trick was to do it good enough but not so good as to be too expensive because the work of building the railroad was all done by hand. There were no shovels; no power equipment was used at all. Horses weren't even used in construction of the railroads. Just Swede power.

All the railroad was built by contract. Each station was a hundred feet and it was so much a station. The station crew were all partners. There would be six or eight men, or ten or twelve, or even three, and they would take a mile at so much a station and split evenly. They would be provided with the tools and the powder and the saws, and these would be charged against them, and the crew would be paid what was left for the construction of the mile or the station.

The rock work was negligible. The little there was didn't amount to anything. The work was really not heavy enough for much wheelbarrow work. There was only a foot or two of fill or cut. That was easier to do, with these strong, husky fellows, just throwing it rather than wheeling it. It was harder work but you could do it a little faster by just shovelling than you could by wheeling it.

Robert J. (Bob) Filberg, president of Comox Logging. He was born in Sweden, moved to Colorado as a child and came to BC in 1909 at age seventeen. After working briefly for David Jeremiason's Anderson Logging Company at Union Bay and International Timber at Campbell River, he joined Comox Logging as chief engineer in the fall of 1909. He was superintendent when the company became part of Crown Zellerbach.

The Walking Dudley, McNair Fraser Logging's climbing locomotive, dragging a turn of logs between the rails at West Vancouver in 1911. The machine pulled itself along two cables laid in either side of the rails, held in place by pegs driven into the ties. It could haul 20,000 board feet of timber down the two-mile railway to the dump near the foot of 17th Street. (CVA Log P50 Neg 34 #4)

During the transitional period in which the logging industry converted from animal power to steam, the BC coastal timber market went through some major upheavals. Following completion of the CPR, there was a gradual shift in the demand for lumber away from the Burrard Inlet mills' export business toward the markets served by railways in the East. During the late 1880s in the U.S., an intense competition among Washington and Oregon mills selling into the California market drove log prices up, although the increases were moderated by downward pressures exerted by the lower costs of steam donkey logging. By 1899, for example, it cost $4.50 a thousand board feet to log by oxen, compared to $2.10 with steam donkeys. Not only did oxen need to be fed and cared for when they were not working, but they did not perform well in wet weather.

In 1887 the Washington and Oregon mills set up a log buyers' cartel, the Puget Sound Brokerage Company, and began to run their own logging operations instead of buying from market loggers. This, along with other pressures, encouraged many of the U.S. companies to look for timber in BC.

Throughout most of the 1890s, the Pacific Northwest suffered a severe recession, particularly the lumber industry. When the dollar-a-thousand tariff on lumber going into the US was lifted in 1894, the BC mills increased their shipments into the California market, hurting the American mills even more. This led in 1896 to the formation of a lumber marketing combine, the Central Lumber Company, which the Hastings Mill and three others in BC joined; Moodyville and VL&MC did not. The latter, in fact, had been closed for three years and re-opened to sell into California by undercutting the combine's prices.

With the advent of the Klondike and Alaskan gold rushes in the late 1890s, the recession eased. As settlement of the North American plains accelerated, the interior markets grew rapidly. BC mills were soon cutting at capacity for shipments to the Okanagan and the prairies. A boom was under way that lasted until World War One.

Initially, the outbreak of war devastated the logging and lumber business. The presence of German warships in the Pacific brought an immediate halt to export trade, and the enlistment of men in the armed forces severely disrupted the labour supply. With a provincial population of less than 450,000 people, BC sent 43,200 men overseas—almost ten percent of the population. Within a couple of years, however, the Allied war machine developed an insatiable demand for lumber, and by the time the conflict ended, logging levels were at record highs, in spite of wartime manpower and equipment shortages.

In sum, from about 1890 to 1920, the province and the logging industry were transformed. In three decades logging went from its horse and oxen days to the mechanized world of steam and beyond. It was as if all the advances and many of the problems of the industrial revolution had been compressed and thrust upon the area at once.

The forms of industrial organization experienced equally dramatic changes. Drawing on the influx of European immigrant workers who flooded in after the war, the workplace changed from the slow-paced kind of camp run by Jerry Rogers or Saul Reamy. Most of the Lower Mainland had been logged by men working out of camps like these, or by handloggers who lived at home. By the end of the war the first of the big upcoast railway camps, employing hundreds of single men operating complex, massive and powerful logging equipment, had been established. But this was not the whole picture. Even though the vast south-coast stands of high-value timber had been acquired by the big companies, there was still a lot of room for smaller, independent loggers, from handloggers on up. They too flourished during this period. Together, all these developments resulted in the opening up of the entire coast to logging by the end of the war.

Around 1885 the Miller family came here. That was my maternal great grandfather. They logged in Vancouver for Hastings Mill, with oxen and . . . you name it. They did a lot of logging down around Clark Drive, Vernon Drive, in the east end of Vancouver, and Hastings Mill wasn't that far away. They worked for them until they ran out of timber.

Then he worked for Hastings as a falling contractor at Rock Bay. He was up there in 1900, 1901, 1902. He was a logger. They had been logging in Sweden. He came out to Vancouver as a seaman, jumped ship and stayed there because a friend of his by the name of Larson was there at the time. Logging was his profession.

The Mosher family had been in the sawmill and logging business in Nova Scotia for a long time. They came to Nova Scotia from Rhode Island a hundred years before the revolution, in 1676, and started a grist mill. They were farmers, loggers—I guess you were a little bit of everything in those days. They got into the lumber manufacturing business, lumber sales and the logging business. Two of the Mosher brothers, my grandfather and my uncle, for whatever reason, decided they were coming to BC, around 1900. My grandfather picked up his thirteen kids and his chattels, with my uncle they got everything and themselves into a railroad car, a whole car. They came out to Vancouver.

As soon as they got out here they got timber rights. My grandfather, with his boys and my great uncle, got the camps running and they built a sawmill. One camp on the Sechelt Peninsula, one around Hardwicke. I remember them talking about Hole in the Wall, so they must have had a camp on Maurelle or Sonora Island. They had camps going in the Jungles.

They were selling logs at first, while they were building their mill. They never got the mill built until 1911 or 1912. They had picked up some timber rights in the Fraser Valley and they built the mill around those timber rights, I guess. It was across the Fraser on the Great Northern Railway, at some kind of settlement called Strawberry Vale, not too far from where Ritchie Brothers has their auctions now.

They had the mill all built and ready to go. They had the logs and had cut a bunch of lumber. The Great Northern was supposed to put a spur into their mill off the line that had just been built to Vancouver. But a depression hit and the Great Northern never built the spur. The bottom dropped out of the market then, the prairie market. They never got to ship any lumber and they ran out of cash. They went belly up.

They continued to log. Eventually the camps closed. My dad had gone overseas during the First World War and my uncles continued to work in the industry. They'd work for someone else for a while, then get a gyppo operation of their own going. Two of my uncles had a gyppo operation going around Pender Harbour at one time. They were trying to use a Cat with one of those pans they lay the front of the logs on.

My dad got into Rock Bay and he was there for a long time. My Uncle Bill too. They all worked at Rock Bay at one time, they'd all come through there—all eight uncles. And five aunts, all of whom got mixed up with loggers. It was sort of a big family organization. Between the wars they worked for themselves and for other logging outfits. They didn't like working for Bloedel, Stewart & Welch, though. They referred to that as Blood, Sweat & Work.

Monty Mosher, logging contractor, co-builder of the Cyclo Crane and vice-president of the Truck Loggers' Association.

This nine-by-ten, two-drum roader was probably the first steam donkey used on Howe Sound. It was operated by Mosher Logging at Hopkins Landing in 1908, after which it was sold to Bill Cook and used on Lasqueti Island. (Elphinstone Pioneer Museum)

The Jungles was perhaps the area most thoroughly populated after 1890. The pioneer Hastings and Moody mills obtained timber rights and established camps, Hastings throughout the area and Moody, for the most part, along the mainland side of the Gulf of Georgia. They were joined first by new mills built on the Fraser and in False Creek—Royal City Mills building in both locations, Leamy and Kyle in the Creek. Leamy and Kyle began logging in Cameleon Harbour on Sonora Island as early as 1888, and was joined in the same area by VL&MC and Pacific Coast Lumber at Owen Bay a few years later. Haslam from Nanaimo was also an early operator in the Jungles, mostly on Vancouver Island along Johnstone Strait.

In the early 1900s, Brooks–Scanlon of Minneapolis, the second largest timber company in the US, acquired timber in the Fraser Valley and south of Powell River under the name Brooks, Scanlon & O'Brien. (They had bought out John O'Brien who had been logging in the area since 1900.) Timber broker Charley Pretty, acting on behalf of a group holding the leases on the water and timber rights for a proposed pulp mill at Powell River, approached the Brooks–Scanlon people and persuaded them to buy these rights, forming the Powell River Company in 1909. This company began logging at the top end of the Jungles, pushing a railway up the Kingcome River the same year.

During this same period another transformation was going on in the Jungles. Many of the handloggers who had worked the area since the early 1880s, along with recent recruits to their ranks, were prospering, and they began to acquire some of the new machines available to loggers. In most instances this meant buying a steam donkey, mounting it on a float and using it to yard logs down the hill, into the water. With the addition of a block hung from a frame, a system called "A-framing" was born. It quickly became one of the simplest and most effective forms of mechanized logging on the coast. A donkey was mounted on a large log float, and the mainline was strung through a block hung from the apex of two raised poles and run uphill. The logs were yarded down the hill and into the water.

As well, these early donkey loggers learned to mount their machines on heavy log sleds and pull themselves up the hills in pursuit of logs. It was a lethally effective method in the hands of the early handloggers. As long as they stayed away from someone else's timber, they were not restricted in the methods they employed or the areas in which they worked. In 1906, they were prohibited from using steam-powered machines and in 1908 an unsuccessful attempt was made to ban them from the south coast altogether. With some justification, they were accused of wasting large amounts of timber and of skimming off the easy-to-reach timber along the shores.

As stated earlier, it was in this region of the coast, more than any other, that the independent loggers found their feet. Some of them started as handloggers during this period of industrialization. One of the most renowned was Hans Bendickson, a Norwegian who came to BC to fish salmon in 1901. Within a year he was ox logging at McNabb Creek on Howe Sound, and in 1904 he received Hand Logger's Licence #1752. Old Ben, as he was called, logged throughout the Jungles, buying a nine-by-ten double-drum donkey built by the Vancouver Engineering Works in 1906 and putting it to

A typical road donkey at work on Hardwicke Island, 1919. The ten-by-fifteen Vancouver Engineering Works machine, built in 1910, had a mainline drum capacity of two miles of one-and-one-eighth line. Louis Larson, engineer, and John Rohan, wood bucker. (CRM 12368 Bendickson Collection)

work on a timber sale on Broughton Island. Until 1918, Bendickson and his growing family—six sons and a daughter—logged throughout the area, including shows at Port Moody, Quadra Island, Pitt Lake and Britannia Beach. Then the family settled into a homestead on Hardwicke Island, establishing gardens, orchards and a herd of cattle to feed the growing family and logging crew.

As the Bendickson boys became teenagers, they were put to work logging, greasing skids and punking whistles. They used a ground-lead system until they moved to Hardwicke, where one of the coast's first high-lead systems was installed to yard logs down fore-and-aft skid roads. In 1936 they started Cat logging, and they began truck logging in 1951. Bendickson Logging was one of the first companies to join the Truck Loggers' Association after it was formed in 1943, and Art Bendickson was a TLA director for ten years between 1950 and 1965. The Bendicksons formed their own logging and towboat companies and in 1952 received one of the few small Forest Management Licences in the province, covering all of Hardwicke Island. In 1971, their ranks swelled by a third generation, they sold this licence to Crown Zellerbach, but retained their logging equipment and went into contract logging on Hardwicke and elsewhere.

Another Scandinavian who came to BC after logging in Minnesota and the Bellingham area was Swedish-born P.B. Anderson. In 1910 he started logging with two steam donkeys—a yarder and a roader—near Pender Harbour. By the

My father bought out Clarence Dougan's father, who had a camp at Flat Island, south of Pender Harbour. I joined him there at the age of fourteen. That was in 1909. I was no student, and along with three other boys had been expelled, in my junior year, from the Bellingham high school.

The first few months were pretty tough. It was a camp of forty or fifty men. They were all the older, rough, old-time loggers. In those days there were no young fellows in the camps.

I was started in the blacksmith's shop. But as soon as my father made his first trip to Vancouver, I went out to look for a job that suited me better. I started following the PF man. That was what suited me.

This was an old skid road camp, with a fore-and-aft flume. The logs came down the fore and aft, dogged up with grabs. The grabs were removed and put in an old-type canoe pig, which was skinned back up again. The fore and aft had two large logs on either side, and a smaller one on the bottom. There was a big roader, or yarder, or donkey with a drum holding over a mile of one-and-one-quarter-inch line. The logs came down the trough, from as far back as a mile. They were bunted in the chuck, and the haulback was taken up to the woods again.

Dad soon put me back in the blacksmith's shop, feeling as he did that the woods were too dangerous. However, it wasn't too long before I was out, blowing whistles. Soon I went on to work in all the jobs in the woods.

Dewey Anderson, longtime BC logger. He was part owner of several logging companies with his father, P.B., and brother, Clay. He was president of the BC Loggers' Association in 1939 and 1940.

In May 1909 I started to look for another proposition. The timber in the northern part of the state of Washington was pretty well taken up so I had to go to some other parts. So I started for Canada and went to Vancouver. I cruised there for nearly six months before I found a suitable proposition. I found a bankrupt outfit forty-five miles northwest from Vancouver at a place called Flat Island, four miles south of Pender Harbour.

There was two donkeys, one a yarder, one a roader, and lines and equipment to do the logging. There was also a cookhouse and bunkhouse for the men but everything was in bad shape so it took quite a bit of money and time to put same in good working order. There was 30 million of timber in the deal. I paid $28,000 cash for the deal. (Timber at that date was worth very little.) This was spring of 1910. I logged here with fair success but had to have more machinery so I bought a twelve-by-fourteen yarder from the Empire Manufacturing Company. This with lines equipped cost $9000.

P.B. Anderson, second from left, with pile driving crew building a railway bridge at Knox Bay, West Thurlow Island, in 1917. (UBC Special Collections, Anderson Collection)

After this addition I was getting along fine 'til summer of 1912. In August a fire broke out in my logging works. I lost all my fell and bucked timber, lot of rigging and one donkey.

This put me in pretty bad shape again. I had not been in British Columbia long enough to have established a bank credit. So I sent a message to my friends Victor Roeder and William McCush, president and vice-president of the Bellingham National Bank. Mr. William McCush came up to my camp and said after looking it over, you are all right but you need some money. So he loaned me $3000 and told me to try to get along with it and don't stop logging.

I was doing fine after this and was doing well. In the fall of 1912 I took a contract to log 50 million of timber for Hastings Sawmills, Mr. John Hendry owner, Mr. Arthur Hendry manager, Alexander (Sandy) McNeir timber superintendent. I now had to increase the output so McNeir loaned me $3000 with an option to later buy in with me. This he did after returning from a trip to California.

I cleaned up Flat Island and moved in to Pender Harbour which was closer to my holdings, but the timber was six to eight miles from the salt water so I needed a railroad. I went to Mr. Arthur Hendry and bought eight miles of steel rails and switches for spurs, etc. Also rented a small ten-ton locomotive and some RR cars. Now I started railroading and got along fine for a while but found out that the ten-ton machine was too small for the grades so I went to the interior of British Columbia and bought a fifty-five-ton geared Climax locomotive. I got through with my contract as well as other timber I picked up in the neighbourhood.

In the meantime I had made another contract with the same company to log 40 million of timber at a place called Skokum Chuck on Agamemnon Channel at the mouth of Jervis Inlet. I picked up another 60 million there and logged successfully 'til I got through.

At the time at this camp I also had a contract with the Gordon Development Company to log 40 million of timber for them near my old camp at Pender Harbour. I put my son Dewey to do the logging there and he done the job in a first-class shape and got through about the same time as I did at Skokum Chuck.

In the meantime I had bought 150 million of timber up Johnstone Strait across from Vancouver Island 150 miles north of Vancouver (at Knox Bay on West Thurlow Island.) So in June 1917 I moved there. I now had a good railroad outfit, twelve miles of rails, twelve new logging cars, six donkeys and rigging to run two sides, a fifty-five-ton locomotive and money in the bank to move and open up the new camp. My output now had a capacity of 5 million per month. It took me three months to move the outfit, build the railroad up to and into the timber, fall and buck timber, build landings, set donkeys and get riggings out to start logging. At that time my oldest son Dewey was called to the army. He being an American went to Fort Lewis. Clay, my youngest son, joined the Canadian Army and went to Siberia, so I lost two of my standbys.

This tract of timber on Thurlow Island was one of the finest quality of old growth fir and cedar on the Pacific coast. We had to build camps in the woods as well as at the beach. There was also a long wharf from the shore to deep water for steamboats to land, and there was a big booming ground to drive. Got this all completed and started to log about the first of October 1917. Finally the boys came back from the army. While they were away I had bought twelve timber claims, twelve square miles adjoining my original holdings so now I had timber to last me eight years. My oldest son was superintendent and Clay was foreman.

I had bought a nice tract of timber from Hastings Shingle Company on the mainland near Grassy Bay on Loughborough Inlet. The boys wanted to go logging on their own so I bought them a railroad camp that was operating at Grassy Bay. This place was only eight miles from Knox Bay. The boys started and logged for about a year when the market on logs dropped so they had to close the camp 'til the market got better and prices for logs came back. The boys then came back to Knox Bay again and stayed for about a year when the market had changed so they could run profitably.

I had appointed another superintendent and kept logging with good results, but in the last part of June it was very dry. I was about to close the camp when fire broke out in our workings. I had 160 men at the upper camp and we fought the fire with this crew two months before we got it under control. It took three months before it was all out. I lost three donkeys, several bridges and the whole upper camp.

I had insurance for $200,000 but that only paid the wages of the fire-fighters. My extra loss was $110,000 over the insurance. From this time I operated this camp without any mishap 'til I got through in 1920. The boys got through with their camp about the same time.

I was in splendid shape financially but I have always been easy and did not value money the way I should. I had backed a cedar saw mill to the extent of $180,000 and it got in difficulties. I had to buy it at a receiver's sale so before I got it I was in $220,000. Now I had a mill on my hands so I decided to run it. I thought I could run it, but milling and logging don't work the same so I got in difficulties. I sold my logging outfit, paid my debts and walked out. The city taxes on the mill site were $500 a year. I had no income to take care of this so I let it go back to the City of Vancouver. This fourteen-acre mill site is today worth half a million dollars. Such is life in the lumber game. I saved my home and about $20,000 out of the mess.

I had to start over again but I was now a little over sixty years old so I had to be more careful from now on. My credit had not been hampered so if I could get a piece of timber I could start again. I had a small piece of timber, only about 8 million, near Greene Point Rapids on Cordero Channel. It was handy to the water and it would give me a start. So in the spring of 1930 I started on the small tract and had wonderful luck. The log market took a big jump and I was ready with logs at the time. I cleaned up the tract in three months, paid my bills and cleared $40,000.

Now I was getting back on my feet and looking for prospects. A timber broker, Mr. Charley Pretty, looked me up so I took my two sons with me and we looked the timber over and bought it on a stumpage basis. This tract had about 350 million on it. It was located on Harrison Lake on the Fraser River watershed. I had in the meantime bought back all my Knox Bay machinery so we now had a good outfit so we moved everything to Harrison Lake. We started to log there in the early spring of 1931. I logged there with great success 'til all the timber was out, taking out 300 million. This brought us up to 1935.

We were in a good financial standing, had cash, a good outfit but no timber. Now again we saw Mr. Charley Pretty, the timber broker. He put us on a tract of timber on Vancouver Island (at Salmon River) twenty miles from my old camp at Knox Bay. I had been through this timber and knew it was first class and of good quality and a big tract. The price for the tract was $1,350,000—$600,000 cash. Now that was more than we could handle alone so we took in a partner, Mr. J.C. Robson of New Westminster (who owned Timberlands Mill).

The nearest timber was fourteen miles from the salt water. There was long tideflats, lots of gullies and three big rivers to cross before we got to the timber. After we had made the down payment on the timber we had $500,000 in the bank. We were getting along fine, building the cookhouse, dining hall, office, driving piles over the flats, grading railroad and three big bridges over rivers. Money was going fast. We had three rivers to cross and had the bridges nearly built and ready to take out the false works when we got a big flood that took them all out. Each of these bridges was from 250 to 300 feet long so we lost a lot on this freshet. But we started right over to rebuild and got them built. But now we needed more money, lots of it. Our own bank turned us down so we went to the Bank of Montreal and got a credit for $1 million. We used up three-quarters of this before we got timber to the market and returns coming in.

P.B. Anderson, pioneer independent railway logger. He was born June 1, 1866 in Sweden and died in December 1959, in Vancouver.

The small steam tug Dart was used by BC Mills Timber & Trading at Saul Reamey's camp on Bear (McCreight) Lake in the 1890s to tow logs to the landing, where they were loaded on rail cars and hauled to Bear (Palmer) Bay. A second, gas-powered Dart was used by the company later on Roberts Lake. (CVA Log P38 Neg 23)

time he retired twenty years later, he had built an extensive railway logging company with shows at Loughborough Inlet, Knox Bay, Granite Bay, Salmon River and Harrison Lake. His were some of the biggest and most successful railway logging camps to operate in the Jungles, rivalling those of well-financed companies from the U.S. and eastern Canada.

Bill Dolmage, who for many years ran the coast's biggest log-towing company, used to claim that P.B. was the most gentlemanly logger on the West Coast, and named one of his new tugs after him. When P.B. retired, his sons Dewey and Clay continued to log at various locations until well into the 1950s. Their operations were eventually bought up by larger companies, with, for instance, the Salmon River camp forming the Kelsey Bay division of MacMillan Bloedel.

A third independent logger to establish himself in this area was A.P. Allison. Born in England, he was raised in Chemainus and went to work for the VL&MC in 1905. He soon went out on his own and by 1908 had a railway camp at Greene Point Rapids, which eventually included one of the largest shingle mills on the coast. During the war he was associated with the Brooks, Bidlake and Whittall Shingle Company. After setting up a second railway camp at Homathko River, he opened a third at Cumshewa Inlet. The latter became the major producer of airplane spruce on the coast and during World War Two was taken over by the federal government and renamed Aero Timber. When he died in 1957, A.P. was president of the Truck Loggers' Association. His three sons took over the family operations, including the Lion's Gate Lumber Company, and logged for many years on Johnstone Strait.

Between 1890 and 1920, a population of more or less permanent residents settled the area known as the Jungles. Several thousand people, not counting those in the railway camps, gave as their place of residence one of the many communities, bays, coves or inlets between Powell River and Allison Harbour. By the end of this period the area had regular steamship service, postal service, schools and church mission boats. In 1905, a hospital opened at Rock Bay to serve the entire area. With few exceptions this entire economy was built on the logging industry.

Elsewhere, a similar kind of activity was evident. Along the mainland, from Powell River south, a number of large operations were established. Brooks, Scanlon & O'Brien opened a camp at Stillwater in 1909, and by 1918 was hauling over sixteen miles of track, making it one of the biggest railway shows on the BC coast. Bloedel, Stewart & Welch opened another large railway camp a few miles north at Myrtle Point in 1911. It was started by Bellingham lumberman Julius Bloedel and his partners. They bought 100,000 acres obtained by timber speculators during the staking frenzy a few years earlier, paying a dollar an acre for it. From this base BS&W expanded into one of the biggest railway logging companies on the coast.

About a dozen smaller railway operations were established throughout the adjacent areas.

In 1906, when I hit the logging camps, they were still using the line horse, but most of the machines had haulbacks. As I remember, the yarders had, even with a line horse rigging on it, they still had a small drum in there with which they could kick the logs onto the road to make up the turns to send them on through to the beach.

Then we also had in those days, even up to 1906, what they used to call hand lines. The old hooktenders used to want a hand line so they didn't have to string their haulback. They would reach out, maybe a hundred feet on either side of the centre of their line to bring in the logs, and then they'd unhook and put the chokers on to the main bull hook. To hook in logs that were out too far to spring the mainline over to catch them. So we would log probably a swathe of about 500 feet wide.

In 1913 we opened up a camp at West Lake, just in from Powell River. There we put the logs into the lake from the shorelines—went around the shore and would reach back with a yarder—the company didn't have enough money then to buy equipment so we tried to get along the best we could—so we used what was called a drop line. A drop line in that case was about 200 to 250 feet long with a connection, you might say—they used to call it a bulldog—and that would fit into the "I" splice of the main line which had the bullrigging on it. The other one had the bullrigging the other end. So, when we cleaned up all the logs as far as the mainline would work, we then would hook on the drop line and go back another 200 to 250 feet.

Then we hauled these logs down so we could hook on to them with the mainline. In the meantime we would be back of there and we would be hooking on to some more logs and we'd drag those down. The process was to bring them as far as we could with the drop line and take them the rest of the way on the mainline. We'd get locations up on a hill where we could put in fore and afts, pull the logs there, and then we'd bunt them for maybe 200 feet. So we were logging in those days, you might say, 1500 feet back from the lake. Of course there was nothing spectacular in the work except that we were getting the logs and making a little money on them, until such time as we could buy more equipment.

The superintendent quit the company in 1913 and I took over as superintendent of the outfit. Everything was ground yarding in those days—there was no such thing as high lead. The highest lead we had was when we put the bull block over the top of a "hot cut" stump. It was cut off high, and coming into the landings it would have the lift.

In 1915, when the camps opened up after the First World War started, we then went to the high leads, and that was a disappointment to the hooktender who, in those days, was what you might call the bull of the woods. He felt that his position in camp had become secondary, and in my own experience they bucked the high lead until you just had to clean house with the whole crew of your old-timers and get the younger fellows in.

The high lead took away the necessity of having good chokermen and rigging slingers and the hooktender, because he used to depend on his roads to come in dodging stumps and windfalls and obstructions that might be in his yarding road. And you had to know how to put the choker on for a roll on a log, or on an undercut on a log, and the high lead took that out of the picture. You used to have in those ground-hauling days a sniper who rounded the end of the log. That part of the operation was thrown out.

Then we stopped building landings. Previous to high lead we had what were called "jump-up landings" or "roll landings," where the log would come in and they would load them from there. When they put in the high lead we would just drop the log and they could get a hold of it with the tongs. Under the old system of ground lead we used to have end hooks that raised the log and put it on. When we went to the high lead, at first we worked with single tongs. Then we got in the duplex system. The duplex system took away the necessity of having levermen that could almost dictate the policy of the loading, so that there was a transition in there that disappointed many men because they felt that their importance to the camp was shrinking.

As I remember it, the first high lead that was used was down in Oregon. I think Myrtle Point was the first high lead in British Columbia, and that was put in there after we opened up after the winter shutdown, in the spring of 1915. We stopped building landings after the fall of 1914.

Sidney G. Smith, born in Michigan in 1882. He started working for Bloedel, Stewart & Welch in 1911. For many years he was in charge of timber operations for BS&W and retired as vice-president of MacMillan and Bloedel.

A Shay wood burner at Lamb Lumber operation at Wolfssohn (Lang) Bay in 1916. The camp and all its equipment had been purchased from the Vancouver Timber & Trading Company the previous year. Left to right: John Reeves; Henry Hall, fireman; Toy Toll, engineer; Bob Grieves, brakeman. (Leonard Frank photo; CRM 60133)

Most of them consisted of two or three miles of track and were short-lived operations, closing down or moving elsewhere when the timber ran out. They included camps on West Redonda, Hernando and Texada islands, and several camps up Jervis Inlet. One of these small companies, Lamb Lumber, established itself in Wolfssohn Bay in 1915 when it bought out the Vancouver Timber and Trading Company's assets, including timber, four miles of track, a locomotive and other logging equipment. Four years later, when the timber was gone, everything was loaded on barges and towed across the Gulf to Menzies Bay, where the Lambs commenced operations for another twenty years.

On the east coast of Vancouver Island, logging operations were established all the way from Victoria to the northern limits of the Douglas fir zone north of Sayward. After the E&N land grant was awarded, most of the logging companies south of Campbell River were large-scale shows, attracted by the availability of high-value timber that could be purchased outright. In the Chemainus–Cowichan area, there were three major companies by the beginning of the war: VL&MC, the Cowichan Lake Lumber Company, backed by Ontario money, and the Empire Lumber Company, financed from Philadelphia. The E&N extended a branch line to Cowichan Lake in 1913, and in 1925 the Canadian National Railway reached Youbou on the north shore of the lake. These lines provided reliable log transportation to the various mills operating and planned on the salt water, and hauled lumber from the mills on the lake to deep sea docks on the coast.

For the first decade of the century, VL&MC also ran some small camps on its land around Courtenay. From a log dump just south of Union Bay the company built a railway line 4.5 miles northwest. In 1903, the legendary timber cruiser and logger Mike King logged along the Tsolum River for VL&MC, using one of the first steam donkeys in the district to yard logs into the river. A dam on Wolf Lake was used to flush them down to the bay, but the twisted river course presented many obstacles to a successful river drive.

Across the island, in the Alberni Valley, the forest industry was coming to life again after a twenty-five-year hiatus following closure of the Anderson mill. A short-lived paper mill on the Somass River, which used rags imported from England and the U.S. instead of wood pulp, had

Comox Logging, at a site near Black Creek in 1912, loading onto flatcars with a single-tong system. The wood-fired steam yarder used here was one of the first skyline yarding skidders in BC. (CDM P200–1001d)

made little mark. Through the 1890s and the early part of this century, a series of small mills cut lumber for local use; the small amount of timber they used was likely obtained from land clearing in the valley.

After the E&N brought a branch line into the valley, the first mill of any significance was built along the railway at Bainbridge in 1916. The Bainbridge sawmill was established, in part, to fill an order for unusually long timbers—up to 120 feet—needed to repair the locks in Ontario's Welland Canal. Logging began around the mill's log pond, using a ground-lead donkey, a slow process because of the length of the logs. As the Bainbridge logging crews moved farther out they added a high-lead yarder, swinging the logs from it to the pond by ground lead. In time a railway was added, along with a two-thousand-foot incline used to lower logs down Baldy Mountain.

By October 27, 1927, the big trees needed to produce the long timbers were running out. The logging crew was working back of the incline with a yarder and an Empire donkey swinging from the yarder and loading incline cars as well. The donkey engineer was pulling in a particularly large log at the same time as the leverman was attempting to load another big log. One of the guylines on the spar tree broke under the strain, and the tree came down, killing the fireman. All work ceased immediately and the mill never reopened.

Meanwhile, just beyond the top end of the Alberni Valley at Comox, the manipulations of remote money managers had launched one of the major logging operations of the era. Years earlier, in 1889, Quebec lumbermen had built a sawmill at Barnet on the Fraser River. After passing through American hands it was reconstituted in 1910 as the Canadian Western Lumber Company, owned in part by the ubiquitous promoters of just about anything that would turn a dollar—William Mackenzie and Donald Mann. The Canadian Northern Railway, which ran out of steam after reaching Cowichan Lake and had to be taken over by the federal government, was a scheme of theirs to run a line up the centre of Vancouver Island, from Cowichan Lake through the Nitinat and Franklin river country, up the Alberni Valley to Comox, and then north, island hopping to—it was hoped—Prince Rupert.

Canadian Western Lumber, through its logging subsidiary Comox Logging and Railway,

One of several Shay locies owned by International Timber Company operating near Campbell River in 1916. (BCFM 4–5)

owned large stands of E&N-granted Douglas fir lands near Chemainus and at the bottom end of the great fir forest running north from Union Bay to Sayward. In a very short time Comox Logging became one of the biggest logging operations in the world, in a league with Weyerhaeuser in the U.S. A year after it began, 500,000 feet of logs a day were going into the water at Royston, hauled over rail lines extending up several drainages behind the main camp at Headquarters, north of Courtenay.

Even at this very early stage, when most other companies were just learning how to use the small ground-lead yarders, and horses and oxen were still in use throughout the coast, Comox Logging was using the latest in big, powerful skyline machines just then coming into vogue in Washington and Oregon. This was the first of the big-time, highball logging shows—all under the direction of superintendent Bob Filberg.

To the north, another substantial logging company was started by Charles Cobb and Nick Haley in 1906 on 30,000 acres of the northernmost E&N lands. International Timber began with a camp in what is now downtown Campbell River, and a railway running south from the dump on the Campbell River estuary to Oyster Bay. In 1914, a slow market year, IT was putting 275,000 feet a day into the water, hauled down fourteen miles of railway extending into some of the best railway logging country in the world. Lagging a little behind Comox Logging in yarding equipment, it boasted one of the best camps in the world, according to a July 1914, report in the *Western Lumberman*:

> Some of the conveniences are white enamelled beds, with springs, electric lights in all the buildings, as well as street lights. The company also runs a store, a reading room, pool room, and bathroom for the benefit of all the employees. The bunk houses have stoves. Fires are started about 30 minutes before the gong sounds every morning, so the men have a good warm room to dress in. During the rainy season the fires are kept going all day, so that when the men come in wet they will feel comfortable. The company assembles more men in one camp than any known company in the same business in British Columbia, the dining-room having a seating capacity of 225 men at one sitting.

An early small logger in Campbell River was Sigurd Hage, who came from Seattle in 1909 to contract log for the Cameron Lumber Company. Starting with horses and graduating to steam donkeys, he logged much of Campbell River's residential area from a fifty-man camp at the present site of the RCMP station. Hage had two elevated log chutes over the Island Highway, with a roader donkey on the beach to drag the logs into the water. As he worked farther back from the beach along the International Timber railway, he contracted to haul logs on the company line. In 1922 Hage moved to Coquitlam and set up a small railway show with a high-lead yarding system to log the Coquitlam highlands. He briefly operated the old Rat Portage camp at Twenty Mile Creek on Harrison Lake in 1924, earning a reputation as one of the best, most innovative steep-country loggers on the coast.

Logging also began on the west coast of Vancouver Island in the early 1900s, first in the south along the relatively protected shore in the lee of Cape Flattery. In 1906 a group of Nebraska businessmen established the British Canadian Lumber Company at Port Renfrew. By 1909 they were logging down seven miles of railway. Their biggest problems began after the logs went into the water and had to be towed to the Cameron Lumber Company's sawmill in Victoria. The exposed waters of Juan de Fuca Strait were not a very secure place to tow the flat booms used at that time. In March 1909, a million-foot boom of fir was lost.

Out in the woods they'd have to burn wood in the boiler. There'd be a wood yard where you'd buck the wood and split it. There's a wood bucker, and a wood splitter, and a fireman. He'd look after the fire. He'd have a poker, is all, to stir up the fire. They'd have a damper in the stack of the boiler. The engineer opened the damper. And the fireman, before he opened the firebox, he'd always take the iron poker and tap on the side of the boiler so you'd know he was going to put the wood in. Otherwise, if the damper was closed, and you opened up the throttle, the steam from the damper would come down through the tubes and blow out the door, and burn the whiskers off the fireman.

Tom Hall, longtime Quadra Island resident and early steam logger. He was born in 1901, started punking whistles as a teenager, worked hand falling and got his steam donkey ticket in 1930. He began truck logging in the 1930s. He also mined and commercial fished before retiring to Gowland Harbour.

G.G. Davis, the camp's logging superintendent, devised a new method of rafting logs first used in 1911. At the company's booming grounds at the mouth of the Gordon River, a line of piles was driven about 75 feet out from, and parallel to, the dump wharf. A long boom with a small steam donkey lifted the logs, one at a time, from the rail cars into the water. A flat boom 120 feet long and 70 feet wide was built with two single sidesticks. Several lengths of 1¼-inch galvanized wire rope were tied to the sidesticks and woven through the logs, under one and over the next, to create a large, flexible woven mat of logs. More logs were stacked on top of this mat until the sidesticks were about to sink. Then more wire rope was laid over top of the logs and attached to the sidesticks, using the donkey to cinch them as tight as possible. The resulting raft, containing 1–1.5 million feet (depending on the species), would draw 15 to 20 feet of water and ride about the same height above the water line. A big tug could tow it at about three miles an hour.

The Davis raft was an immediate success and was able to withstand storms in the open Pacific. One of them was lost at sea and sighted several weeks later, 3,000 miles away in the north Pacific, still intact. This simple device had a profound influence on coastal logging: it opened up the entire coast to the loggers, and played a key role in two distinct developments in logging.

One was the crash program to log Sitka spruce on the Queen Charlotte Islands during World War One. The British had discovered that Sitka spruce was the only species of wood suitable for building aircraft frames, and BC was the only place in the British Empire where it grew. The best stands were on the Charlottes, a remote area of the coast virtually untouched by loggers because of the difficulty in moving logs across Hecate Strait. In 1917 the federal government assigned Major Austin Taylor the task of mounting the biggest logging operation in history, with H.R. MacMillan, BC's chief forester, seconded as his assistant. Although a much smaller spruce operation was set up in Quatsino Sound on northern Vancouver Island, the main assault was farther north.

Most of the best spruce stands had already been claimed, so an act was passed by the province allowing the confiscation of cutting rights if owners would not accept the prices offered. The logging was contracted out to a small army of loggers with a production target of 3 million feet a month. Two collection camps were established, one in Masset Inlet on Graham Island and the other at Thurston Harbour on Moresby Island.

A three-pronged approach to logging was

Cables are woven through logs to form a mat for the base of a Davis raft at an unidentified location, c. 1920. (Link & Pin Museum)

Logs are loaded onto the mat. When loading is completed, more cable will be tied over top. These rafts were built at several locations on the west coast of Vancouver Island, in the Queen Charlotte Islands and on the North Coast before the development of log barges. (Link & Pin Museum)

taken. About 125 handlogging contracts were issued to small-scale operators who combed the inlets for suitable timber they could get into the water without machinery. Another 40 riving contracts were let. Riving entailed bucking trees into fourteen-foot logs and hand-splitting clear spruce segments, which were easier to move to the beach where they were loaded on barges. The remainder were contracts let to a number of larger-scale loggers who set up camps and ground-lead shows employing, in all, about 100 steam donkeys. The Masset Timber Company had fourteen camps in Masset Inlet, employing about 800 loggers, and another six independent operators, including J.R. Morgan, started up.

Some of these logs were milled at a handful of small sawmills already in existence, and Masset Timber opened a larger mill at Buckley Bay just before the war ended. The lumber was barged to Prince Rupert and sent east by rail for shipment to England. The logs were made into Davis rafts and towed to Vancouver by a large fleet of tugs acquired by the government. The rived cants were barged to a cut-up plant in Vancouver, which milled up to 1.5 million feet a day.

In the final months of the war, airplane spruce production reached almost 7 million feet a month, with another 1.4 million feet of airplane-grade fir being produced as well. About 60 million feet of logs were towed to Vancouver in Davis rafts, with losses of only two percent, according to C.A. Kohlman, who had developed the rafting system with Davis at Port Renfrew, and superintended the spruce effort.

Davis rafts also played a significant role in the early evolution of pulpwood logging. In the early years of the twentieth century there were no pulp mills on the BC coast and, therefore, no market for the huge volumes of hemlock, balsam and lower-grade fir logs passed over by loggers. In 1901 a group of influential BC financiers persuaded the government to amend the Land Act to grant pulp leases. Several companies were formed and granted leases on almost 400,000 acres of timber. None of the original companies ever built a mill; all ran into financial troubles and were granted extensions by the McBride government. When the original recipients of the leases began to sell them, there was an immediate protest from coastal loggers, represented by the BC Loggers' Association. They claimed, backed by timber cruise reports they had commissioned, that the pulp leases contained very little wood of pulp quality, but did have large volumes of saw logs. The government, anxious to attract investment to the province, ignored them.

Eventually, however, several pulp mills did get built. The first to open was at Swanson Bay in 1909. Later that year the second mill opened at Port Mellon, on Howe Sound. In 1912 mills opened at Powell River, Woodfibre and Ocean Falls, followed by the mills at Port Alice in 1917 and Beaver Cove in 1919. These mills, and others like them in Washington State, created a ready market for timber the loggers had until then been bypassing. But, because of the long tows involved, the heavy weight of the hemlock logs, and the rough waters that had to be traversed, towing pulp logs to these mills was a risky proposition before the development of the raft.

Early on, spruce was rafted from the Queen Charlotte camps across Hecate Strait to the Swanson Bay mill. The Powell River Company towed hemlock rafts from Kingcome Inlet and the Charlottes, and the Barnard Logging Company regularly towed rafts from Bute Inlet to the Washington Paper Company mill at Port Angeles. The coincidental development of pulp mills and Davis rafts had a profound effect on the logging industry in several areas.

Logging for the mill at Ocean Falls began four years before start-up, when company logging crews cleared the millsite. Later there were two components of log supply. The company opened a series of camps throughout the area, yarding with steam donkeys, first by ground lead, then switching to high lead. At some point,

In the meantime logging operations had been started. As early as 1908, handlogging operations had been carried on chiefly to obtain piling for wharves and other construction work. In the spring of 1910 the first logging camp was started on South Bentinck Arm. Two small donkeys together with lumber and entire equipment and camp outfit were loaded on a scow which was towed to its destination by a small gasoline launch named the *Pulp*, the only power boat then owned by the company. The whole camp crew of about thirty men also took passage on the same scow.

Next spring Camp No. 2 was established across the inlet from Camp No. 1. In 1912 Camps No. 3, 4 and 5 were started, so that year five camps were operating, using thirteen donkeys. In addition to these main logging camps, handlogging was carried on quite extensively, the amount of logs derived from this latter source almost equalling the combined output of the large camps.

Mark Smaby, who helped establish the Ocean Falls pulp mill. Later he had a general logging contract and sublet hand logging contracts until 1927 when he retired to the US.

probably about 1920, the company closed its camps in favour of contractor-run camps. There was some early development of A-framing in this area, and as the timber that could be reached by this method was logged, cold deckers yarded themselves up the hills and into the more remote stands. Very little of this country contained flat enough ground or large enough stands to warrant the use of railways.

Almost half the logging for the Ocean Falls mill was done by a large, colourful assortment of handloggers who worked the channels and inlets and along the shores of Link Lake, the mill's power and water source. The handloggers were known throughout the country as Smaby's Lambs, in reference to the general handlogging contractor, Mark Smaby, who sublet contracts to others. Since there was no other market for logs within reach of these small operators, and their output was too small for the construction of Davis rafts, they were stuck with Smaby's prices. He became a legendary character, viewed by some as little better than the man-catchers who prowled the streets of Gastown shanghaiing loggers for the big railway camps.

It was out of the ranks of these handloggers that many of the larger contractors emerged, slowly acquiring the equipment to become A-framers—all under Smaby's watchful eye. Some, like G.H. "Doc" Gildersleeve, built substantial operations and established the first truck logging camps on the mid-coast. Others plotted their escape. The only alternative for a contractor was to load his equipment onto his floats or a log boom and tow the whole camp south, across Queen Charlotte Sound to the Jungles, where small-market loggers could get and sell timber. One of the few who succeeded was Oscar Soderman, who relocated to Blackfish Sound in the 1930s.

In the first decades of the twentieth century, logging on the North Coast was a relatively insignificant concern. The most valuable timber was east of the Coast Mountains, in the Interior, and until the expansion of the pulp and paper industry after World War Two, most North Coast logging was for local consumption. An exception, which itself was primarily an Interior-oriented operation, was Olaf "Tie" Hanson's company. Hanson was an Alberta farmer who spent his winters logging and hand-hewing ties for the Grand Trunk Pacific railway, which was then working its way to Prince Rupert. Realizing there was more money in timber than farming, Hanson struck off ahead of the railway builders to locate and secure timber stands suitable for ties. He soon dominated the business, and by the time the railway was completed he had a thousand men working for him, most of them cutting ties from hemlock stands around Terrace. After the line was completed Hanson began cutting replacement ties from interior pine and cedar poles along the interior portion of the Skeena River, all of which were shipped east by rail. Hanson went on to establish several other businesses in northwestern BC and was elected Member of Parliament for three terms.

I think it is generally conceded that Douglas fir weighs in the log on an average of eight pounds to the foot, making a weight of four tons to the thousand feet; so a log scaling 5000 feet would weigh twenty tons. If a flume is of any great length it is essential that little of this weight be allowed to come in direct contact with the surface of the flume. The logs must at all times be almost entirely supported by the water in the flume.

The flume my company built at Clowholm Falls [head of Salmon Inlet] is 935 feet long. The grade varies from zero to eight percent. The cross section of the flume shows it to measure six feet wide on the bottom with a six-foot wall. It measures nine feet across the top. The intake is eleven feet wide. In each corner is placed a fore-and-aft log, averaging eighteen inches in diameter. This keeps the large logs from rubbing on the bottom.

At the end of the flume the ground dropped off to about sixty percent to the salt water. An ordinary log chute was built here. We placed our bents eight feet apart and used twelve-by-twelve-by-fourteen caps with eight-by-eight posts and braces, with four-by-twelve bottom linings and three-by-twelve side linings, battened with two-by-fours. If ordinary pile bents could be used I would estimate a flume could be built for about $10 a running foot. The cost was high in our case as we had to haul everything in over a very bad road and tight-line it across a river twice. Where the flume was built in or near the bed of the river we used concrete footings, and on account of unusual high water had a lot of expensive delays putting these in.

Three men can put over 800,000 feet down in eight hours at a cost of operation of about two cents per thousand.

A.P. Allison, founder of Allison Logging. He was raised at Chemainus and began logging in 1905. When he died in 1947, he was president of the Truck Loggers' Association.

Loading with ends hooks off a gin pole at BC Mills Timber & Trading's Rock Bay camp, c. 1918. These hazardous loading devices, which often broke loose, were eventually replaced with tongs. (VPL 5809)

Two small saddle-tank rod engines used by BC Mills Timber & Trading at its Rock Bay operation, 1919. (VPL 5798)

At the southern end of the coast, logging moved up the Fraser Valley, mostly to clear land for farming. By the turn of the century, loggers, as opposed to land clearers, were beginning to make serious inroads up the valleys feeding into the Fraser. By 1910, both the BC Electric and Great Northern Railways had laid track on the south side of the Fraser, almost to Chilliwack. A number of small mills were built along these lines, while the earlier mills along the river continued to cut logs towed from distant sources. Logging and milling throughout much of the 1890–1920 period was dominated by the Trethewey family, which built the Abbotsford Lumber Company mill in 1909 and logged most of the lands surrounding the town. Numerous other small mills were scattered throughout the area and small logging operations, most using horse or oxen power, set up to feed them.

Developments around Chilliwack lagged a few years behind Abbotsford. Beginning in the 1890s, a large number of small lumber and shingle mills drew their timber supply from the lower slopes of Mount Cheam, and the bottomlands of the Chilliwack and Columbia valleys. In this area horses, and even the occasional ox team, were used well into the twentieth century. One technique used in the Popkum–Rosedale district was to load rail cars, or sets of rail trucks, and let them coast down the gentle grade to the mill. Horses hauled the cars back up the hill. The Graham & McNair Company, which built a mill near Rosedale in 1915, took this system one step further and erected a boiler and steam engine on a flatcar, using chains to drive the wheels. In 1919 this company was the first in the area to put in a high-lead system, and it was one of the first outfits in the Rosedale area to use trucks when it purchased a 3.5-ton Federal in about 1923.

Developments on the north side of the river took a different turn. A narrow band of easily accessible land between the river and the steep mountains provided insufficient timber for the early growth of a logging industry. But with the advent of steam power, several big logging shows were established up the side drainages. In the area between Haney and Mission, three big railway logging camps were operating by 1915.

The first was developed by the Heaps Timber Company, an offshoot of E.H. Heaps and Company, probably Vancouver's biggest and most progressive machinery manufacturing and distributing firm. When the Heaps company took over the assets of the Ruskin communal sawmill in 1899, it found itself in the logging business. Having access to the most up-to-date equipment, Heaps replaced the horse logging show with a ground-lead steam donkey yarding setup, feeding into a railway running north of Ruskin, up Kanaka Creek. It lasted until 1913, then folded.

At Mission an earlier, small-scale operation was acquired by a logger from Grays Harbor named Henry James, known throughout the coast as Jesse, after one of his ancestors—the notorious U.S. bank robber. In 1915, James set up Keystone Logging, with a railway extending from a dump on the Stave River in behind Mission. Over the next six or seven years James established a reputation as the hardest-working, fastest-living logger in the Pacific Northwest—a role model for successive generations of coastal loggers. By 1921 his timber was gone and, taking the smaller of his two Climax locomotives, a steam donkey and other equipment, he moved to Cowichan Lake and set up a camp at Cottonwood Creek to log in association with established mill owner C.C. Yount.

In 1915, on the other side of the Stave River, the biggest railway company in the area went into business. Abernethy Lougheed used the Western Canada Power Company's chartered railway, running from Stave Lake around the falls and six miles to the dump on the lower Stave River. In the same year the company started A-framing on the lake, and built shingle and sawmills there as well. The lake operations were wound down in the early 1920s, and Abernethy Lougheed went on to establish one of the biggest railway shows in the province between Pitt and Alouette lakes.

After the false start with the mill at Port Douglas in 1858, developments in the Harrison area progressed slowly until the early 1900s when the Tretheweys took over the mill at the junction of the Harrison and Fraser rivers. The Rat Portage Lumber Company took over the mill a few years later and increased the cut to 150,000 feet a day, creating an enormous demand for logs. A shingle mill across the river, the largest in BC, contributed to that demand. The result was a number of attempts to establish logging operations up the Chehalis River and on Harrison Lake, with initial efforts concentrated on oxen logging on the Harrison River flats.

In 1900 Loathium Logging used steam donkeys to yard in the Chehalis Valley, with oxen skidding logs to the river. But the log drive down the Chehalis was a failure and the company folded. Another pioneer steam logger in the area was Bill Hanna, who took a vertical spool rig up Harrison Lake in 1904.

The first company to log the lake seriously, with a view to towing the timber out and milling it down the Fraser, was Brooks, Scanlon & O'Brien, which was establishing a railway camp at Stillwater at the same time. In 1910 the company used steam yarders to log into the Lillooet River, at the head of the lake, and built a 124-foot sternwheel tug to tow logs to New Westminster. The Rat Portage mill brought another sternwheeler to the lake around 1920 and operated it until 1924, when the company's logging camp was sold to Sig Hage.

Yale is usually considered to be on the border between the coast and interior forests. The growth of the logging industry in the lower section of the Fraser Canyon, below Yale, occurred in two phases. The first was the very early

They're a peculiar class of people, loggers, generally speaking. I used to be one; of course, I didn't stay long at it, thank goodness. But in the early days these people would walk from Vancouver, Westminster, wherever it might be, and they'd go clear across to Puget Sound and that country, follow the logging camps from one place to the other. They lived in the logging camps, they worked in the logging camps, they knew nothing but the logging camps, and they would gamble everything that they had. That was their amusement, gambling. They would gamble in there, and I have known men that didn't get down to see a city for three solid years because he never had the price to pay his fare. He didn't have money to pay his fare because he lost it.

I've seen men, many a time, just sign over their cheque in a poker game. They were always fed, but they didn't have anything else. They couldn't buy a thing. Only through the quartermaster stores that they had there you could buy anything you wanted, pretty well, a commissary they called it. You would buy what you wanted and they would put it against your bill. At the end of the month you just got your cheque minus that. So they were not in debt, but they didn't have anything.

There was one young man came to me one day, and he said "Drinky, I'm going downtown." I said, "You're going downtown; what's the matter with you? Can't you work any more?" He says, "It's over three years since I was downtown but I made good last night." I says, "You did? What did you do last night?" He said, "I made $900 last night." Now, it had been three years. He'd lost three years' pay, but he got $900 this particular night and he just went and he pulled the pin and went down to Vancouver. He said, "I'm going to have a holiday."

Albert Drinkwater

development which began during the gold rush and culminated with the construction and early operation of the CPR. The steep terrain of the canyon and its side drainages retarded the development of major logging operations until the advent of good trucks in the 1950s. But a considerable amount of timber was logged during the 1880s to build the railway and to supply trains with fuel during the early wood-burning years of locomotives. Most of this logging was done with oxen, skidding logs down the creek valleys flowing into the canyon.

The second phase occurred in the early 1900s to supply the small mills that were cutting lumber on both sides of the river for the local market, mostly around and below Hope. But even here the steep ground delayed logging until the easier-to-get timber of the upper Fraser Valley and Harrison Lake areas was depleted, and more modern equipment was available.

One other area on the southern mainland coast that experienced a major prewar development was the Squamish Valley. The Howe Sound, Pemberton Valley and Northern Railway, built independently of logging interests, was completed in 1910. It hauled timber for several logging camps, including the Squamish Timber Company, which in the same year was shipping 60,000 feet of logs a day to tidewater. In 1912 this railway was incorporated into the Pacific Great Eastern Railway, which was then in the first stages of its long, drawn-out journey to Prince George.

In the brief span of about twenty-five years the industrial order of the BC coast went through a radical transformation. It became industrialized, in much the same way other parts of the developed world abandoned the old ways of doing things and adopted new, mechanized means. What occurred was a technological explosion, and it was manifested most graphically in the development of logging railways. In the 1890s there were only a few primitive locomotives used to move timber, most of them working in conjunction with oxen or horses. Steam-powered equipment was still the exception, animal power the rule. When the twentieth century dawned there were probably less than a dozen miles of logging railway on the BC coast. Yet by 1912 there were 227 miles of track, built by twenty-two logging companies. In 1919 *The Timberman* magazine published a survey of the western logging industry:

	Rail mileage	Locomotives	Donkeys	Loggers
Washington	2,000	430	1,700	27,500
Oregon	825	162	825	12,500
BC	370	75	390	7,800

The mechanization of logging on the BC coast was not a sequential process. There is a conventional view of logging history which suggests that first there were oxen and horses, which were replaced by ground-lead steam donkeys, which in turn were supplanted by high-lead, then skyline yarding systems. Similarly, trains replaced the animals on the skid roads and were themselves replaced by trucks. In fact, by the early 1920s, all of this equipment was in use at the same time.

By the end of World War One, many loggers still thought only about getting a little two-drum steam roader to replace the horses hauling down their skid roads. Even some of the bigger camps, such as Hastings' Rock Bay camp, were still ground leading. But at the same time, Comox Logging had been working for years with state-of-the-art Lidgerwood skidders that had been used by Louisiana loggers for many years before they came to the Northwest.

Likewise, in 1918, there were BC dealers

In 1909 this Baldwin locomotive was used by the Comox and Campbell Lake Tramway Company, which later became Comox Logging, on its extensive rail network near Courtenay. (Leonard Frank photo; VPL 6044)

selling Republic, Garford, Duplex, Federal and Little Giant trucks, with many in use at various camps around the coast. By the end of the war, there were 150 three-and-a-half-ton Standard trucks working in Washington state spruce logging camps alone, leading to complaints in BC that these models were hard to get.

Once the mechanization of logging began, it happened very quickly. As markets developed for northwest timber, manufacturers began adapting machinery from other parts of North America to logging conditions in the Northwest. So, rather soon after the first yarders were in use, rapid advances in the development and manufacturing of enormously more sophisticated and powerful yarders followed. Internal combustion engines did not merely succeed steam engines; to some extent they developed along with them. It was market factors more than anything that retarded adoption of the new equipment flooding onto the market between 1900 and 1920.

At the end of World War One the coastal logging industry was poised on the verge of dramatic growth. The extent and location of the timber resource was known. Much of the required infrastructure, such as a coastal steamship service, was in place. The equipment was available and the machinery companies needed to repair and improve it were open for business. A sizable workforce of skilled loggers was either at work or waiting to be hired out of the bars and brothels of Gastown.

The so-called glory days of logging were just around the corner.

IV Yarding Them Out

Previous page: One of the Lidgerwood steel spar skidders used at Franklin River between 1935 and 1955. (VPL 6119)

The end of World War One plunged most of North America into a severe but short-lived recession. In the logging camps of coastal BC, this slump passed unnoticed, primarily because of a rapid expansion of lumber sales to Pacific Rim countries, particularly Australia, China and Japan. Tied in with this export boom was a huge expansion of shipbuilding activity in BC—some of it for deep-sea lumber carriers, the rest for vessels in the coastwide trade—which created a big demand for premium-grade Douglas fir.

That A-frame we had at Sproat Lake would burn eight cord of wood a day. That was a big machine. It held 2000 feet of inch-and-a-quarter line on the drum. Big machine, fast. That thing was fast. I can always remember there when the hooktender and the rigging slinger and Box Car Pete, they were all sitting side by side on the log. They'd set their turn. I was on the A-frame, I'd done some working on the A-frame, and they were sitting over on the side. I guess they had a hang-up there and they skinned the rigging back. That thing was so fast when it come back on them, it up-ended that log, it swung the line way out and it just picked that Box Car Pete right up out of the three guys and killed him dead. That night when old Bob Swanson asked me what I had for the book today, well I said, we just finished off Box Car Pete today. He wrote a poem about that.

Swanson's day and mine were right on the tail end of the steam. Steam was just going out then, eh? They were going out just in my time. I was sorry to see them go, you know, they were a good thing to work on in the wintertime. You could come in there for lunch and go behind the boiler and you could get warm, if it was soaking wet out. Every day logging, it was interesting. There was never a dull moment.

Sam Telosky, who began logging in the early 1930s. He worked at various locations on Vancouver Island and the Queen Charlotte Islands until his retirement as logging supervisor at Elk River Timber in 1978.

Two new lumber marketing establishments vigorously pursued the export trade and quickly took most of the business of selling BC lumber abroad away from the San Francisco brokers who had dominated it. The first was a cartel of coast mills, the Association of Timber Exporters of BC, later called Seaboard Lumber Sales, organized in 1919. E.J. Palmer of VL&MC was the chief proponent of this group. The second was a company involving H.R. MacMillan, who had left his job as BC's chief forester in the early stages of the war and gone to work for the federal government promoting Canadian timber in Britain and Europe. In that capacity he dealt with the British government's wartime timber buyer, Montague Meyer, one of the most influential timber importers in the world. In 1916 MacMillan went to work for Palmer at the Chemainus mill, but they didn't get along and MacMillan left to assist in the aero spruce operation. In 1919 Meyer visited BC and formed a partnership with MacMillan in a timber exporting business, H.R. MacMillan Export, in competition with Seaboard.

Both firms began aggressive marketing campaigns and through the 1920s were successful in wresting business from the American brokers. The export trade grew steadily, matched by a growing domestic market. The stock market crash of 1929 was followed by a $4-per-thousand duty on Canadian lumber going into the US, however, and within two years coastal log production was halved. The British also imposed a tariff, but under the preferential terms granted Commonwealth countries, the British market opened up. By 1938 almost half the lumber milled in BC was going to Britain.

When war broke out again, the new trade patterns established for BC timber products were critical to the Allied military effort. Coastal log production increased to record levels in 1940 and remained firm throughout the war.

Working conditions in the woods went through a series of rapid changes following World War One. The growth of the big railway logging operations—some of which, like Comox Logging, had started before the war—introduced many changes into the working and living situation of loggers. Some of these changes were welcomed, others resisted. It was a turbulent period, regarded quite differently by the various participants in the industry.

The adoption of railways and, particularly, high lead and aerial yarding systems, accelerated the pace of work. The leisurely pace afforded by the use of horses and oxen had hardly increased at all in ground-yarding shows, but it was now a thing of the past. Yarding and loading machines ran faster and faster. There were always rail cars waiting to be loaded. Highball logging had arrived.

The new equipment that came into widespread use after the war had many profound and subtle effects on working conditions. It extended the logging season, making it a year-round operation and creating a need for a more stable work force. Likewise, the new technology required a higher degree of teamwork. Efficient logging crews took longer to meld into a trained unit, so it became desirable to keep a well-functioning crew together.

Parallelling the increases in production were rates of injury. Some companies seemed to view their workers as expendable and easily replaceable items of equipment. They encouraged competition between crews, often with disastrous results. In his memoirs, Harper Baikie recounts one of these competitions at Comox Logging when the managers provoked the crews to increase production from 390 to 750 logs a day.

Bloedel, Stewart & Welch acquired one of the worst records for accidents. During the 1930s, for example, the Menzies Bay camp offered a daily keg of beer to the crew loading out the most cars of logs. In one large, difficult setting known as the Glory Hole, as many as fifty men were said to have died. The corpses and the injured were transported to the Campbell River hospital in Carl Thulin's ambulance—a flat-deck truck. Loggers who succumbed were buried in a special cemetery, which in later years became the site of an apartment building. Loggers, it seemed, were not quite human.

The first accident compensation legislation, passed in 1891, allowed injured workers to sue their employers. The courts became clogged with claims, and in 1916 a new Workmen's Compensation Act established a government-run, compulsory premium, no-fault plan—the most advanced in Canada. It also eliminated the workers' right to sue the employer.

Probably the most dramatic postwar changes were in loggers' living conditions. In the previous era, workers crowded into bunkhouses holding up to fifty or sixty beds, stacked in three tiers. There were few or no windows, no ventilation and a central wood heater, over which hung the smelly, wet clothes of hard-working men. There were no mattresses or bedding provided for what was, in reality, a transient, migrant and brutalized work force. Owners and managers like E.J. Palmer considered themselves enlightened if they provided these primitive camps with saloons and whorehouses.

The need for more reliable, skilled, stable workers led to a number of significant changes. Comox Logging had built one of the most luxurious camps in BC before the war; it was followed by most of the other companies, large and small, after the war. Smaller, steam-heated, electrically lit bunkhouses quickly became the norm, with individual rooms for two to four men. Bedding was provided, and bullcooks did the cleaning. Dining rooms and meals improved.

From a broader social perspective, the most significant change was the establishment of durable logging communities, including family housing. Comox Logging encouraged its employees to buy company land and build houses. Others, such as Merrill-Ring-Wilson at Rock Bay, provided family quarters. Some of these camps—Englewood, for example—survived as permanent communities. Most of them lasted for decades before succumbing to forest

There were places that always had a bad reputation, they wanted you to know that they never missed a turn for anything. The old saying was always that if a man got killed at the back end, put him in a choker and send him in so they wouldn't waste time taking him in. And the boss raised hell because they didn't send a turn in with him. That was typical of those big outfits at those times. Body count didn't mean anything. I'm talking about bodies, not live men. That didn't mean anything in those days, you know, they could care less. There was lots of men to fill in, so they could care less. I don't say there weren't good outfits. There were, and they were really interested in the men, they didn't want them to get hurt. But there was all kinds of them that didn't give a damn about them at all.

Al Hendrickson

policies geared more toward corporate expansion than community stability.

It is likely that the most renowned settlement of this era was James Logging's Camp 2, established on Cowichan Lake in 1921. Its bunkhouse was a large, two-storey structure with thirty-two two-man rooms, eight hot showers, two recreation rooms and dry rooms—all electrically lit and steam heated. Mattresses, sheets and pillows were supplied for the beds, which were made daily by a bullcook. Workers were charged fifteen cents a day for accommodation. The spotless dining room served up meals with several choices of main dishes and desserts for another $1.20 a day. The well-stocked commissary sold merchandise at town prices or less.

James had other camps in the area and at one of them installed a saloon and whorehouse, both of which he frequented himself. As well as providing some of the best living accommodations and after-hours activities on the coast, James was famous, or infamous, for the hard-working, fast-living way of life he was more than willing to share with his employees, both in camp and on legendary forays to Victoria and Vancouver. On one of these expeditions, in 1926, he choked on a steak at a restaurant and died, depriving the coast of one of its most colourful and well-liked owner-operators.

A debate about the source of improved living

The contrast in loggers' living accommodations is apparent in these two photos, taken about the same time in camps a few miles apart. In the Powell River Company's 120-man camp at Kingcome Inlet in 1916 (above), four men shared a steam-heated room equipped with lights, desks and windows. Work clothes were kept in a separate drying room. Nearby, at the Clark & Lyford camp on Broughton Island in 1917 (right), forty men shared a bunkhouse with only one wood-burning heater and no windows or ventilation. Even this camp was thought to have good bunkhouses, compared to many that existed at the time. (John Cress photo; author's collection)

Roy Olson, who was the superintendent, had gone out to see where the fallers had fell some trees on a bridge out on the mainline. He grabbed a peavey to roll something off and the peavey came out and he went over the bridge, and got killed. So then I jumped up to general foreman. Harold Bronson, who was general foreman, he went to super. We didn't have managers; supers and general foremen. And then I stayed in management. In those days you practically had to run everything because you didn't carry spare men. If the locomotive engineer didn't show up, the foreman ran the locie that day. Lots of times, if the engineer on the skidder was sick, I had to go out and run the skidder all day. You had to know how to do everything, pretty well, because you filled in the key jobs. The union would take a dim view of that today, for the superintendent to go out to run the logging truck or something for the day.

Jack Bell

and working conditions in logging camps between the wars runs through the historical and contemporary accounts of the era. There are those who maintain that the new-style camps, the first attempts to use safe working procedures, and other changes that improved the lot of the working logger were initiated by far-seeing, intelligent and compassionate company owners or managers whose struggle for profits was moderated by respect for their employees. There is little doubt this was true in many cases; Bob Filberg, Dwight Merrill, P.B. Anderson and others come to mind. But it is also true that in some cases these changes were wrung out of reluctant and hostile owners by the efforts of angry workers who organized themselves into a powerful union to fight on their behalf. This group and its supporters interpret the improvements as attempts to manipulate the working class into a docile flock of obedient, more subtly exploited workers. And there is no shortage of evidence that this attitude existed. Hank Phelan at Campbell River Logging appears to have earned his notorious reputation, as did Tom Lamb of Lamb Lumber. It is no coincidence that possibly the best logging novel ever published, *Timber*, was written by a Campbell River magistrate, Roderick Haig-Brown, who was ideally positioned to obtain a clear and compassionate view of the loggers' living and working conditions.

The organization of labour in the woods began well before the war with the formation of the Industrial Workers of the World, the Wobblies, in Chicago in 1905. The Wobblies spread across the continent, particularly in the US, and within a couple of years had established themselves firmly in northwestern logging camps. As much a militant political movement as a labour union, the IWW was resisted strenuously by employers and politicians. During the 1915 shingle weavers' strike in Everett, Washington, an armed mob fired on a shipload of Wobbly supporters from Seattle, killing several and wounding many more. Four years later another gun battle between Wobblies and local authorities occurred at Centralia. IWW members and sympathizers scattered throughout the Northwest, with more than a few finding refuge in camps along the BC coast.

By the end of the war, a previously unexpressed militancy was evident among BC loggers and millworkers. Some of it came from south of the border, some from returning veterans and more from the ranks of immigrant workers fleeing war-torn Europe, with news of the Russian revolution. In 1918 the BC Loggers Union was organized, with Ernest Winch as its secretary. The next year the loggers were joined by millworkers, creating the 15,000-member Lumber Workers' Industrial Union. It organized several strikes and walkouts with no great success, and by 1926 had all but disappeared.

On a winter day in 1928 a small group of loggers, all former LWIU members, met in a Hastings Street rooming house in Vancouver to revive the Lumber Workers' Union. At this time unions had no legal status. There was no legislated certification process whereby workers in a mill or camp were free to form a bargaining unit

The Capilano Timber location close to Vancouver, in 1917, was unfavourable. The One Big Union in Canada and the Industrial Workers of the World were just becoming powerful. We hired from Vancouver, and the men were close enough to attend all the meetings. Our strikes were continuous. The grievance committees would come in and Dave Scott, our timekeeper, would listen to them. In the meantime I would be writing out their cheques. When they were through we would hand them their cheques, and tell them they were fired. That method cannot be used today.

Lloyd Rodgers, who grew up in Vancouver and began logging in 1915. He was a partner in H&R Logging at Harrison Lake and owned the Consolidated Timber Company and Spring Creek Logging. He managed Aero Timber during the war, and later was logging manager for the Powell River Company.

that was accepted by management. The new union focussed on recruiting members in the bigger camps, who responded by organizing themselves into the BC Loggers' Association. Some camps hired thugs to keep union organizers out, and physical confrontations increased. Some companies fired union sympathizers, who found their names on the notorious "blacklist." Opposition to the union and its aims was not universal. Smaller employers tended to accept the union and negotiated acceptable wages and living and working conditions. Lots of blacklisted workers found jobs in the smaller operations. But many of the larger camps run by hired managers and superintendents took a hard-line position.

By the Christmas shutdown in 1933, feelings were running high on both sides, and Vancouver was a hotbed of militant talk during that holiday season. On January 26, a minor dispute at the Bloedel, Stewart & Welch Menzies Bay woods camp led to the firing of 64 loggers. By breakfast all 500 loggers in the four BS&W camps in the area had gathered at the beach camp to present their demands. These were rejected, and the men walked out.

In earlier strikes, when the workers walked off the job they usually headed to Vancouver to announce their grievances to the world from the barroom tables of Gastown. When they were broke, they found another job and the strike died. This time the union was prepared. The Bloedel strikers moved into a picket camp in Campbell River and began organizing. There were several large railway camps in the immediate area—Lamb Lumber, International Timber, Campbell River Timber, Comox Logging—and soon their workers had walked out too. The union launched a massive publicity campaign which reaped a harvest of public support, including that of the Vancouver City Council and many local businesses. Donated food was trucked in from Fraser Valley farmers, and tag days sanctioned by city council raised money on the streets of Vancouver.

Over the next two months the strike spread from camp to camp, until practically every large logging operation on the coast was shut down. The BC Loggers' Association fought a losing battle on every front, and when it attempted to rally public support by arguing the strike was Communist-inspired and Moscow-led, public sentiment shifted even more to the strikers. Some Association members, such as P.B. Anderson and J.R. Morgan, settled with the union and met its demands. The Association divided between those willing to deal with the union and negotiate an agreement, and those wanting to destroy the union.

In April, with public sentiment firmly behind the strikers and the wider business community demanding the provincial government end the strike, BS&W decided to reopen its camp on Great Central Lake with strikebreakers flown in from outside the Alberni Valley. The union organized a march on Alberni. Hundreds of strikers assembled at Parksville and began the fifty-mile walk to Great Central Lake. By the time they reached the lake their ranks had swelled to 500 men. There they were confronted by a large contingent of provincial police who stopped them from entering the camp. The trekkers set up camp in the woods and for two weeks they harassed the strikebreakers, preventing them from moving many logs.

Meanwhile, in the outside world, demands were increasing for the government to intervene. On April 26, it raised the minimum wage in logging camps and sawmills to what the strikers had demanded, $3.20 a day, effectively ending the strike. The Lumber Workers' Union came

The speed-up was simply making you work at the utmost limit. They had these efficiency experts who would come into industry in those days, into the logging camps too, would study a given job and a given operation and would suggest to the boss that an operation could be done in X number of minutes less, you see, and within a few days you would find yourself working at the pitch suggested by this efficiency expert. This extended throughout the industry.

It got so bad that men didn't walk on the job, they didn't walk between various little jobs that they were doing, but they ran. Indeed, it became almost a matter of pride to loggers, that somehow or another the employer was able to, in the absence of unions or in the absence of any opposition at all, the employers were able to almost establish a tradition that the logger was a special breed, he was big and tough and he could therefore work much harder and faster than anybody else and spit in everybody else's eye, sort of thing. And as I said, the result was that on the job you would find men, not walking or even walking quickly, but literally running. And of course this led to many accidents too.

It became so absurd that old-time loggers finally began to take notice of the fact that when the crews would come in at night on the railway crew cars, the crummies, the cars would hardly be in the outskirts of the camp before men would be swinging off and running for the wash house. Running, first of all, running to their bunkhouse to drop their lunch kit, and then running for the wash house to get washed up and then running back to their bunkhouse to part their hair, you know, slick it down, and finally running to the cookhouse for their meals. I'm not exaggerating, this became almost a custom.

Al Parkin, an IWA organizer in the 1930s.

Merrill–Ring–Wilson's Rock Bay woods camp in 1931. Located near where the Island Highway now crosses Amor de Cosmos Creek, this was one of the coast's major railway camps, using three Climax locies, two Shays and a Baldwin. The speeder on the tracks hauled loggers to work in the woods and the small, neat houses on the right were company-owned family dwellings. (Monty Mosher collection)

out of this first major labour action as a recognized representative of loggers. From there it went on to organize sawmill workers, and in 1937 joined loggers and millworkers from the northwestern states to form the International Woodworkers of America (IWA). Later that year the province passed the Industrial Conciliation and Arbitration Act providing workers with the legal right to organize and protecting them from employers who tried to prevent them from doing so. In 1943 the IWA staged and won its first major strike when it shut down the Queen Charlotte Island spruce camps. The following year the union negotiated the industry's first master agreement, and emerged from the war as the undisputed industrial union of woodworkers.

On the technical side of things, the yarding systems that came into use in BC and other parts of the Pacific Northwest around the turn of the century originated in the logging camps of Michigan and Wisconsin and the cypress swamps of the southern US. A northeastern logger named H.H. Butters began by converting three-drum steam-powered pile-driving engines into primitive overhead yarding machines, on which he received a patent in 1883. Because wire rope was not yet available, these early yarders used manila rope, which severely limited their effectiveness. At about this time northeastern loggers were beginning to migrate to other parts of the continent, including the South, where some of the most valuable timber was found in swamps. Butters and Spencer Miller, an engineer with the Lidgerwood

Opposite: One of the Lidgerwood skidders purchased by Comox Logging in 1910. The twelve-by-twelve steam-powered unit was skyline rigged and had a separate loading engine operating an eighty-two-foot-long hayrack boom that could swing in a full circle. These units were mounted on a frame that was moved on rail trucks to trackside wooden spars 150 to 200 feet high. (Mauno Pelto collection)

Manufacturing Company of New Jersey, developed the "pull-boat" logging system, which consisted of a pair of steam-powered winches mounted on a scow, on which was erected a mast or spar. A fixed skyline was strung from the top of the spar to a tail tree several hundred feet in the woods. The skidding, or main line, ran through a block at the top of the spar, through a block on a carriage hanging from the skyline and out to the logs, which were hooked to it with tongs. A receding or haulback line went from the second drum through a block on the top of the spar and another idler block on the carriage, and through a third block on the tail tree, then back to the carriage.

When it became practical to build railways into the higher portions of these regions, Butters installed his machine—which he called a "skidder"—on a railway car, and used a convenient tree for the spar. Other refinements were added by Butters and other innovators to meet the varied conditions encountered in different forest sites.

Critically important was the 1899 invention of the interlocking main and haulback drums. Devised to keep the turn suspended, they replaced the previous inefficient practice of using the brake on the haulback to keep the mainline tight and to provide lift. As rigging became stronger and wire rope heavier, it became increasingly difficult for the hookers and chokermen to drag the mainline off to the side of the yarding road to reach logs. To alleviate this problem, which required frequent moving of the skyline, a slack-pulling line and drum were added to pull a desired length of mainline out from the carriage so that chokermen could easily reach logs on either side of the skyline.

Engines with compound gears were developed for low-speed, high-power application on the mainline, with a higher, direct gear on the haulback drum for rapid return of the rigging to the woods. Additional drums were added to operate loading devices, to move the rail cars, or to move the skidder car to a new location. The Lidgerwood Company acquired the patents to most of these devices, including Butters', allowing it to dominate the logging equipment business for many years. It wasn't until these

The first time they set Skidder Three up, the crew went to work in the morning and the fallers were working up beyond it, about a mile or two beyond where the first setting was. So they set the steel spar up, and it had removable trucks underneath it. You jacked it up, and it had big hydraulic jacks. You put blocking in, big twelve-by-twelve, sixteen-by-sixteen blocks, and then you lowered it down. Then you pulled the trucks out from under it.

So they jacked it and raised the tower and jacked it up there and in the late afternoon. They had one of these little Casey Jones speeders, just a little bit of a thing, carried about eight men, no roof. When they took the trucks out the speeder could go under it, it was high enough. Going up he went under it, and he picked the fallers up. Coming back down, he never said anything, of course, and the crew at the skidder had gone home. So they come down, around the corner and these fallers see this skidder sitting on one track and the trucks on the other. They didn't know they could go right underneath it, and they all bailed off. Couple of broken legs and arms. All skinned up. The speeder sailed right through underneath it.

Jack Bell

I started in 1928, right up here back of Nanaimo at the foot of Mount Benson, in the old Nanaimo Lumber Company, and they had a railroad show. The mill was on Comox Road. They went and logged up Mount Benson, right at the foot of Mount Benson. And I started there in '28 blowing whistles on the steam yarder—high-lead steam yarder. I worked there until the Depression came, and then my two brothers and myself went out to Nanaimo Lakes and built a log cabin. We stayed there for four years and lived off the land. In '34 my brother, younger than myself, we went over to Alberni, because we heard they were building a sawmill over there, and they got work. And we did, we got a job in the sawmill for a week stacking big planks. Four by twelve and forty feet long in the middle of the summer, oh, it was a killer. We worked a week and then they laid us off. Then I went to old Camp Three up at China Creek, APL. And we got a job up there setting chokers. They had slacklines up there. From there the claim worked out, so I went down the canal and got a job from a gyppo by the name of Phil Welsh. He was right across from the Franklin River, and Franklin Camp started in '34. I went over to Franklin in '35.

Bloedel, Stewart & Welch started in 1934, started to build a camp in the spring of '34 and started to log in October. So I went over and started to work there building Davis rafts. They were shipping all the big hemlock down to Bellingham, the Bloedel Donovan mill there. See, they didn't want the hemlock up in the valley, there's no pulp mill. All they wanted was fir or cedar up in Alberni. So they made Davis rafts and shipped it down to Bellingham.

Well, then they wanted me to go up in the woods and go high rigging, which I did. I high rigged for, I guess, about ten years. Then I went on supervision at Franklin. I was at Franklin for thirty-five years.

Jack Bell

patents expired that western manufacturers were able to start producing machines designed specifically for the conditions of the Pacific Northwest.

Consequently, when the Comox Logging Company was established in 1909, the only available yarding equipment that was compatible with the railway equipment then available were the Lidgerwood skidders, which were an enormous advance beyond the ground skidding equipment commonly used. However, there were disadvantages to the overhead or skyline yarders. The large amount of rigging equipment they required was expensive to buy, maintain and set up, and they were not very efficient at moving big logs over long distances. As a result northwestern loggers backed off to a simpler system combining features of the overhead and old ground-lead methods. Around 1909, northwest loggers began moving the bull blocks of the ground-lead system higher off the ground, but still used the ground-lead donkeys being built by a number of western manufacturers—including Washington Iron Works in Seattle, Willamette Iron & Steel Works in Portland, and the Empire Manufacturing Company in Vancouver. Initially these systems were called "tree rigs," but they soon became known as high-lead systems. In their early stages, some of them still employed line horses for the haulback, and a single skidding or mainline passing through the raised block to pull the logs in.

The location of the first high-lead system in BC is another one of those points on which there is no agreement. A logger named Gardner used a high-lead system in Washington in 1906, although a report that he used this machine along Discovery Passage, on Vancouver Island, seems doubtful. A 1915 article, "written for Western Lumberman by an expert," claimed the first high-lead system was at Comox Logging in 1913. Matt Hemmingsen is also credited with being the first to hang a block up a tree after arriving in Chemainus in 1906, and there are several other reports of the "first high lead show in BC." Most likely many of the loggers who acquired the original drum yarders began experimenting with a raised bull block. Once they started building yarding units with enough weight to prevent the machines from lifting themselves off the ground when pulling in a big log, and had mastered the art of climbing a tree to hang the block, high-lead logging was an established system.

In fact, high-lead logging was so efficient in some conditions that overhead logging units, such as those produced by Lidgerwood, were often rigged for high lead by foregoing use of the skyline. It was particularly useful on sites where

I worked dragging those tongs, what they called hooking, for about a year. Then they insisted I start learning to rig, to do high rigging. In 1921, that's when I started to climb trees. I thought I was going to die climbing trees—I think I climbed trees fifteen years. That was quite a long time, especially skidders, you were up in the air so much. There was so much aerial work. It wasn't like the old high lead where you just go up a tree about twice a month.

There was more to do with skidders. Quite often you were moving the lines around a bit so they wouldn't cut. All that sort of thing—so many lines on a tree and so many blocks on a tree. Later on I had to rig all the trees, and that's all I did was rig, for many years. Just sort of up in the air every day, for hours and hours.

Normally each spar tree had just one skidder. We have worked two skidders on them, in an emergency, one working each side. I don't suppose they'd allow it these days, the compensation. But with six skidders it took a lot of rigging. I also had to take it down. Of course, it didn't take too long to take it down, not near as long as to put it up. Later on, when the big trees got scarcer, we used to haul our trees. We'd get them where we could—good big trees—then we'd haul them to the place where we wanted to put them up. Then we could put the logging jewellery on the tree down on the ground, before we raised it. Then we raised it up, all it meant was a little adjusting, here and there, tighten the guylines and all that. You never knew it was raised, of course, unless you looked. That way you got away from the swelled butts, and you got them exactly where you wanted them, close to the track. Of course, everything was getting more efficient, no fooling around, no guessing. Like any other business, it was right down to brass tacks.

Joe Cliffe, who started logging for Comox Logging in 1920. He retired from its successor, Crown Zellerbach, in 1961.

railway spurs could be built easily, where there were numerous big logs of more than two to three feet in diameter, and where it was not necessary to yard more than 700 or 800 feet. Beyond that distance, especially if the logs were smaller and the ground was too steep and rough for economical railway construction, overhead systems were favoured. And, even after the war, ground-lead systems were still the most economic on uphill yarding over distances of 600 to 900 feet.

In 1915, logging on the easy ground around Campbell River, International Timber operated five high-lead yarders—three 10-by-11 Willamette compounds, one 12-by-14 Willamette compound, and an Empire 11-by-14 four-drum machine—on settings of forty acres each. Only four settings were worked at a time, the fifth engine being rigged ahead by the rigging crew.

The introduction of high-lead logging provoked an enormous controversy, both within and outside the industry. High lead was resisted by loggers for several reasons. It replaced chokermen and hookers, skilled in snaking logs through the stumps and standing timber, with yarding crews which could work fast enough to

---

We had two Lidgerwood steel spars at Franklin, and Great Central had one, they had Skidder Number Two. It was a little smaller, it had a ninety-foot tower, Franklin's were a hundred foot. Skidder Five was the largest. It weighed 252 ton. And Skidder Three weighed about 225.

Franklin got the first skidder, Skidder Three. They bought that in July of 1935. That was the first steel spar skidder. They were going about nine months before they got the first skidder. They brought the equipment down from Great Central Lake and from Menzies Bay. And they were using units and cold deckers, trackside, 'til they got the skidder. They were not a big change 'cause we did a lot of skylining anyways. But it was a self-contained unit.

Once that thing was set up, the locie would bring up the empties and push them through, under the skidder and up behind. They had the spotting line, a big drum with a heavy line. They'd hook on to the cars and they'd lower them as they kept loading. So the loads just kept going ahead of it, as they were loading with the heel boom.

It was fairly fast once you got onto it. We finish a setting, and move, and get logging again in six to eight hours. Because they only had four guylines on it, and they went on the corners, and then they had what they called heel blocks. And the top of the tower had a huge donut, a steel ring, went right around the top, two halves with a big bolt through, and the heel blocks and that's seven purchase blocks. And they had what they call block and anchors, two big straps that you put on the stumps behind, two-and-a-half-inch lines, and they went up and tied to the heel blocks. So when you put the tower up, you just had four guylines. And then these heel blocks took all the tension. The guylines were just to steady it when you raised it.

It had a big steam pump. You got steam from the boiler. It pumped hydraulic fluid to jack up the skidder and then you could pull the trucks out from under it.

On the yarding function of the machine you had your receding line, it was the back one, your skidding line and your slack pullers and then your utilities, and it sat right under the tower. And they went up through the tower. And then you had the transfer line and you had two heel lines for tightening up your heel blocks, one in front, one behind, two drums there, then you had your strawline drum and a pass line drum, that was the utility line.

Then on the loading end you had the spotting line, the drum to move your rail cars, and then you had swing drums, to swing the boom either way, and one to raise the boom. When you moved you just lowered the boom and lay the tower on top of it.

To load, you had both hands for your swing drums and taking up your boom. You worked the throttle with your knee. It had a head loader and a second loader. If they were yarding they used to average about anywhere from twenty to forty carloads a day. If they were taking out of a cold deck pile then they would get fifty or sixty. If it was a rough show they might only get twenty loads, if it was really canyons. The loader had no problem keeping up.

They were cumbersome big brutes. I had five locomotives pushing Skidder Five up a nine percent grade, taking it up into a setting. We had two in front and three behind, pushing. They were heavy, hard on the tracks. They were used at Franklin 'til 1955. Then we switched to trucks. Those Lidgerwoods were there as long as the railroad was there, then scrapped. We kept one and put it on a sled and we used it for, oh I guess until about 1957 or '58. Just the sled, not the tower, we just put the two sections on, the yarding section and the utility section and did away with the loading end of it.

These steel spars, they weren't welded, they were all riveted. Built like the same as they used to build the old boilers, and the steel bridges. Before they got all the welding, they used rivets. The rivets, they held up all right. They'd pop rivets every once in a while. In fact, we pulled over those steel spars two or three times. The tail-hold stumps would pull out or that big donut used to break every once in a while. Just the tower would go over.

Jack Bell

At one time the whistle punk had to put up his own wire, just common wire, and there was quite an art to it. You cut saplings, fastened them to the donkey roof and attached wire so that they were rigid—would stand the weight of the wire, extended in the air 200 or 300 feet before you could hook it to another tree. It had to take this weight without blowing the whistle, you see.

You didn't want kind of sloppy whistles—you wanted them snappy—so you had to have this spring attachment fixed just right. Then the guy would hold the whistle and play it like a telegraph, for all the different signals.

When Comox first got their skidders, for several years—in fact for a lot of years—they used hand signals, and sometimes in foggy weather we might get three different men out there giving signals, because with the skidder you logged farther between the tracks. Comox used to log a 1100-foot circle. The whistle punk used to be quite an important guy, but they've gone now, pretty well.

Joe Cliffe.

One of two Lidgerwood steel spar skidders used in BC was acquired by Bloedel, Stewart & Welch for its newly opened Franklin River Division in 1935. These 250-ton units had 100-foot towers and heel-boom loaders. Mounted on rail trucks, they could be set up on a spur or mainline and raised with hydraulic jacks, allowing rail cars and speeders to pass underneath. They were used until 1955, when trucks replaced railways at Franklin. (VPL 6120)

Below: One of the earliest high-lead operations on the coast worked off the first spar raised at Powell River Company's Kingcome Inlet show about 1911. The engineer on the sled at the left operates the main and haulback lines, holding the two logs in the rear. The loading leverman, also on the sled, operates a duplex loading system utilizing two tongs that hold up the log in the foreground. This was a standard-sized crew for such a setup. (BCARS 83515)

keep up with the new machines. Opponents charged that the logger was superseded by brute machine power. This power, when it was used to yard logs through unwanted standing trees and snags, introduced an additional element of danger to the work. The use of high-lead equipment in a manner that destroyed the unmerchantable portion of the forest brought criticisms of waste and wanton destruction from outside the industry. Fallers didn't like it because they had to fall all the trees, including snags, within about 300 feet of the yarders—a task for which they often were not paid additional rates. And finally, the destruction of the remaining forest, combined with the large tops and limbs left in the slash, created an enormous fire hazard. Sparks thrown by the various steam boilers caused numerous fires that spread from slash to surrounding forests, consuming far more timber than was logged in many years. But when they were stacked up against the economic advantages of such an effective, efficient system, these objections counted for little. By the early 1920s, high-lead was the dominant logging system on the coast, a position it maintained in various permutations for the next half century.

Mauno Pelto, at age seventeen, preparing to rig a spar tree at Caycuse, 1935. (Mauno Pelto collection)

The first skidders to come in were Lidgerwoods and the Lidgerwood company sent two men here to show how to use them. One was Herman Matson, he was a machine operator who ran the steam machine, and the other was a high rigger by the name of Henry Norman. They were the ones that really showed them, up around Cowichan Lake and, I believe, Comox before that, how to operate those skidders.

They were both Finlanders. They came from Washington State but they'd both worked in Louisiana before that. Henry stayed here, lived in Caycuse and died in camp. During the Depression years, when there was so much talk about Russia and their five-year plans, they were looking for help, for experienced people to show them how to log and one thing and another, and Henry got a chance to go back there. He went back and took his family—there wasn't much work here anyway so I guess he thought it would be good. He got back there and there was no machinery, a very bad state of affairs.

Ken Hallberg, born at Ladysmith in 1914. He started logging when he was fifteen, and retired in 1974 as general manager of BC Forest Products' Renfrew Division.

By 1920, a wide array of cable-yarding systems had all but eliminated animal-powered logging operations, and for a time cable systems were the only ones used in the primary yarding phase, from stump to roadside. The most widely used systems employed cable yarders in combination with other hauling systems—mostly railways, although trucks were also coming into use at this time.

The rapid and widespread adoption of cable-yarding systems spawned enormous activity. In the first place they required a huge array of rigging gear—blocks, hooks, carriages, chains, clevises, swivels, tongs and chokers. Miles of wire rope of various sizes and types were needed. Throughout the Northwest, numerous manufacturers built foundries and factories to meet the demand. In BC most of this activity occurred in Vancouver, mostly in and around False Creek. Fill was dumped on a sandbar in the creek to create Granville Island, where several firms were established to service the logging industry.

In the woods, loggers had to learn the new skills of rigging cable systems. When the first

Lidgerwood skidders arrived they were accompanied by experienced crews from Louisiana who taught local loggers how to set up the machines. The Lidgerwood steel spars were equipped with a rigging engine that hauled line through a set of heel blocks to raise the spar. But when tree rigs were used, a convenient tree had to be prepared, or a spar cut elsewhere and raised at the desired location. In the bigger camps, specialized rigging crews of four to six men put up the spars and the rigging gear.

Rigging a standing tree (a Douglas fir if possible) required the high rigger—or high climber as he was sometimes called—to don a set of spurs and climb the tree, chopping the branches off as he went. At about the twenty-four-inch diameter mark, anywhere from 100 to 200 feet up, he used his axe and a small falling saw to take the top off the tree. Then he or another member of the rigging crew lowered a light rope and pulled up a rigging block through which a "pass line" was threaded and used to haul up the rest of the rigging—guylines, the various blocks or "tree shoes" and other paraphernalia, depending on the nature of the system. Once the rigging block was in place, the rigger would be raised or lowered with the rigging donkey, or "goat."

In 1929 the Industrial Timber Mills operation at Camp Six had somewhere around 250 or 300 men working. They used to run four skidders. They used one for cold decking and built big piles away from the rail line for winter logging. Some of those piles had up to 14 million board feet in them. When the winter would come they would keep the roads open and swing these piles down.

There was only one power shovel used to make road, one for the whole operation. They reached out a long ways from that road down at the bottom of the valley. You could be out 1800 feet from the road with the trackside machine, and the second machine could go out another 1800 feet so you wound up 3600 feet away from the railroad track. Today, the whole country would be full of roads. They just had that one shovel.

They were steam skidders except for the one on the cold deck. It had a gasoline power plant in it, a 300-horse Waukesha gas engine. I can remember on wet, cold days the whole skidder crew would go into the motor house and have their lunch—fifteen or sixteen men.

It was quite a big affair to move that thing up those mountains. Lots of times it was so steep you could barely get up there yourself, let alone move that thing up. It must have weighed 80 or 90 tons. It came with the Waukesha, built by Washington Iron Works. It was one of the first that came with a kind of power other than steam. A very good machine.

It had a winch capacity something the same as a twelve-by-fourteen steam skidder. I remember it burned seventy-two gallons of gas a day. It had an inch-and-three-quarters skyline. As a kid I wound up with a job as second rigger on that skidder, I used to rig the back end on it.

The tail end was usually rigged between seventy and ninety feet. They were always standing trees; they used to mark them before they felled the timber. They chopped an X in them with an axe, for the back spar. The engineers did that. When they became exposed out in the fell and bucked they took the tops off them so they wouldn't blow down.

The crew was divided. There was a rigging crew, which was my domain. We used to be four men and the second rigger. I was the second rigger. That was our job totally, rigging back spars. It would take us about five to six hours, sometimes a little more, seven hours, to rig a back end up.

That entailed putting seven-eighths guylines up, three of them. There was a hook on the end to choke the tree up. You coiled your strawline up, half-inch line, and pull that double on your new road. The strawline would take what they called the transfer line. It was seven-eighths and would pull the skyline cable out. There were two cables, one that you were working on and the one on the new road. That process went on continually. When a road was finished you dropped that rigging and came into the new road. When they started to work on the new road we would get the next road ready. It would take them, on average, half an hour or thirty-five minutes to change road. So the machine never stopped very long. They would work up to three days on a road.

The head spars would be 130, 140, maybe 150 feet. I've rigged head spars many times that were almost 200 feet. Then you'd have four sets of guylines

There was a tremendous amount of thought had to go into these things because in those years your major production came from maybe three or four sources, settings, instead of maybe ten or more, like there is today. It had to work, work well, otherwise the outfit would go broke.

There was up to eleven miles of wire rope on those rigs when they were totally rigged—the big ones. About eighteen guylines, about 300 feet each. One at Chemainus had 6500 feet of inch-and-thirteen-sixteenths haulback line on it.

Mauno Pelto, who came to Cowichan Lake from Finland in 1929, when he was eleven. Four years later he started punking whistles at Camp 6. He retired in 1981 as logging superintendent at Crown Zellerbach's Nanaimo Lakes operation.

Right: A high rigger limbing a tree that will be used as a spar. With the use of climbing spurs and a steel-cored rope, the rigger works his way up, chopping off branches as he goes. (BCFM 3–3)

Far right: At a height of 100 to 200 feet, he tops the tree. At this point the greatest danger is that the tree will split and crush him with his rope. He has loosened his spurs and is ready to descend if this happens. (BCFM 4–4)

A character we called Big Mike was the head rigger at the International Timber camp for many years. Mike was primarily concerned with raising and rigging the track tree for the yarding and loading operation—very seldom a standing tree was in the right spot. As a general rule, it was the second rigger's job to top all trees that were used in this way, pretty hard work cutting all the limbs, after which he had to cut through twenty-four to twenty-six inches of the treetop with an axe. Mike, however, liked a change, and one day he asked me just where a certain cold deck tree was on a felled quarter. I told him and said, "I will head you in the right direction."

A little later I thought, maybe I should just quietly sneak over there to watch Mike. Well, I got to where he was as he was just starting to put in the undercut for the top, with no one in sight but myself, and he had no idea I was there. The treetop was finally severed and fell to the ground, and Mike just rested in his belt for a few minutes. Then without show or fuss he leisurely climbed up to a sitting position on the top. Then he stood up on this twenty-four-inch diameter, 160 feet above the ground, and calmly and leisurely relieved himself, and again sat down. There were no other witnesses to this but myself, and he never did know of my watching.

Archibald Kerr, who worked at many jobs in the woods over a sixty-year period. He retired as a professional forester.

Although rigging was not considered a dangerous job, it was not unknown for riggers to fall out of the tree, either when the top went off and it was swinging wildly through the sky, or as a result of cutting the climbing rope, or by any one of a number of other possible mishaps. Some riggers were killed or maimed in this manner, and there were several who fell a hundred feet and survived to work again. Colin Brooks was a high rigger almost 60 years old in the early 1960s when he fell ninety feet out of a tree on Quadra Island. When he returned to work a few weeks later, held together by a handful of stainless steel pins, he found that in his absence the crew had finished the setting, felled the spar and used their axes to carve the stump into a remarkable likeness of his craggy features.

If there was no suitable tree in a convenient location, the workers found one elsewhere. They limbed and topped it before falling, to avoid breakage, then skidded it to the new landing. If the distance was short, 100 feet or less, the tree was "jumped" into position. Guylines were fixed to the topped tree, which was cut off at the butt. The donkey then pulled the butt into place, with adjustments made to the guylines to keep it in a vertical position. If the spar had to be raised from the ground, a "gin tree" or

Funny things happened to green people, me too. I remember we had moved this machine up, got the tree all rigged up and then started rigging a high-lead setup. My job was punking whistles, I was pretty young, maybe fifteen. It was raining pretty hard so the engineer pulled the slack off when the last turn was going in before lunch. The hooktender said to me, "Don't you leave until all the crew's on the machine." So I watched the crew go down and the last turn as it went in, they landed the logs on the pile. So I thought, that's good enough. I put my leg over the haulback, and you know what happened then? The haulback picked me up and I don't know how high I flew, but I was up there a long time. I landed in a logical spot. I was fine and looked around. I found my hat and as far as I could tell nobody saw me. I didn't say anything. After lunch the hooker says to me, "What are you going to do when you get the kind of vocation you'd really like to have?" "What would that be," I said. "You know, I kind of like this logging, I think I'd like to log." "Well," he says, "I thought you were practising for the circus. Just before lunch I saw you flying through the air."

Mauno Pelto

The top has broken free and the rigger holds on tightly as the spar whips through air. (BC Forest Museum 3–3)

"dummy tree" was left near the location and rigged with guylines and a multiple-block purchase to haul it upright.

If the setting required high-lead rigging, a series of sturdy corner blocks were installed across the back edge of the setting. A light strawline was pulled out by hand, through the corner blocks and back to the yarder, and attached to the haulback line. The strawline drum then pulled out the haulback line through the corner blocks and back to the spar. The mainline was pulled from its drum, threaded through the mainline block on the home spar and attached to a device called the "butt rigging," which consisted of swivels and some sort of hook to which chokers were attached. The haulback line was attached to the other end of the butt rigging. Once the drums were lined up, "in lead" with the main and haulback blocks on the spar to prevent the lines from binding when reeled onto the drums, the operation was ready to start logging.

This rigging-up procedure was considerably more complex if an overhead system was being used. In addition to raising and rigging the head spar, a smaller tail or back spar was also needed, along with a strong tail-hold stump to anchor the skyline. In the bigger camps this task was

If a suitable spar tree was not available where it was needed, one was felled elsewhere and raised at trackside. This spar at Bloedel, Stewart & Welch's Menzies Bay camp in 1927 is being raised by the rigging goat on the rail car, with the aid of the short spar, or gin pole, in the background. (BCARS 73729)

With the spar topped or raised, the riggers attach the guylines, and yarding and loading gear. The yarding donkey has been moved into position and, with the use of a small pass block carried up the tree by the riggers, hauls up the lines, blocks and other gear. (BCARS 73730)

My first logging job was in 1939 up on the Malahat for a fellow named Harry Jones. He had two portable sawmills up there. That was a school for loggers. They had very light equipment and they needed a lot of people. They were forever hiring people to work in the sawmill. As soon as you were subjected to the slavery of working in the sawmill you wanted to get out of there into the woods. I made my way from the sawmill to the logging end of it.

These were bush mills they had, and they used to drag the logs in with a little Fordson-powered donkey to a spar tree, and then the logs would roll down the rollway to the mill. The sawmill and the loggers were working very close together all the time.

A fellow by the name of George Locke was logging on the Malahat just across the road from where you turn into Bamberton. He had some timber and he didn't want the poles, so we got going with the poles. By a strange stroke of fate there was another fellow who had attempted to start in the logging business. He bought a donkey engine and one thing or another but he got disheartened because he didn't understand logging and he couldn't seem to get anyone to work for him that understood it. He was on the verge of going broke so he said he would sell us this logging donkey if we'd just pay him what he had in it and then take over the payments. There was some felled and bucked timber too.

It was an original Skagit donkey with a Fordson engine on it. Basically, what they did with the thing was to build a sled and bolt the frame of the winch onto the sled. Then you took a Fordson tractor and took the wheels off it, the back wheels, and put sprockets on. Then you put sprockets on the end of the haulback shaft. Great big sprockets on where the wheels had been, and little wee sprockets on the winch. Then you bolted your Fordson tractor down to the sled and you had a donkey.

There were just two drums, a mainline and a haulback. But it's amazing what you could do with that thing, combined with a whole lot of blood, sweat and tears, out in the woods. It would hardly pull the hat off your head so you were having hang-ups all the time. You practically had to walk the log in, rolling it, turning it and one thing and another to get around the stumps.

The particular bit of timber this fellow had got had no spar tree for this donkey engine to work through. In those days there were no rubber-tired skidders and bulldozers weren't in general use in this area, so the only way you did it was with high lead. So here was a place where the timber was all felled and there was no spar tree anywhere very close by.

He got us to help him. So we found a tree and managed to drag it to where we wanted the spar tree to be. In order to raise one of these spar trees we had to rig what they called a gin pole, which can be something smaller, a forty-foot log, say. That's not too difficult if it's only forty feet. So then you hang some guy wires on it and then you use it to raise the spar tree. You have the guy wires on your gin pole and the spar tree is lying on the ground. You anchor the butt end and get a hold on the spar tree at an appropriate height, so that when it comes up it's going to be about the right height for the top of your gin pole. You hang three guy wires on this spar tree before you pull it up.

You pull the thing up and you have to guide it by having people with the guylines at each of the three stumps that the guy wires are going to go around. As it's coming up it can't swing too far this way or that way. You constantly have to let slack go on the guy wires as the tree comes up. When the tree is up there then you anchor all the guy wires and it's going to stay.

But, he had a bunch of greenhorns. When the thing was on the way up, one of the guys let all the wraps off the stump and when the tree started to swing, the whole works came down. I was running the donkey. I saw that guy wire go slack and I knew it was a disaster, so I just bailed off that thing and ran as fast as I could. The tree came down right alongside where I'd been standing.

That's what precipitated the demise of this fellow. He said, "That's too tough for me. There's no way I want to get exposed to the dangers. You guys can take this thing over. I don't want anything more to do with it." That was our start on our own.

Frank Norrie, one of the founders of Norrie Brothers Logging, and now president of Vancouver Island Helicopters.

A typical skidder setup that could be found on many of the big railway shows from the mid-1920s until after World War Two. This one is rigged with a North Bend tightline system, and the logs are being loaded with a hayrack, single-tong or heel-boom system. (BCARS 81745)

performed by a separate back-rigging crew. Although the variations in overhead rigging systems were limited only by the imagination and ingenuity of the loggers, two basic systems were utilized: tightline and slackline. Both utilized a carriage, or "bicycle," hanging from blocks running on top of the skyline.

In a tightline system the skyline was run through blocks, or tree shoes, at the top of the head and back spars and secured to convenient stumps at each end. One of the great drawbacks of this system was that it had to be re-rigged as each corridor of a setting was logged, a time-consuming and expensive task. The tightline system did not require a dedicated drum on the yarder, and the mainline drum was used to raise the skyline. Various tightline systems were devised to deal with the different conditions encountered in the woods. One of the most popular was the North Bend and its various modified versions, which was particularly favoured for swinging logs from cold decks to the landing. The Tyler system, which required an additional mainline drum, was capable of lifting the turn off the ground and was used when yarding over canyons or handling brittle species such as cedar. It was primarily a swing system, far more costly to rig than a high-lead system,

Opposite: This skidder spar used at Victoria Lumber & Manufacturing's Copper Canyon operation in the early 1940s utilized an enormous complex of guylines, skyline, main and haulback lines, and loading lines. The expense of rigging spars for this system was one reason high-lead systems were often preferred. (Jack Cash photo; UBC BC1930/546/869)

The Columbia Block & Tool Company's new plant on False Creek, c. 1900. This was one of many manufacturing companies established to provide logging equipment. It later became Opsal Steel and moved to its present location on West 2nd Avenue in Vancouver. (Ted Dunn collection)

In '44 I came to Copper Canyon and rigged on the first skidder, the big Willamette skidder they had there. It was on railway wheels. This machine weighed 225 tons. It had two-inch cables on it and they were 2500 feet long. This is the machine with 6500 feet of one-and-thirteen-sixteenths haulback. The mainline was 3500 feet.

They would put a little spur in, off the railroad track, to the spar tree. You'd have about a ten-foot jill poke and square off one end with an axe. Then you'd make a square place in the bottom of the tree and jill poke the machine to this, with a wire rope from the front of the machine right around the tree. You'd tighten it tight to the jill poke.

There were hydraulic jacks on each corner. You'd get behind the drum and line up on the blocks up the tree. They'd raise or lower on the jacks and when the blocks would line up on the edge of the drum you would be in perfect lead. It had to be right on so you wouldn't chew your lines up.

Mauno Pelto

and was used extensively in bridge construction.

Slackline systems used a special donkey with a drum large enough to hold the required length of heavy skyline—perhaps as much as 3,000 or 4,000 feet. It needed sufficient power to raise the turns and strong brakes to keep the skyline in the air. In this setup the skyline was pulled off the drum, through the head block, through the carriage and tail-tree shoe, and tied to a tail-hold stump. When yarding, the skyline was lowered and the chokers attached to the logs. Then it was raised to lift the front end of the logs, and the carriage was hauled in with the mainline. The slackline was better at yarding than most tight-line systems, especially on rough ground and downhill.

The need for different rigging configurations led to the production of a variety of yarding engines. Once the Lidgerwood patents ran out, machine manufacturers from San Francisco to Vancouver began building them. In 1925, for instance, the Willamette catalogue of logging machinery listed a dozen basic yarding machines that could be used in various combinations with

Westminster Iron Works plant in New Westminster, possibly in the 1940s when diesel-powered yarders were being built. The company was established in 1874 and operates today as Lamb–Cargate. (CVA Bu P337 Neg 7500 #2)

each other, different boilers and loading systems, either on sleds or railcars. This was a uniquely innovative period in the industrial development of the Pacific Northwest. Several highly competitive manufacturers could respond quickly and creatively to the needs of their customers—loggers with access to an untapped resource and selling into an expanding world market.

It was during this phase of BC's history that most of the industry's major manufacturers were established or grew into prominence. In 1877, John Reid came to New Westminster from Ottawa and went to work for a local blacksmith, W.R. Lewis. Within a few years he bought the business, along with partner William Currie, and expanded to build horse-drawn carriages. By 1892 the company, now called Reid and Currie Iron Works, had established a foundry and early in the twentieth century began manufacturing yarding and loading machines under the name Westminster Iron Works. Until the 1960s it was one of the most prominent logging equipment manufacturers and dealers, producing a full range of steam machines and selling a variety of other machines, including trucks and gas-powered locomotives. In 1962, two local engineers, Charles Carcross and Norman Springate, bought the company and manufactured logging and mill machinery. Four years later the Lamb company, a logging and milling equipment manufacturer founded early this century in Hoquiam, Washington by logger-inventor Frank Lamb, bought the operation, and the name was changed to Lamb–Cargate. It is still in business under that name, manufacturing pulp mill equipment.

There was some construction of steam equipment, mostly boilers, in the last century by the Victoria Machine Depot. About 1907, the Empire Manufacturing Company was opened on West 6th Avenue in Vancouver by J.F. Smith. Sometime later it moved to 2nd Avenue and Cambie, where it also produced a full line of

Granville Island was dominated by manufacturers of logging and milling equipment in 1925. Gibson's, Vulcan Iron and Wallace Foundry all produced yarding and loading equipment, while Britannia provided wire rope. In the 1970s the island was transformed into a shopping, theatre and restaurant complex. (CVA 393–16)

yarders and loaders. Over the next few years several other firms were established to sell, repair and manufacture logging machines. The Vancouver Iron and Engineering Works and the G.B. Murdie Company, established in 1914 in Victoria, were two of the more prominent companies in the late 1930s and 1940s, producing gas- and diesel-powered donkeys.

In the late 1930s, one of the BC logging industry's greatest mechanical geniuses, Frank Lawrence, came out with his first yarder, the Lawrence 36. It was known none too affectionately as the Nutcracker because of hazardous positioning of the strawline drum between the operator's legs. It had a small gas engine and was replaced in about 1942 with one of the most popular yarding engines ever made, the famed Lawrence 10-10. It was a three-drum high-lead yarder holding 1,200 feet of one-inch line on the main drum. Usually it was powered with a six-cylinder Chrysler gas engine. Lawrence had begun designing and building machinery in the 1920s; he later went on to build a whole range of yarders, some in conjunction with the local Hayes truck company, and also developed and built the popular Gearmatic winches and transmissions and a portable steel spar, the Sparmatic.

In 1922 a Scottish engineer, Jack Wilson, founded Tyee Equipment on Granville Island to build steam yarders and loaders. He came to the Northwest in 1907 and worked for Washington Iron Works and Willamette, designing the latter's first high-lead yarder, their famous 13-by-14 two-speed yarder and their duplex loader. From its inception, Tyee was the BC dealer and main repair depot for Shay locomotives, as well as manufacturing, selling and repairing a wide array of logging equipment. In the industrial heyday of Granville Island there was a steady stream of locies rolling on and off barges via the tracks that ran to the Tyee shops.

The first major manufacturer of rigging equipment, Columbia Block & Tool, was founded by Andrew Opsal on False Creek about the turn of the century. In 1908 the company moved to West 2nd Avenue and changed the name to Opsal Steel. It began producing the OSCO line of blocks, hooks, jacks, swivels and other cast items required for cable logging. In

A Federal truck on Granville Island, 1925. The building occupied by Britannia Wire Rope Company at that time is now the home of the Emily Carr College of Art. (VPL 604)

1942, Ted Dunn bought the company and has operated it ever since, with his son Tom. Opsal is the oldest logging equipment company in BC, and one of a handful of surviving Pacific Northwest equipment manufacturers which began at the turn of the century.

The establishment of these and other companies in BC had a significant impact on the logging industry. The people who made the machines were close to those who used them, so they were able to respond quickly to changes in demand. A logger could get a machine modified by its maker or have a new one made, making him a more efficient and competitive producer in the global marketplace. In turn, the local manufacturers were able to sell top-of-the-line equipment throughout the Northwest, easily competing with the bigger manufacturers in Washington and Oregon.

At Sproat Lake in around '38 or '39 when they were going pretty good, they had a log chute. That was Manning Timber, they had a big sawmill there on the lake. They had a camp right down on the float there on the lake. They truck logged right down into a dump at a town called Kleecoot, they had a post office, that was about all, before you got to Sproat Lake. That's been so long ago that I kinda forget what kind of trucks they had. They weren't big trucks. They had about six, seven trucks there.

We had a cold decker going all the time and a skidder working. Everything went out over a timbered road, down to the lake. Seven miles. I can remember moving the donkey out of there to the beach. When we got that donkey, we couldn't move her down the road, it was all timber roads and it was too wide for the roads. So we moved her cross-country and down over the rock bluffs. There wasn't hardly any sled left on the donkey.

There was quite a knack to moving them old steam pots. I even took one up the mountain cold decking one time. That's a lot of work to take one of them steam pots out into the woods. You gotta have water all the time and it was quite a deal in the wintertime. It was hell to keep one going. Your water would freeze up. And you had to have good wood, if you didn't have good wood, you couldn't get steam. So our best logs went in for wood. Whenever a good log would come in, the old woodbucker, he'd be just, "Get that good log in here for me," he'd say. We'd tightline the log in for him to buck up.

Sam Telosky

Comox Logging's A-frame on Comox Lake, 1933. The yarding engine, in the foreground, was a twelve-by-twelve Lidgerwood, and the smaller engine in the rear was used for moving the raft along the shoreline. Crew quarters were on floats at left. (Mauno Pelto collection)

In BC, one of the innovative uses of cable systems was on A-frames. At one end of the scale, these consisted of the small A-frame setups used by gyppo loggers in the Jungles and farther north, around Ocean Falls. But there were also some enormous A-frames installed, using the biggest yarding machines available. Comox Logging built a big A-frame unit on Comox Lake in the early 1930s. A million feet of timber was stacked in three layers to build a 150-by-200-foot float, on which the A-frame was erected, built of two huge spars lashed together at the top. A big steam skidder, rigged with a skyline, reached 2,000 to 3,000 feet up from the lake, yarding logs down into the water, from where they were loaded and trucked to the dump at Royston. Bloedel, Stewart & Welch operated a similar A-frame from an even bigger float on Great Central Lake during the same period. The yarding engine on this rig was a Washington Flyer with a skyline drum holding 4,000 feet of line. When the yarder was eventually dismantled, the drum was used on a Wyssen skyline in Sechelt Inlet, and the mainline and haulback drums were used to build a winch on the tugboat *Polaris*.

A gas-powered Clyde cold-decking yarder at Bloedel, Stewart & Welch's Menzies Bay camp, 1927. (BCARS 73733)

At some point in the late 1920s or early 1930s a method of supplementing the reach of the overhead yarding systems was developed. It consisted of a lightweight yarding machine mounted on skids that could winch itself up the hill, over the felled timber, to a position near the tail spar of the roadside or "home" yarder. This machine, high-lead or skyline rigged, reached even farther, along a ridge, up a side valley or down the other side of the hill, skidding logs into a pile or cold deck, within reach of the home skyline. The logs in the cold deck were skidded quickly to roadside with the main machine. In some locations several cold-deck settings fanned out behind the tail tree; in others a succession of two, three or even four cold-deck settings worked back from the tail tree. Logs were swung forward in stages, from cold-deck to

A Comox Logging crew moving a cold-decking yarder through the slash at Camp 3 near Black Creek in the 1920s. (CDM P200–829c)

A Bloedel, Stewart & Welch cold-deck pile after the yarder has been moved. The spar will be used as a tail spar for a skidder that will yard the logs out to the railway. (CRM 7704)

In 1933 I went as foreman for Thomsen & Clark, with George Bell as my head rigger, and opened up Rainy River. This was where the Port Mellon pulp mill is now, on Howe Sound. One morning I was running the donkey and everything was going along very well, but at 11 a.m., George picked up his lunch pail and walked off. He had a habit of just taking off from the job, and for no apparent reason. Perhaps for this reason, he was nicknamed, variously, Panicky Bell, Calamity Bell or Flighty Bell.

One day George was taking our old gasoline Cat up the skid road, into the woods. There was an old bridge and I told him not to go over it, but around it. He went over it and down, settling on a rock. Before I could hear about it he got some powder and put it between the rock and the Cat, and blew her. He was lucky, he blew the Cat right out, and not right up. When I came along he was continuing down the road as if nothing had happened. That was typical of the way in which he worked.

Another time George and I were working together, moving our skidder overland for two miles. It was a very heavy piece of machinery and the work was slow. Each night our timekeeper, who was the brother of the boss, would ask us how far we had gone that day. Then he'd call up the boss and give him a report of the progress for the day. The timekeeper was called Buckfist because he had one hand amputated.

One morning we started out up a slope of about 400 feet. When we got to the top, we decided to change the line. We didn't snub the line very well and the donkey went right down the hill, 100 feet farther back from our start. The lines were tangled all over that hill. It took us all day to get these untangled and back on the drum. At dinner Buckfist asked George how far we had gone that day. George told him we had gone 900 feet. Buckfist got on the phone and told the boss how well we had done that day. The next morning, first thing, the boss roared up in his motorboat. He was more than surprised to see we were farther back than we were two days ago. George was quick to explain that no one had asked in which direction we had gone 900 feet.

Another time George and I were going over to our booming grounds in an old boat. There were no panes in the windows and the weather was bitterly cold. George threw a bucket of gasoline on the deck, and threw in a match. The whole ship looked as if it had caught afire, from the shore. They thought we were gone for sure. But we were warm for a little while, anyway.

Lloyd Rodgers

A Fordson Little Tugger, used to ground skid. The slack cable to the right is the haulback. (BCFM L-4)

cold-deck, eventually ending up at roadside, or in the water if an A-frame was used. This entire method became known as cold-decking.

Cold-decking with steam-powered equipment was often a difficult proposition. The big, heavy machines were hard to manoeuvre over rough ground. After oil replaced wood as boiler fuel, it had to be hauled to the cold-decker, or pumped up through pipes laid for that purpose. It was often difficult to get water to these machines: typically they were located on hilltops or ridges. When a fire started, it was hard to move them to safe ground.

When they became available, gasoline engines were used extensively on cold-deck machines, being lighter and not requiring a water supply. The fire hazard of gasoline was a drawback, though, and the early gas engines had very little power, so when diesel engines came on the market in the late 1920s most cold-deckers were equipped with them.

In the 1920s, when gas engines were first available, many gyppos tried using them to replace the engines on steam donkeys, but with little success. In 1921, Skagit Steel and Iron Works of Seattle came out with a machine called a Little Tugger, consisting of a Fordson gas-powered farm tractor, the back wheels replaced with sprockets linked through a chain drive to a set of winches. Many were used throughout the industry for loading and, occasionally, cold-decking and yarding. Abernethy Lougheed bought the eighth one of these machines produced and used it near Mission. It was later used by Wellburn Timber at Cowichan and now rests at the British Columbia Forest Museum in Duncan. The significance of this odd-looking machine is that it was the first even mildly successful gas-powered donkey used in the Northwest.

The evolution of overhead and high-lead cable yarding systems required a means of loading the logs hauled into the home spar onto the rail cars or trucks used in the next phase. It was a standard practice with animal and steam-powered ground-lead skidding to haul the logs up onto a landing, from where they were rolled onto the vehicle that took them away. But building landings cost money, and with this system it was difficult to select logs in order to build a well-balanced, optimum-sized load. It was

First and second loaders with hayrack boom setting tongs on a big fir log at Copper Canyon in the 1940s. (UBC BC1930/546/870)

rarely possible with elevated cable-yarding systems to drop the logs parallel to the rail or truck road, and the output of the new yarders overtaxed the capacity of this primitive loading method. Therefore it was necessary to develop loading machines powerful and fast enough to keep up with the yarding equipment that fed them and the capability of the rail systems to haul away the loaded log cars.

The solution involved a number of methods of lifting logs and lowering them onto the carrier. With ground-lead logging it was necessary to erect a separate loading device. A common version consisted of two short spars, or A-frames, with a set of lines running between them, crossing both the landing and the railway. A spreader bar with a set of tongs hanging from each end was lowered to the landing, hooked onto a log, lifted and swung over the car, then lowered.

The first Lidgerwood steel spars to arrive on the West Coast came equipped with a swing boom and engine. The yarder dropped logs in a pile where the boom could reach them. A single tong was hooked on the log, by the head or second loader, at a point slightly behind its centre of gravity. When the tong line was hauled in, it lifted the near end of the log up against the bottom of the boom, raising the far end of the log off the pile. The boom was swung over the rail car and the log lowered. This method became known as "heel-boom" loading, and has been a basic principle of log loading until the present time.

With the advent of independent wooden spars, loading mechanisms were incorporated into the rigging. Guylines with blocks hung from them were positioned so that tong lines could be used to pick up the logs and load them onto the cars. Initially, this was done with a single tong

A Comox Logging crew loading flatcars with a two-tong hayrack boom. (CDM P200–829a)

hooked onto one end of a log. The raised end of the log was lowered onto the car and the tong released, then hooked onto the other end, which was lifted onto the car. Alternatively, a log was hooked at its centre of balance and lifted onto the car in one pass. It was a clumsy method and hard on the rail cars.

The next stage was the development of the two-tong or "duplex" loader. This system required a specially built loading donkey equipped with two reversing drums. Several variations were possible with duplex loaders. When loading rail cars it was common to first hook one end of the log and drag it onto the car, holding it in position with the tong. The other tong was set and the lower end of the log loaded. With logs weighing as much as twenty tons, the machine onto which they were being loaded had to take a considerable amount of abuse. Rail cars were capable of taking this rough treatment, but when trucks came into use they had to be loaded more carefully. When used to load trucks, both tongs were set and the entire log lifted and lowered onto the trailer.

Duplex loading, particularly onto rail cars, required more precise teamwork than any other

When I worked in the woods I ran a duplex, loading. It was said to be brought out by Riley of Bloedel and Donovan in the States. A Riley loader is what it was first known as. A duplex used steam like nothing else. And a duplex was a noisy bastard but, Christ, could you log with it. If you ran a duplex you'd be out on the nose of the machine, on the front. Usually they had a yarder and loader on one sled, so a duplex man had to have his eyes open. You watched your loader. He'd take the tongs and put it on his shoulders, up behind his head. He'd run down that log and you'd feed slack. He ran down and he'd duck his head and throw the tongs. If you were a good leverman you'd do it right and catch the log right. Clunk! He'd run up the bloody log and get away. You'd bring it up and put it across the car. And the other loader would pick the other tong up and run to the other end, plunk it on and he'd get the hell out of the way and you'd pick the log up and swing it over the car and put it where he wanted it. That was loading.

Bob Swanson, who grew up at North Wellington, near Nanaimo, and started logging at age fourteen, firing a steam donkey. He earned an engineering degree in 1942, became chief inspector of BC railways and was a key figure in keeping the Aero Timber operation going in the Charlottes during the war. He has published four books of the finest logging poetry ever written. For many years he has been designing and building air whistles for all the major North American railways.

A wood-fired Alberni Pacific hayrack loader picking logs out of Great Central Lake in the 1920s. (W.J. Moore photo; Finning Tractor collection)

phase of logging at this time. The loading crew consisted of the leverman, or engineer, who ran the donkey, and as many as three teams of two loaders, each consisting of a head loader and a second loader. The loading procedure, in its high-ball form, began with the head loader hoisting a set of tongs, which might weigh as much as ninety pounds, onto his shoulder and carrying them out onto the log pile, with the leverman carefully feeding him line by easing ahead on one of his levers. The head loader selected his log, determined where to set the tongs and, in a move requiring skill and agility, bent forward, ducked his head and threw the tongs at the log. He then ran clear. At the moment the tongs came down around the log, the leverman hauled back on the lever, setting the points of the tongs firmly into the log and lifting one end up and onto the car.

The second loader was ready with his set of tongs and ran to set them in a similar manner at the other end of the log. The leverman lifted that end on and gave some slack in both lines so the next team of loaders could repeat the process, while the last pair could recover their energy and nerves. This was high-ball logging at its most hazardous. A split-second miscalculation by a leverman could maim or kill a loader. A misstep by a loader hustling across a log deck carrying a heavy set of tongs could, and often did, lead to serious injury.

At a fairly early stage, an alternative to guyline loading was devised—boom loading—which also had a number of variations. The most basic boom-loading systems used on wooden spars were the double-tong boom and the single-tong or heel boom. One double-tong system, known as the "hayrack" boom, consisted of two parallel logs from forty to eighty feet long held about three feet apart by spacers. One end of the boom straddled the spar and was held up on a short line, while the other end hung from a guyline, all arranged so the boom swung in a wide arc over the log deck and the road. Two loading

A tracked shovel at Victoria Lumber & Manufacturing, equipped with tongs. Although the primary use of the shovels was to build roads, and tongs were first used to move logs aside while clearing right-of-way, this use soon evolved into a loading method. (Jack Cash photo; UBC 1930/547/#53 VL1487)

lines on separate drums raised and lowered the tongs, and a third drum swung the boom in one direction. A counterweight, usually a chunk of log or some old rail car wheels, was hung from a guyline or nearby tree and pulled the boom the other way. In some places this counterweight was called a "squirrel" because it was constantly going up and down the tree. The hayrack boom was capable of lifting logs and gently lowering them onto the carrier.

Construction of a similar boom was made easier with the use of cast-iron fittings patented by Fraser Valley logger Claude McLean and manufactured by the Columbia Block & Tool Company in Vancouver. The McLean boom used a square beam in place of the hayrack, with tongs hung from the patented fittings. The McLean boom and homemade versions of it were used extensively throughout the Northwest.

The single-tong boom was similarly constructed, with the addition of heel rails, usually railway track, attached to the underside of a hayrack-type boom. Requiring only one loader, it could heel a log and, keeping it under complete control, lower it onto the load. The boom systems, along with duplex loaders, were used in the vast majority of cable-logging systems for twenty-five or thirty years after World War One.

There were some methods of loading used only occasionally until after World War Two.

My company operates a skidder outfit and we have been logging almost exclusively with skidders since 1910. I'm nearly ashamed to admit that we produce all of our logs with skidders that were purchased twenty-two years ago.

Each skidder and loader is mounted on a steel wheel car with swivel trucks. The overall length is only twenty-seven feet. The loading engines are ten by ten with four drums. The skidding engines are twelve by twelve with four drums. The boiler is sixty-six inches with a seven-foot-high firebox. Total weight of each machine is about 150,000 pounds.

Our booms differ from any others I know of in that they are eighty feet long and we log entirely around the setting without moving the skidder. Perhaps I'd better explain what I mean. We set the skidder at right angles to the track; our first road is usually straight out from the skidder, then we go right around the tree. Some of the roads come behind the boiler but the boom is long enough and hung high enough to swing over the boiler and pick the logs up. We use a block if the logs are large but usually a straight line with a crotch and two tongs, one drum swings the boom to the log and a counterbalance swings the boom to the load. This rig works well and has the advantage of permitting a full setting to be logged without moving the skidder around the tree. Frequently we have averaged 300 logs per side per day for a month and our best day was 856 logs loaded by one side in eight hours. In skidding we try to make 1250-foot roads our maximum and seldom go out farther than that, usually about 1100 to 1200 feet.

Robert J. (Bob) Filberg

Victoria Lumber & Manufacturing's Chemainus shop crew in 1939, after rebuilding the Lidgerwood skidder in the background. The chief engineer, Robert Swanson, is fourth from right in front. This photo was reproduced as a mural in Chemainus. (J.M. Howe collection)

They were moving around men quite bad in those days. There was no compensation, inspectors or safety or anything like that. It was pretty bad. I remember one day while I was working I heard the whistle blowing on the machine over a little ways, and something had happened all right. The brakeman got killed. And they just took him off to the side, put a blanket over him. Kept on logging.

It was not a good time at all. They like to talk about the good old days, but now they have safety committees, and they have their union—we didn't have a union in those days. The IWW had been there and the OBU was organizing at times, but when I arrived in '24 there was no sign of a union around to be had. I didn't see anything like that at Bloedel's.

Skate Hames, who logged at many locations around the Gulf of Georgia beginning in the 1920s.

These were different types of self-propelled loading machines, some of them adapted from machines designed for different or additional tasks. Locomotive cranes, used for constructing railways or lifting derailed machinery back onto the tracks, were employed in picking up logs along the tracks, either as part of the logging operation or to recover logs that fell off rail cars. When track-mounted shovel loaders made their appearance, they were occasionally employed in loading by equipping them with a longer boom and, usually, a single tong. The first track loaders—Marion being one of the popular brands—were, by modern standards, fairly cumbersome machines and were expensive loaders if dedicated to that task. Their importance was that, unknown to those who first used them, they were the precursors of an entirely new logging system which would dominate the coastal BC industry forty to fifty years later.

The initial-phase logging system that dominated the industry by the early 1920s—essentially, high-lead yarding—profoundly changed the nature of the job. What had always been hard work for the men on the ground was now also exceedingly dangerous. The power and speed of the machines employed, particularly by the big companies, made them capable of smashing through practically anything that got in their way. Few young trees survived to grow the new forest, and accident rates escalated.

Logging wasn't all like this. There were smaller operators, camps where the owners worked in the woods with their crews. Not all of these small, independent loggers were angels; there are plenty of stories about gyppo operations that were real horror shows. But for a small logger to prosper he had to provide decent working conditions or he couldn't keep a competent crew together. And he couldn't afford the big powerful machinery capable of destroying forests on the scale of the big railway camps.

This era of big-scale logging has often been characterized as the Glory Days of Logging, primarily by those who did not work in it very much. For those who were actually out in the woods, and got rain in their lunch buckets, it was about as glorious as war. Logging then, as now, was a job—hard, dangerous, often wet and miserable, and for the most part conducted in remote camps, far from family and friends. The people who worked in it all their lives—the fallers, buckers, riggers, yarding and loading crews—did not think it was glorious. But, being neither whiners nor snivellers, they did their jobs without complaint and, more, with a pride and dignity that outside observers often tried to romanticize as a grand and glorious calling.

What occurred in the Pacific northwestern forests over a period of twenty years or less was an accelerated process of heavy industrialization, in which the powerful technology and the methods of managing workers that had been developed in the factories of Europe and eastern North America were adapted and applied in the most valuable forests in the world. The major players in this transformation were, as always, non-resident financial interests. But there were others who came to work and live, to raise their families, to build and become citizens of the communities springing up along the coast. These were the people who created another sector of the industry, the "citizen business" concerns as H.R. MacMillan called them.

Through the period 1920–1950, there was lots of room and timber for all these players. They coexisted, with a great deal of movement of individuals between them. What allowed and fostered the diversity of logging firms in this era, apart from the availability of sufficient timber, was the development of different types of machinery. Some of it, like railways and the enormous logging machines required to keep them running, suited the big companies. But there was also equipment coming on the market that individuals and small, enterprising firms could use to advantage in a free and active timber market. The development of the internal combustion engine and its application to the business of logging in the 1920s soon had a profound effect, not only on the nature of logging itself, but also on the social, economic and political character of coastal BC.

PACIFIC
COAST

V The Railway Era

Previous page: The Pacific Coast Shay was the most popular geared engine on the coast. This one, shown working at Franklin River in 1947, belonged to Bloedel, Stewart & Welch. (UBC 1930/188 51451)

Over the twenty-year period that ended with the Great Depression, there was a rapid and major increase in the use of logging railways on the BC coast. The markets and mills created a demand for logs, and capital was available to establish large and expensive operations. Also available was a wide range of logging railway and related equipment, designed and manufactured for conditions in the Pacific Northwest.

Specifically, a type of locomotive had been developed to meet the particular needs of railway loggers. During the early, tentative stages, locomotives were either small engines such as "Old Curly," used at mines, on construction projects or in mill and factory yards, or they were larger engines originally used on main-line railways throughout North America. Neither type was particularly suitable for logging. Either they lacked the pulling power required on steep logging grades, or they were too big to negotiate the sharp corners and cramped quarters of a logging show.

These deficiencies were overcome by a Michigan logger, Ephraim Shay, who designed, built and used the first geared locomotive. Previously, locomotive power was transferred from the steam cylinder to the drive wheels by a rod connecting the piston to the wheel. These were known as rod engines. Shay's first engine, which he patented in 1881, consisted of a single vertical cylinder with a powerful piston, which turned a driveshaft, which was connected by a set of bevel gears directly to the axles of the drive wheels. Loggers, who like to rename everything they use, called Shay's engine the "stemwinder." Shay soon added a second cylinder to his engine, and then a third, a design retained as long as they were produced. He sold the patent to the Lima Locomotive Works in Ohio for $10,000 and went back to logging. By 1893, 450 Shays had been built.

The Shay was a distinctive-looking locomotive. Its boiler was offset to the left, with the three-cylinder engine set just in front of the cab on the right side. It had a short wheel base for operating in confined areas, and was mounted on separate trucks with no frame, making for a flexible machine that could stand up to the rough conditions of crude logging railways. The gearing system allowed enormous power to be

---

I started work firing a locie in 1933 for McNair Shingle Company. They were hauling shingle bolts from Lois Lake to Stillwater. They took over the railroad when Brooks, Scanlon & O'Brien left it in 1928. They used it to haul bolts out themselves then. Brooks Scanlon took all their locies out except one, and McNair got that, the smallest, the small forty-five-ton rod engine. They used that. She stayed here until 1954, when they got another locomotive.

My dad worked for McNair, up in the woods driving a team hauling bolts out of the woods. They had a jack ladder in a chute, with the bottom end down in the water. The guys would bring the bolts in a big bag right up to it and there was a platform each side of the bottom of the chute. The guys would stand there and poke them in. This chain had dogs on it, put a bolt up in that, the dogs would grab them and carry them up and dump over the top end into the railway cars. Haul them out and dump them in the chuck.

Then they had another jack ladder down there, like they had when they took them out of the lake, only it lifted them into a crib. They had quite a rig for dumping them off, too. Each car had big stakes up at each end, heavy timbers, twelve-by-twelve timbers up at each end, a big framework, and they were well braced. They'd load these cars up from one end to the other, just round them up on top. They had a cable laying on the deck of the car and it went up through these walls, through sheaves, and they hung out over the end so they could grab ahold of them with the engine. They'd bring them down and hook them to an anchor at the end of the dock. Then they'd get ahold of the other end of the cables with the engine and take a rare on them with the engine. It would tip them right over as slick as anything.

Woodrow Runnells, who grew up and worked at Stillwater as a locomotive engineer.

applied to the drive wheels, enabling the Shay to operate on hills two to three times as steep as those which could be climbed by rod engines. Because of the way power is applied to the drive wheels on a rod engine, they tended to pound out the light tracks used in logging. The Shay, however, applied steady power to the wheels through the gears, thus avoiding this problem.

Shays quickly became the most popular logging locomotive in the Northwest, making up forty percent of all the steam locies used in BC. In 1945, when the last Shay came out of the shop, 2,761 of them had been built. They came in several models of varying sizes, and were so commonplace in the Northwest that one of the biggest models, introduced in 1927, was called the Pacific Coast Shay. This ninety-ton engine was built under licence by Willamette Iron and Steel Works in Portland. When the patents held by Lima expired in 1922, Willamette came out with its own version of the Pacific Coast Shay, almost a direct copy. One of these engines, the only one in BC, was bought new by McDonald Murphy in 1928 and retained when the company became Lake Logging and, eventually, Western Forest Industries. But the Willamette was not a success, and after building thirty-three engines the company quit producing them.

This classic wood-burning Shay locomotive was acquired in 1925 by the Theodosia, Powell Lake and Eastern Railway, a subsidiary of Merrill & Ring Lumber Company, for its Theodosia Arm operation. It was used there until 1937. This view shows the Shay's distinctive vertical cylinders. (PRM)

Scottish Palmer Logging's Climax at Mile 54 on the Canadian National Railway's Cowichan Lake line in 1926. (H.W. Roozeboom photo; Royal British Columbia Museum B93 984.3.472)

Merrill & Ring's fifty-ton Shay, "One-Spot," was brought to Squamish new in 1910 by Norton & McKinnon. After working at Cracroft Island and Duncan Bay, it returned to Squamish in 1926. Its last owner was Comox Logging. George Percy, Merrill & Ring superintendent, is second from left in this photograph of Squamish camp, 1927. (Ed Aldridge photo; Ed Aldridge Collection #15)

Above: Abernethy Lougheed Logging bought this Heisler in 1926 for a mainline locomotive on its line into what is now Golden Ears Park. It could haul more than thirty loaded cars at a speed of twenty miles an hour. (Leonard Frank photo; VPL 5558)

This Climax locomotive, shown at Salmon River Logging in 1939, was one of the biggest models built. Left to right: George McDonald, brakeman; Scotty McCampbell, fireman; Percy Stacey, engineer. (CRM 11136)

In 1888 the second geared engine, the Climax, went into production in Corry, Pennsylvania. The Climax company had sold an earlier crude model of locomotive, mounted on a wooden frame. Shawnigan Lake Lumber's *Betsy* was one of these. The new model Climax had a single engine on each side of the boiler, mounted at a forty-five-degree angle. The pistons drove a crosswise crankshaft, which transferred power to a longitudinal shaft and through bevel gears to the wheels. The action of these two cylinders, thrusting down and back, produced what has been called a "pile driving action" which made the Climax one of the roughest-riding locies in the woods. But they were powerful and rugged engines, and accounted for about twenty-five percent of the locies on the BC coast. For about twelve years Climax buyers had the option of purchasing engines with double-flanged and corrugated wheels for operating on pole roads. By the time production ceased in 1928, almost one thousand Climaxes had been sold.

A third gear engine, the Heisler, was introduced in 1894 in Erie, Pennsylvania. It too had a single engine on each side of the boiler, but they were angled down under the boiler and connected to a central driveshaft, which was connected through enclosed gears running in oil, to one axle on each truck, or set of four wheels. Connecting rods went from the powered wheels to the other set on the truck. The Heisler was faster and quieter than the Shay and Climax. It was not as popular as the other two, and only seventeen of them were used in BC before production ceased in 1945. Heisler also made a Northwest version, the West Coast Special, and one of them came to BC, first to Malahat Logging, then to BC Forest Products.

The fifty-odd rod engines employed in coastal logging operations had certain advantages over the geared engines. They were smoother, faster-running machines and cheaper to operate and maintain. But they cost fifteen to twenty percent more than a comparable gear engine.

The makers of most rod engines—the most popular model among coastal loggers was the Baldwin—introduced modifications in their designs to meet the challenge of the geared engines. Two drawbacks of conventional rod engines were overcome by shifting the water tanks from a tender that trailed behind the locie to side or saddle tanks positioned beside or over the boiler. This shortened the unit and put more weight on the drive wheels, improving traction.

In practice, most of the larger companies utilized both types of engines to their best advantage. They used Shays, Climaxes and Heislers along the spurs and in the steeper, rougher ground at the back end of settings. They also used them to move the big skidders, which weighed as much as 130 tons. Hauling this much weight over poorly built spurs with light steel rails was a hazardous task for which the more powerful, slower, geared engines were better suited.

The faster rod engines were used on the relatively well-built, flat mainlines. Some sort of transfer point was established, to which the geared engines hauled loaded cars from the landings. The mainline rod engines then took over and hauled the cars to the dump. This system normally required one mainline locie for every three or four geared engines. Rod engines were used on grades of about two percent or less, with geared engines handling slopes up to twelve percent. If the mainline had high adverse

I was bull bucker at International Timber and on New Year's Day 1928, Pete Harambourne came up to Camp 7 in quite a sweat to take an inventory of cedar poles piled away up the track. It was quite amusing. He looked for everyone in the office and scaler's shack for someone to do this work, passing me by with scarcely a glance. Pete just could not find anyone that was able to do this work. It was a Sunday anyway, and most of the guys with a touch of the morning-after were not interested. So there was nothing to do but ask me. I said okay, how do I get up there. He said, "I'll send you up on the Baldwin mainline locomotive, and he will stay with you until you are finished." So, away I went.

The work was some distance from camp on a newly laid spur road. There was just myself and Bill Johnson, the engineer. Going over this newly laid spur road that had practically no traffic on it, the big locie, for some reason, spread the tracks and settled right down in the mud, all wheels off the rails. We were not far from where I had to be, and Bill said, "You go ahead and I will try to get her on the tracks." So I went to work out of sight of the Baldwin. I guess I was away about two to three hours, and when I came back Old Bill was sitting in the cab with the fire door wide open, and the locie still mired down between the tracks.

Bill said, "It's no use Arch. This old girl is settled down for the night, and we might as well stay with her. It will take a section crew half a day to put her on the tracks, and when we don't show up Pete will be up here raising hell." Well, Bill was a guy with quite a sense of humour and not given to getting excited over anything. So there we sat, with the windows shuttered off with the side flaps and the fire door open. Since it was New Year's, Bill had fortified himself with a mickey of rum. I heartily commended him on his foresight. We stayed there till past suppertime and, sure enough, Pete himself came looking for us. After his usual blast when things went haywire, Bill poked his head from behind the curtain and said, "Pete. This is one hell of a railroad you have—it is not the longest in BC, but it sure is the widest."

Archibald Kerr

Brooks, Scanlon & O'Brien's big Baldwin rod engine, shown here at Stillwater in 1926, spent only a few years in BC before being shipped to the US. The water tank is at the rear of this model. (H.W. Roozeboom photo; VPL 1598)

Bloedel, Stewart & Welch's Baldwin Four Spot at Menzies Bay in 1926, a year after it was built. The water tanks on this model were located alongside the boiler to improve traction, giving it an appearance similar to the Porter in the photo on page 138. (H.W. Roozeboom photo; VPL 1516)

Shown at the Kelsey Bay log dump, Salmon River Logging's 2-6-2T Porter was built in 1923. It was later used by O'Brien Logging and Canadian Forest Products until it was scrapped in 1961. (Percy Stacey photo; BCARS 82115)

grades, one of the larger geared engines, such as the Pacific Coast Shay, was probably used.

There were several other rod engines available, although most of the ones used in BC were either Porters or Montreal Locomotive Works engines. One Vulcan, a 45-ton model, worked at Island Logging, and Alberni Pacific acquired an Alco, a 135-ton monster, in 1920. For a time it was the largest logging locie on the coast.

During the 1920s, a variety of gas locomotives appeared. Their development parallelled that of trucks, as gasoline-burning internal combustion engines became available. There was, in fact, some crossover of machinery at this stage, with attempts to put railway wheels on trucks. Most of the relatively small gas locies were used

The Humdergon was composed of two sets of boxcar wheels bolted to a 10-by-20 timber frame with a wooden deck and a truck bunk just ahead of the rear wheels. A drive chain connected the sprockets on the wheel axles so she drove on all wheels, and another sprocket and drive chain was connected to the rear end of a Fordson power unit. The trailer was another set of wheels and bunk connected to the Humdergon with a piece of railway steel. The braking system was something unique. The front wheels had a regular boxcar brake set-up which the driver could control by turning the wheel and winding up a chain. Then we had a cable running all the way back to the trailer—to apply the brake on the trailer the driver had to leave his seat, cross over to the other side of the machine, and pull a big lever which had a ratchet which would hold the required tension. The whole log hauling system cost $500.

After we had been hauling logs for several months with the old Humdergon, we decided that it needed a new body or framework. The old wood was getting all rotten and the holes that the bolts went through were all worn and enlarged, and she weaved along the track like a loose jointed bobsleigh.

We bought some new 8-by-16 stringers for the main frame, several cross timbers, etc., even new bolts and nuts and new 2-by-12 decking and bunks. Caesar Scott who was pretty handy with tools was the chief carpenter and jack knife engineer. We fitted all the joints together and bolted them all down tightly onto the two sets of railway wheels, moving the Fordson engine onto the frame and she was ready to go. We decided to take her on a trial run. Harper grabbed the controls and away we went up the track. There was six or seven of us aboard.

The old frame was waterlogged and heavy but with this new body which I guess was lighter, the new Humdergon was sailing right along making 15 or 20 miles per hour.

We came to a curve about a mile up the track and all of a sudden she jumped the track and bounded into the ditch. No one was injured and we spent a half hour jacking her back onto the track.

It didn't take Caesar long to figure out what was wrong. He grabbed the monkey wrench and loosened all the bolts on the corners of the frame as well as the bolts holding the wheels in place. This done, we went merrily on our way weaving down the track just like the old one had been doing for years.

Wallace Baikie, a partner in Baikie Brothers Logging, in his memoirs, Rolling With the Times (Campbell River, 1985).

by small loggers, particularly along the east coast of Vancouver Island. Often they used lines built by bigger companies, or rented steel rails and installed two or three miles of line along an old railway grade into small patches of timber.

Between San Francisco and Vancouver there were many manufacturers of these early non-steam locomotives. Most of the ones used in coastal BC were manufactured here. One of the most widely used was Westminster Iron Works' Tugaway, which first appeared in the mid-1920s and ran on either steel rails or pole roads.

The Reliance Motor & Machine Works of Vancouver built a Fordson-powered machine that barely rated as a locomotive. Baikie Brothers Logging acquired one of these contraptions in the early 1930s. They called it a "Humdergon," a name used by loggers for just about any unusual machine lacking a suitable label. It, or a similar machine, was built by Reliance for the Dawson and Taylor logging company, which logged the same area in 1924.

This machine and others like it, haywire as they now appear, played an important role in launching the Baikie brothers and many other small loggers struggling to get and stay afloat in the tough years of the Depression. They couldn't move a lot of logs in a hurry, but they could keep up with a small yarder and take care of the hauling phase of a four- or five-man operation. In this regard they were a kind of bridge or transition between railway and truck logging. If the ground was flat enough and some rails could be scrounged—not too difficult a task in those days—it was cheaper than building a fore-and-

Engineer Ed Aldridge at the throttle of Merrill & Ring's fourteen-ton Plymouth gas locie, at the Squamish dock, 1926. (Ed Aldridge Collection)

I brought the new gas locie into Squamish. They had this little seven-ton one at Duncan Bay. I operated that for them over there. The old man—that was old George Moore, the General Manager—could foresee diesel power coming in, although we never had one. They finished logging before they got to that. That's one reason I stayed with the little gas locie. The money was a little less than the steam locie engineer, and I figured he'd be getting the bigger one pretty soon, but they never did get one. This was a hell of a nice machine. It had a Climax engine. Nothing to do with the steam Climax. It was built in Clinton, Iowa, I think. The locomotives were built in Plymouth, Ohio, they took the name of the town for the name, Plymouth Locomotives. The story that goes along with this one, this is the one they got for the Theodosia Arm show. Like I say, the old man, he come and talk to me about these gas locies and everything, so I knew he was getting this twenty-ton Davenport for Theodosia. I was still at Duncan Bay then. I heard that they got this machine there, it was all written up there someplace. The old man come over to Duncan Bay and I said, "How's the new gas locie, Mr. Moore?" And he says, "A prettier machine I have never seen, nor a more useless sonofabitch." He was sending her back. He says, "It ain't got enough power. I been trying to tell them, it ain't got enough power in it, but them engineers won't listen to me."

He did get a twenty-ton Plymouth after that. I asked the old man one time, he was talking about it, he was over in camp one time. "How much gas does she burn, Mr. Moore?" He said, "I think a gallon rings the bell once."

Ed Aldridge, a locomotive engineer who went to Squamish with the first bargeload of equipment to establish the Merrill & Ring Lumber Company camp in 1926. He worked for the company until it left in 1940.

Cheeseblocks, used to hold these big logs on Abernethy & Lougheed Logging's flatcars at Alouette Lake in 1926, were later replaced with stakes as the logs got smaller. (H.W. Roozeboom photo; VPL 1486)

aft or plank road for a truck, which was also a pricey piece of equipment for a small logger. Before bulldozers were readily available to build truck roads, these little gas locies filled an important role for small operators.

There were a few bigger camps that also utilized gas locies in the 1920s. Skagit Iron and Steel marketed a four-wheel, chain-driven locie powered by a six-cylinder Buda gas engine, and at least two of these were purchased by O'Brien Logging in 1928 for its camp in Simoom Sound. The Maritime Motor Car Company of Vancouver produced gas locies in the two- to five-ton range, including one for Brooks, Scanlon & O'Brien in 1917 and another for Merrill, Ring & Moore's Duncan Bay camp. They were used for switching cars and road maintenance and construction. George Moore, MR&M manager, was an enthusiastic proponent of their use, citing the fire safety they provided, along with the fact they only required one man to operate. The Maritime company did not remain in business for long, and in the 1930s the most popular gas-powered engine was the US-made Plymouth.

The period of greatest expansion in the railway logging sector was during the decade between the end of World War One and the onset of the Depression. As described in a previous chapter, several of the big railway camps were established before 1914: Comox Logging at Courtenay, Victoria Lumber & Manufacturing at Chemainus, Bloedel, Stewart & Welch at Myrtle Point, Brooks, Scanlon & O'Brien at Stillwater, International Timber at Campbell River, the Squamish Timber Company at the head of Howe Sound, Abernethy Lougheed in the Fraser Valley and the Powell River Company at Kingcome Inlet. During the boom that followed the war, several more came into existence, while those already in operation expanded by building additional camps or adding more equipment.

Near the end of the war, in 1917, the Robert Dollar Company opened a 200-man camp at Deep Bay to supply its sawmill in Dollarton. A second camp was taken over a few miles north, at Union Bay, in the following year, and by the time the two camps were consolidated in 1925,

The speeder at Brown & Kirkland's Pitt Lake camp was used as a crummy to transport men to and from work, rain or shine. 1928. (Leonard Frank photo; VPL 5589)

it was one of the largest shows on the mid island, with three locomotives and fifty-seven cars on thirteen miles of track. By 1931, a billion and a half feet had been cut and the camp closed.

Also at Deep Bay in the same period, Thomsen & Clark bought out a small railway logging firm, Essery Timber, which had about four miles of track running south to Bowser. They had two Shays working the spurs and a Baldwin on the mainline running to their main camp at the northwest corner of Horne Lake. In 1936, they shut down their Harrison Lake railway show and used the equipment and steel for a fifteen-mile line up Rosewall Canyon. The entire operation closed in 1940.

In 1927, the Canadian Puget Sound Lumber Company built a railway camp at Jordan River which eventually ran four locomotives, including two of the biggest Shays produced. Established to supply the company's mill in Victoria, this camp later operated under the subsidiary Island Logging Company, employing 150 men.

Merrill & Ring, a Seattle company, began logging land purchased by the Merrill family in the 1880s. The Duncan Bay camp started in 1919, and over the next seven years they logged by railway there and in Menzies Bay. In 1922, the company took over the Rock Bay operations of BC Mills Timber & Trading, one of the few companies to experience a decline through the prewar period. Merrill & Ring logged leased land over an extensive rail network reaching back to the headwaters of the Salmon River until 1933. In 1924, a camp opened at Theodosia Arm behind Powell River, and it ran until 1938. Another camp was opened at Squamish in 1927

In 1922 Moritz Thomsen and E.B. Clark of Seattle decided to back an Everett logger, J.D. Essery, in the acquisition of some timber on Vancouver Island for a logging operation. They bought about a billion feet of timber, mostly from the E&N, some from Canadian Collieries and some from the Victoria Lumber Company. I had been working for Clark for ten years in Montana and in 1923 he asked me to come out and take charge of their office. Essery's logging donkeys and rigging were brought up from his worked-out show near Mount Vernon and Thomsen & Clark bought him out in 1924. That year our line was completed into Horne Lake, around which lay our principal holdings.

Maximum grades were two percent on the main line and six percent on spurs. Once we did build a short spur with ten percent grade for 150 to 200 feet. The highest elevation our railroad reached was 1300 feet, and we surely thought we were up in the air at that point. We had no skidders and although we used a slackline skyline from a raft along the south side of the lake, which reached out 1600 to 2000 feet up the hill, we never did reach the top timber. In the thirties we didn't consider it worthwhile to build the railroad up the mountains, and we thought we had too much invested in railroads and equipment to consider trucks. Bert Welch logged that area with trucks after we sold out in 1939 to MacMillan.

Our first locomotive, bought in 1922, was a small forty-ton Shay, geared locomotive, purchased somewhere in the interior of BC. Next we rented a locomotive for about six months from the PG&E. It cost us a few hundred a month, and we returned it when we acquired a new Baldwin saddle tank rod locomotive. It cost us about $23,000. We later bought a large Shay and operated using the three locomotives.

John Burke, who left Thomsen & Clark in 1939 and became manager of the BC Loggers' Association in 1942.

P.B. Anderson's Green Point Logging camp and dump on Harrison Lake, c. 1935. It was built in 1931 and employed 200 men; in 1937 it was closed and the equipment was moved to Salmon River. Married quarters are at the right, with bunkhouses for single men in the centre and the shop on the left, beside the dump spur. (W.F. Montgomery photo; UBC Anderson Collection, Box 1)

and operated steadily until 1940. The company also owned Vancouver Bay Logging, which operated a 140-man railway camp under manager David Jeremiason. All these were substantial railway camps, together logging close to 1.5 billion feet of timber over a twenty-year span. Unlike most other companies, Merrill & Ring held on to the privately owned lands it had acquired in the previous century and began logging them again in the 1980s, making it the oldest continuously operating forest company in British Columbia.

At Beaver Cove on Vancouver Island two Seattle lumbermen, E.G. English and Fred Wood, started the Nimpkish Timber Company in 1917. By 1922, Nimpkish was running five camps, employing 300 men and operating five locomotives with 100 cars on twenty miles of track. This operation grew steadily, adding a large sawmill in 1924 when the name was changed to Wood & English. Brown & Kirkland took over the operation in 1942, under the name Beaver Cove Timber and, later, Canadian Forest Products. In 1944, controlling shares in both these companies were acquired by Poldi Bentley and John Prentice. This company became the longest-lived, one of the largest and, with its numerous big bridges and trestles, the most spectacular railway logging operation on the coast. It is the only operation in BC to continue railway operations into the 1990s.

Further south, at Kelsey Bay, Dewey Anderson moved in rail equipment from Harrison Lake and opened the Salmon River Logging Company during the economic recovery in 1937. This was one of the biggest logging operations on the coast, with forty miles of track, three locies and 150 logging cars dumping 8

Two Shays sit in Merrill & Ring's first camp at Squamish, 1927. (Ed Aldridge Collection #20)

Bloedel, Stewart & Welch's Northwest gas shovel being hauled across the 1100-foot-long bridge over the Campbell River, a few miles above the Glory Hole, late 1920s. (BCARS 73737)

Lamb, he liked to try and farm, log, everything. I guess they had one of the first strikes by a group of loggers at Lamb's camp. He had a big cookhouse, it overhung a gully. There was quite a space under the front of the cookhouse. He had some pigs so he put the pigs under there for the winter. Well, springtime come and things warmed up. The stench was something godawful. Finally the crew asked him if he couldn't move the damn pigs out of there. Oh, no. The pigs had a perfect right, they were going to stay. So everybody went on strike. That night he had news for them, he canned them all. He let them have a week or two in Vancouver, then he hired them all back again. They did move the pigs.

One engineer come up out of Mohun Lake one night with a big load of logs. Had her wide open trying to get over the hump. A damn cow was in the middle of the road. I guess the light confused the cow and it tripped and fell down. He finally ended up running over it and killed the cow. Lamb never said anything. But when the engineer got his paycheque at the end of the month, it was less one beef. The engineer went in to see old Lamb to see what the hell he was doing. "Did you think I was going to stop the damn train just for a cow? Do you think we're logging or ranchin'?" Old Lamb said, "You could have stopped. It'd just take a bit of time." The engineer says, "Yeah, about two hours to go all the way back down and start over again with that damn old Climax."

Glen Duncan, who runs the Link and Pin Logging Museum at Roberts Lake. He grew up and logged for many years in the Sayward area.

million feet of logs a month at Kelsey Bay. In 1946, the Andersons sold it to Westminster Shook Mill.

In 1925, Bloedel, Stewart & Welch opened a big camp at Menzies Bay, utilizing several locomotives—two Baldwin rod engines, four Shays and a Climax. Five years earlier the company had made its first move out of the Powell River area when it set up a camp at Union Bay. It was sold within a few years because the timber, averaging twenty-four inches in diameter, was considered too small to be profitable. The Union Bay equipment, along with the equipment from the logged-out camp at Myrtle Point, was moved to Menzies Bay. This operation worked far back into the Sayward forest and in 1942 opened one of the biggest camps on the coast, Camp 5, at Brewster Lake. It housed 550 men and 63 families. By the end of World War Two, the Menzies Bay camp was working on seventy miles of track. In 1953, two years after BS&W merged with the H.R. MacMillan Company, the railway phases were converted to trucks. In the same bay there were two other railway camps through the interwar period, Lamb Lumber and Campbell River Timber.

The area experiencing the most rapid development of logging activity during the 1920s was the Cowichan Valley, where several railway shows were set up by the companies already there, as well as by newcomers who joined them. Going into the war there were, apart from many small loggers, three major companies logging on the lake—Cowichan Lake Lumber, VL&MC and Empire Logging. The latter was by far the largest with four camps. During the war the Genoa Bay Lumber Company contracted to log Empire's timber. The logging systems had evolved from oxen and horses to steam-powered ground-lead yarders and roaders, all working within easy reach of the lakeshore. After 1913, when the E&N Railway completed a branch line into the lake, most of the logs cut around the lake were hauled out to Crofton by common carrier.

The first substantial addition to the Cowichan Lake logging fraternity after the war was the Canadian Puget Sound Lumber and Timber Company in 1920, which purchased 991 acres of timber from VL&MC to supply its mill in Victoria. The following year Henry "Jesse" James moved his railway camp from Mission to Cottonwood Creek on the north shore of the lake where he joined C.C. Yount to take over

A duplex guyline loader loading skeleton cars at Hillcrest Lumber's Sathlam division above Cowichan Lake, 1940. (Wilmer Gold photo; IWA Local 1–80)

I'd been fooling around with the speeders a little bit, then I get a job running the speeders. That was for Cowichan Lake Logging Company at that time, they went in there in '27. They managed to stay afloat until '31 and I run the speeder until '29 until they shut down and everything went haywire then. It was a gas car, runs on the railroad. They were just small in them days. Probably five feet wide and twelve feet long. You could haul ten, twelve men. In them days it was different. They used to load them down with twenty, Christ, you couldn't stop them coming down them grades there, whooo, when I think of it now. Some guys sitting down on the running board with their arms around each other to stay on—you couldn't see any speeder, just guys hanging on all over. There was lots of guys hurt on them things, I tell you. I don't know how they ever got away with it.

It had a gas engine. Continental Red Seal, I think they called the engine. Four cylinders, about the size of the Model T Ford, forty horsepower. The big crews, in the morning when the crummy would pull out, the locie took them. They were just in the boxcar with seats lengthwise inside them. Camp 6 they had, well if all the fallers were going, the section crew and everything, it took two boxcar loads to haul them all. But they had two different lines, and usually whichever lines had the most men would take two and the other line would take one. They had two locomotives there. I had to get up early because the firemen and engineers had to go out early and get steam up on the donkeys you know, an hour ahead of the crew. And that was one of my jobs, taking them out in the morning. And during the day moving the section crew around, I had to do all of that. There they used to bring in gas in barrels at that time, and oil. They come in on the scow and I had to unload all that stuff and take them and put them up into the oil house. Wasn't too bad until they got a gas skidder in there. Brand new one, I think that was 1929. Well that thing used to eat a drum of gas a day. Then I had more work to do.

Cliff Hallberg, born at Ladysmith like his brothers Ken and Archie. He logged through his working life.

Lake Logging's Willamette geared locomotive at Rounds in 1939. Only thirty-three of these locomotives were built in Portland after 1922, and this is the only one used in BC. It was purchased by McDonald Murphy Logging in 1928 and ended up with Western Forest Industries, before it was scrapped in 1956. (Wilmer Gold photo; IWA Local 1–80)

Empire's timber and mill. Yount was the former managing director of Empire.

James pushed the lake's first logging railway four miles up Cottonwood Creek and built a camp about halfway up it. His crew yarded and loaded with two Lidgerwood units working off wooden trees. They dumped logs in the lake, and those not used by the company's Medina mill, at the site of the present Youbou mill, were towed to the foot of the lake and hauled out on the E&N. A third camp opened across the lake, with a roader skidding the logs down a skid road into the water.

Part of the strategy of the Cottonwood Creek operation was its anticipation of the long-heralded CNR line along the north shore of the lake, en route to Port Alberni, which would preclude the need to dump the logs, tow them down the lake and reload on the E&N. A few months before this finally occurred, in 1925, James sold the entire logging operation, including the logging contract with Empire, to Campbell River Mills of White Rock for a reported $5 million, a staggering sum in those days. The new owners brought in an eighty-ton Climax and one of the first steam shovels on the coast to build roads. But Empire cancelled the logging contract and within a year Campbell River Mills was broke. James took over again, but in the spring of 1926 he died. Empire Logging took over his company and changed its name to Elco Logging, making it the largest logging company on the lake.

In the meantime a number of other companies were established on the lake or at other locations in the valley. In 1922, Channel Logging set up at Mile 70 on the CNR, about five miles south of the lake, while Scottish Palmer built a camp another five miles south, connecting its four miles of rail to the CNR main line. The following year the McDonald Murphy Logging Company moved into the area from Campbell River, Genoa Bay Lumber opened another camp at the head of the lake, Lake Logging started Cat logging at Craig's Landing and VL&MC pushed a line from the E&N line into the Robertson River drainage that connected with the McDonald Murphy lines. In addition, Matt Hemmingsen left VL&MC to set up his own logging company. He logged up Wardroper Creek and also contracted to Elco on the north shore of the lake. It was a record year for the coastal logging industry, and the Cowichan area was probably the major centre of activity.

Through the 1920s the Cowichan Valley was

Caycuse in 1944, when it was still know as Camp 6 and was operated by Industrial Timber Mills. Several companies used this site beginning about 1900. (Wilmer Gold photo: IWA Local 1–80)

the scene of unrestrained logging activity. In 1928, combined production around the lake was probably as high as 300 million feet. The many large camps, two main line railways and a couple of small sawmills provided lots of jobs, which in turn financed the settlement of land around the lake and down the valley to Duncan. Many families lived on float houses, which were easily moved from camp to camp if a worker was fired or quit, or the camp closed. As early as 1926 or 1927, it was apparent that the intense assault on the Cowichan forests was going to create a timber shortage. As this dawned on people there was a reshuffling and consolidation of the companies at the end of the decade.

Early in 1929, a few months before the bottom fell out of the world economy, a second mill opened at what was by then called Youbou. It was built by Frank Beban and his partner, Don Hartnell of Hammond Cedar Mills on the Fraser River, operating under a lease from Elco. This mill and the old sawmill were combined by Beban and Hartnell into a new company, Industrial Timber Mills (ITM), which obtained its timber from Cowichan Lake Logging's Camp 6 across the lake at Nixon Creek—a site later known as Caycuse. The expansion included building a townsite at Youbou to house workers from the big mill complex which was now able to ship lumber out on the CNR at prices competitive with other Vancouver Island mills.

By the end of the year it was apparent the great Cowichan logging boom was over. Lumber consumption throughout the world fell steadily, and in 1932 its price was less than half the average for the previous seventeen years. That year Britain began buying large amounts of lumber from the Soviet Union, and the US Smoot–Hawley tariff restricted access to the US market. In 1933, McDonald Murphy went broke and Lake Logging took it over. This was one of many casualties, as only ten of the nineteen companies operating in the area in 1929 remained in 1933. Most of the timber owned by the failed companies was folded into the newly created companies, particularly ITM, which was in a good position to profit from improved timber prices and markets that appeared in late 1933 and through 1934. In 1933, Hammond Cedar took over ownership of ITM, and two years later ITM bought the Canadian Puget Sound Company's Island Logging Company at the west end of the lake, a site known as Camp 3 and the final end of steel for the CNR. Cowichan-area log production doubled between 1935 and 1936 and remained steady for the rest of the decade and through the war. In 1946, ITM, which had acquired more timber and swallowed a few more smaller operators in the district, was taken over by BC Forest Products.

The between-wars boom also established an extensive rail logging presence in the Alberni Valley. After the failure of the Anderson mill in the last century, it was observed that a logging railway was needed to bring out logs from the distant reaches of the Alberni Valley, beyond Sproat and Great Central lakes. The extensive timber stands that were available along the thirty-five-mile Alberni Inlet were of little value without a mill in the Alberni area because of the difficulties encountered getting logs down the west coast of the island to a market. In 1905, the Barkley Sound Cedar Company built the first mill of any significance in the valley. It was expanded and, after passing through several owners, became the Alberni Pacific Lumber Company in 1915.

In the early 1920s, the E&N Railway ran branch lines in to Sproat and Great Central lakes to encourage the sale of timber lands acquired in the original land grant. BS&W, in partnership with the King-Farris Lumber Company, took 200 million feet of this timber and in 1926 opened a mill on Great Central Lake, with a railway logging operation to feed it. At about the same time, BS&W acquired extensive timber leases at Franklin River and in the depths of the Depression decided to build another sawmill, this one at the head of the inlet, to be called the Somass mill. A few months before the mill

When I was fifteen I went to work at Bainbridge up here, six miles up the railroad from Alberni. They had a logging camp there, Bainbridge Lumber Company. They built that to get long timbers. They wanted that certain kind of timber. They started logging in 1917. Six of the Wests worked in Bainbridge. Five boys and the old man. I went firing locomotive there. Fired a woodburner, believe it or not. It was no snap, that one. It had a water tank, and then you piled the wood all over the top of the water tank, wood slabs you got at the mill. You'd dump a couple loads of logs, and fill up with slabs.

We didn't go that far out. The railroad ran right along the bottom of the mountain. They had several spurs off it, up or down the hill. My job was to keep the steam up on the locie. When you needed more wood in, you wouldn't be opening the firebox door when they were pullin' hard because all the wood would go out the smokestack. We had quite a hill to pull up from down at Beaver Creek, up the hill to where Bainbridge was. A heavy grade, so you had to prepare for that, and get a good fire and lots of steam up, and lots of water in the boiler, so you wouldn't have no stoppages, and then fog it to her all the way up the hill. Sometimes you wouldn't make it, you'd have to stop and get the steam up again. Or you'd have to stop for water. You could run low on water and if you put more water in, it brought the steam down right away, of course. There was no playin' about it, it was a hard job, you had to be pokin' in the wood. And you had to get up and help the brakey, too, when you dumped logs. It was busy. All the Shay could handle was two cars, because they were long timbers, 120 feet long. They had trucks, they called them logging trucks, not a truck, but just a set of wheels with a bunk on them for the logs to set on. And they coupled up when they were empty, all coupled together. But when they were going to load the logs on, they spread them out to the length they wanted. Sometimes the trucks would be eighty, ninety feet apart. One end under one, the back end under the other. The belly would bend down and drag on the ties. A hundred and forty feet long, a hundred and forty-two feet, some of them.

They had a swing donkey right beside the track. It pulled them in so they come right alongside the track and up onto a landing, and then they rolled them onto a car. But sometimes, when we had no area for the landing, we'd have to yard them right to the track, and we had a spreader bar arrangement on the guyline—a heavy guyline, heavy rigging—and lift the whole damn log up, believe it or not, lift those big logs up with a spreader bar, to get ahold of it in two places about thirty or forty feet apart. We'd lift them right up, and swing them around and get them on the car.

I pretty near got killed, a log dropped right on top of me, doing that. It pulled the spar tree down, and then the whole works come down, and I was underneath it. I knew things were going down, I could see the fire flying up in the rigging, and I knew something was broken, and I dove underneath. We had a log on the car, and that saved me, we had one log on the cars, and were putting another one on. So when things started to happen, there was no other place to run but underneath that other log. I dove underneath that and the whole rigging come down. The engineer dropped the log he had up in the air, well he couldn't do anything else. It come down and laid right on my back. I got outta there quick, it had come down and touched me on the back, and it killed the one man. That finished the Bainbridge Lumber Company, they never logged after that.

Al West

opened in early 1935, the Franklin River camp, which would soon become the biggest railway camp in the world, was opened. Within a year it was producing 100 million feet of logs a year. At this point, BS&W was probably the biggest, most modern and aggressive company on the BC coast. With Prentice Bloedel as president and Sid Smith as manager, it was ready to expand even more.

At this same time, 1936, another major player appeared at Alberni. H.R. MacMillan had been building his lumber export business steadily since World War One, and now sold about forty percent of the lumber shipped out of BC. The mills that supplied him were becoming restive and joined the rival export agency, Seaboard. MacMillan's response was to move into the sawmilling business himself, and in 1936 he bought the Alberni Pacific mill, including the company's railway logging equipment, and by a hair's breadth beat Sid Smith to the purchase of the Rockefellers' 17,000 acres of prime fir in the Ash Valley, beyond Great Central Lake. If the 1920s had belonged to the loggers of the Cowichan Valley, the MacMillan and Bloedel interests were about to turn Alberni into the centre of the BC coastal logging industry.

These were the island's major railway shows, scattered from the Nimpkish to the Alberni valleys. They were the biggest companies of their day, having expanded by pioneering new holdings and by buying up other operators and consolidating. Their size and their rapid growth was facilitated by the availability of powerful steam yarders and compatible railway equipment. But by no means did they comprise the entire logging business, or even all of railway logging.

During the same time span, dozens of small, independent railway logging companies emerged, particularly along the east coast of Vancouver Island, most of which sold their logs on the open market rather than delivering them to their own mills. Typically, they used small steam or gas locomotives, usually castoffs from other operations, on a few miles of secondhand or rented steel. Most did not last long, as the patches of timber they were built to log were not large.

The northernmost of the small Vancouver Island railways was B&D Logging's two-mile line at Hyde Creek, east of Port McNeill, which used two gas locies from 1945 to 1949. A few miles south, Beaver Cove Lumber and Pulp built a six-mile railway in the early 1920s.

At Elk Bay, north of Campbell River on Discovery Passage, Elk Bay Timber had a three-mile line with a Climax and fifteen skeleton cars. In 1924, the Elk Bay manager, Fred Brown, joined F.W. Kirkland and took over Elk Bay under the B&K Timber name. In 1926, a spark from the Climax set fire to their biggest trestle so they pulled up stakes and moved to the Pitt River. At this time the Scottish Palmer company, burned out of their show in the Cowichan Valley, moved into Elk Bay with a bigger Climax and ran a line up toward Stella Lake, under the names of Stella Lake Logging and, in 1931, Discovery Passage Logging. This line was taken over again by B&K Timber, which ran it until 1937 before selling the steel to Campbell River Timber.

Across Discovery Passage on Quadra Island, a number of small logging railways operated, in addition to the Hastings outfit, which used two locies out of Granite Bay from 1894 to 1914. Comox Logging had a small, short-lived show in Open Bay in 1927. Abbott Timber used a Climax in Drew Harbour from 1912 to 1915. The Wilson & Brady company used the same Climax and may also have worked two Shays pulling disconnects out of Heriot Bay from 1915 to 1919.

From 1923 to 1927, Craig–Taylor Logging operated with a gas locie at Oyster Bay. The Gwilt Lumber Company used a railway to log for its mill on the Puntledge River at Courtenay in the early 1920s, and also built a pole railway up Supply Creek, on which they used a Day–Elder truck equipped with bell wheels.

The nearby Royston Lumber Company also used a railway to supply its mill, operating on a twelve-mile line up the Trent River valley with two small steam locies. This was one of several small railway companies owned by Japanese Canadians that operated from the early 1920s until 1942, when the government seized them all in the midst of wartime hysteria. Deep Bay Logging, run by a man named Kagetsu, had ten miles of main line running from Fanny Bay up Cougar Smith Creek. A twenty-five-ton Climax was used until 1932, when it was replaced with a sixty-ton Shay, prompting the railway inspector to shut the line down until bridges were strengthened. This camp, including timber, was also seized in 1942. The H.R. MacMillan Company acquired the timber, and in 1944 Jack Fletcher's Tsable River Logging started cutting it under contract. There were as many as a dozen such railway logging camps on the coast, including a 150-man camp in Port Neville, operated by loggers of Japanese descent. These camps were all seized, and their owners and crews interned until after the war. During the same period there were a number of camps on Vancouver Island owned by Japanese companies, logging privately owned, exportable timber and shipping the unprocessed logs directly to Japan. Unlike the fishing industry, to which many of the internees returned, Japanese Canadian loggers were never compensated for the seizures, and it appears none of them returned to the logging industry.

A few miles south, at Mud Bay, Sing Chong Logging used a small gas locie and a single trailer on a two-mile line up Wilfred Creek from 1922 to 1924. On Denman Island during the late 1920s and early 1930s, Henry Bay Logging used seventy-horsepower gas Tugaway locies on four miles of line, in conjunction with the Climax once used on Quadra Island. This well-travelled Climax started out with VL&MC and also was owned by Frank Beban, Timberland and Rounds Burchette.

Two railway outfits worked at Qualicum—Lake Lumber, owned by H.G. Johnson, which ran for three years in the early 1920s, and a small mill and logging show near Hilliers that utilized a thirty-five-ton Climax from 1919 to 1936. In the same area, Locham Singh used a small Shay at Virginia Lumber Company near Coombs during 1919 and 1920. At Nanoose, Superior Lumber used a small gas locie on two miles of track from 1922 to 1924.

South of Nanaimo, near Cassidy, in the early 1900s, the Ladysmith Lumber Company bought a small British-made Manning Wardle steam locomotive called *Nanaimo*, which had been brought to BC in the 1870s for mine work. In 1914, *Nanaimo* was replaced with a twenty-four-ton Forney locie acquired from the New York Elevated Railroad. In the 1920s, the company bought a thirty-five-ton Shay, and in 1927 was reconstituted as the Nanaimo Lumber Company—adding a seventy-ton Climax at about the same time. It went bankrupt in 1931.

At Extension, Frank Beban began railway logging in 1922 to feed his mill, which had a contract to supply the Wellington coal mine with timbers. He used a 32-ton Heisler he bought from Capilano Timber and two small Shays during the ten-year operating period of this camp. Beban's line was tied in to the coal company line, the E&N line, the Nanaimo Lumber line and an extensive main line and spur network owned by Timberland Development

I started up on the '38 fire up at Elk Falls. We were riding out of Courtenay then, fighting fire. They shut everybody down, and everybody was fightin' fire. Two bits an hour. We'd get on the boxcar right at the old Riverside Hotel there, and oh, hell, we'd ride all the way up. We were right near the Quinsam River, where we started on that fire. Every day we'd be two miles farther down the track, two miles down the track. The wind was just blowin' her towards Courtenay all the time.

We couldn't do much. We'd cut fire breaks there, maybe a couple miles ahead of the fire and figure maybe we could stop her. I remember we were at Oyster River, and all the Cats that Comox had, they had quite a few Cats in them days. And they had big fire breaks, wide fire breaks there. The wind died down, and boy, the fire laid right down too. I can remember reading the Vancouver paper, "The fire was under control, it was all quiet there at Oyster River, they had it stopped at Oyster River."

The next day the wind come up and we were ten miles back down the track. Just like that. The winds controlled the fire. They never burned slash then. See, we had lots of slash behind us there and that's what kept her going. You get a wind there and boy. We were at the town of Headquarters. That's where Comox Logging had their main camp there and they had twenty to thirty married quarters there. We were in there at night and they figured the fire was going to take out the town. We had everybody on flatcars there, all ready to go. And you know, when that fire come with about a forty-mile-an-hour wind, and she just blew her over that mountain right behind Headquarters, that Constitution Hill, she just went over that mountain there just like a bunch of jets. Next morning she was all black, just all black for miles there.

Sam Telosky

Loading flatcars at Capilano Timber Company in North Vancouver, 1927. (Leonard Frank photo; VPL 5978)

Company of Ladysmith. The latter started with a 25-ton Climax and later owned the biggest model Climax, a 115-ton engine, as well as a Porter rod engine. Timberland was sold to VL&MC in 1929. Another small line operating in the Ladysmith area, Sorain Singh's Eastern Lumber Company, acquired one of the Lenora, Mount Sicker Railway's 26-ton Shays from 1919 until 1924. At Sooke, Kapoor Lumber operated a railway in what is now the Victoria watershed.

Throughout the interwar period there were dozens of similar small railways scattered throughout the southern coast. Most of them did not last much beyond the late 1920s, though, because by then trucks were more readily available and could get into the small patches of timber available to independent market loggers.

Apart from the large east coast Vancouver Island railway camps and the numerous small operations, there were a few more significant railway camps established after World War One.

The most visible logging railway on the coast was started near the end of the war by the Capilano Timber Company. Its main line ran from the dump at the foot of Pemberton Street in North Vancouver up the Capilano River, with spurs up the mountains on either side. It was powered with Heislers, Climaxes and, in later years, a gas switcher engine. The Capilano show was a substantial one, with peak production at 150,000 feet a day from two high-lead and one skyline sides employing more than 200 men. Through the 1920s, the company fended off attempts by municipal governments using the watershed to put an end to logging in the valley. The company was in financial trouble by 1931 and sold out to Sisters Creek Logging, which continued working for two years until the Greater Vancouver Water District succeeded in banning logging. The equipment was piled onto fifty-five rail cars and shipped to the Industrial Timber Mills at Youbou.

One final logging railway company worth noting is Bernard Timber, which logged at several locations during the 1920s, including Orford Bay in Bute Inlet and Port Neville. Bernard owned at least three Climax locies and a Shay.

Barnard Timber & Logging Company's 100-man camp at Port Neville, 1926. This was a typical operation of a larger independent logging company of this period. When the timber was exhausted, which usually took four or five years, the steel rails were pulled up, the buildings hauled onto a log boom and the camp moved to another location. Barnard moved here from a similar operation at Orford Bay in Bute Inlet. The oil-burning Climax locomotive was serviced in the shop on the right. (H.W. Roozeboom photo; BCARS 67990)

The extensive development of railway logging created the need for a variety of related tasks and equipment. The most basic of these was the construction of road grades. Throughout the West, from the late 1800s, contractors employed a large, transient population of railway labourers building main lines as well as industrial lines for logging and mining companies. In many cases these crews were Chinese, unorganized, underpaid and discriminated against by exclusionary laws preventing them from engaging in higher-paying work or acquiring logging rights. Most of the work on logging lines was done by hand, using picks and shovels to level out a roadbed on which the ties and steel rails were laid. Commonly, labourers contracted to build "stations" consisting of 100 feet of roadbed for an agreed-

Hand grading in a rock cut south of the Mamquam Bridge at Merrill & Ring's Squamish operation, 1927. To build the grade, the rock was blasted with a mixture of stumping powder, gelatin and dynamite. A twenty-four-inch railway was laid; the rock was loaded on small, hand-pushed cars and dumped where fill was needed. (Ed Aldridge photo; Ed Aldridge Collection #32)

Building railway grade by hand at Brooks, Scanlon & O'Brien's Stillwater camp, 1919. The workers hand bucked windfalls into manageable-sized chunks and rolled them aside, then used picks and shovels to level the grade. (Link & Pin Museum)

upon rate. Blasting powder, packed in hand-drilled holes, was used on rock cuts when they could not be avoided. The primary objective was to keep grades on main lines below two and a half to three percent, and spur lines below about six percent.

Some larger companies used scrapers—Bagley and Washington Iron Works made two popular models—operating off a high-lead rigged spar to move large volumes of dirt or gravel. The haulback line dragged the empty scraper back to where the cut was made, and the mainline pulled it ahead, filling it with dirt and dragging it as far as 700 or 800 feet along the grade to where a fill was needed. However, equipping, setting up and running such an operation was an expensive proposition.

Only the biggest companies could afford a much more expensive piece of equipment, the steam shovel. Few BC logging camps listed

Industrial Timber Mills wood-burning steam shovel building grade at Camp 6 (Caycuse) in 1935. It was a track-mounted machine operated by Clarence Whittingham of Youbou. (BCARS 90216)

Wood & English's Marion steam shovel at Englewood, 1926. It was mounted on a rail car, with a turntable. (H.W. Roozeboom photo; VPL 1666)

shovels in their equipment inventories in the 1920s, when steam-powered versions were the only ones available. Capilano Timber bought a Marion steam shovel in 1918 to move blasted rock. Nimpkish Timber also had one, and in the late 1930s Industrial Timber Mills was using a steam-powered shovel to build grade at Camp 6 on Cowichan Lake. In the 1930s, when gas-powered tracked shovels became available, and rail lines were being pushed into rougher country, most of the larger logging railway operations had at least one, and often two or three. Marion and Northwest brands were probably the most popular. After about 1935, bulldozer blades began to appear on the small crawler tractors that some companies used to skid logs to railside landings. But it was not until World War Two and later that these machines became critically important in road building.

In many situations it was cheaper to build a trestle than to make extensive fills of low spots. All the railway companies employed bridge crews, either on their own payroll or under contract, to span the numerous creeks and rivers found on the coast. In the 1920s and 1930s, large numbers of skilled axemen, most of them Scandinavians, were employed building trestles from round wood found on site. As logging operations moved into rougher country, they required more elaborate bridges, and engineers skilled in their design and construction were hired by some of the larger camps. The essential piece of equipment was a pile driver, a versatile machine using a small steam engine that could raise and drop an iron driver to embed pilings in the ground, or lower pre-constructed sections into position. Larger, more complex bridges were usually constructed out of milled timbers.

A standard bridge consisted of a series of "bents" made of four or five poles or piles driven into the ground, with the outer two sloped in toward the top at a slight angle and held together with cross-bracing. These bents were usually placed fifteen feet apart. The pile-driving engine

Green Point Logging's Northwest track shovel at work near Harrison Lake, early 1930s. This was the most popular of the early shovels; at a later date many were also used as heel-boom loaders. (W.F. Montgomery photo; Frank Rustad collection)

Merrill & Ring Lumber's Mamquam bridge under construction in 1927. The eighty-nine-foot high bridge was built on contract by Sam and Bill Culliton. The pile driver on the north side was skylined across the canyon during construction. (Ed Aldridge photo; Ed Aldridge Collection #35)

had a drum with a line to yard the piling into position and hold it while the driver hammered it into the ground. When all the piles in each bent were in place, they were spiked together with braces, cut off level at the top and capped with a twelve-inch square timber or hand-hewn log. All of this work was performed by agile bridge crews, who clambered over the structure and rode up and down on the pile driver line.

Longitudinal braces held the bents in position, and six stringers, three under each rail, were laid longitudinally across the top of the bents. The thirty-two-foot stringers covered three bents, and were spiked into place. Ties were laid crossways on the stringers, and guard rails spiked along both ends of the ties. When the bridge was completed and the pile driver removed, the steel rails were laid and the structure was ready for use.

These old railway logging bridges were some of the most impressive structures on the coast.

Construction of the Vernon Creek trestle, International Timber Mills, 1933. At right is Larry McMullan, BC Forest Products' chief forester from 1946 to 1972, working on a summer job while attending forestry school at the University of British Columbia. (BCARS 90208)

There was usually five or six men on a bridge crew, and the engineer. And there was a spooltender, he had ropes for moving the pile driver into place. I would do that, I would wrap the rope around the niggerhead on the driver and the engineer would crack it—it was just ropes, with four blocks. You had fore-and-aft, and starboard and port. And you could turn the driver, pull it back or ahead or any way you want. You always got the leads right in place, you see.

The driver was mounted on skids. There were twelve-by-twelve caps up top on the bents. On them we had a two-by-six with tallow on it, we'd grease it. The skids we were awful careful with, we never got them on gravel or anything. Keep them nice and slippery so they'd slide easy. Then you had batter wedges. The outside piling had what they called a batter, a slope on it. We had a wedge that we spiked on the cap alongside this two-by-six we had. Then we pulled the driver over on this wedge to give us the right slope. Later on, the leads that guided the driver were free, so you didn't need the wedges any more.

The pilings and everything were just trees out of the woods, any I worked on. Most of them bridges that were built at Camp 6—they had an old mill at Youbou, long before the one that's there now—they cut all timbers. Caps were twelve-by-twelves, and the braces were three-by-ten. You put the caps on top of the pilings, and then you put the stringers on top, they were nine-by-eighteen. You put five stringers, one under either side of where the rail went, just to either side, and then one in the middle. Ties on top of that again, and then your rails.

If you got into some real big ones, like the Bear Creek Bridge, that was a little different. They had 180-foot spars, four or six of them, I forget now, for pilings in each tower. At the bottom they built big cement buttons to set them on. Altogether, from the creek level to the top, right in the centre, it was 247 feet. The spans were 90 feet between them towers. Then there was about 25 or 30 feet of truss work on top of the towers.

What they did there, they rigged a big skyline across, a two-inch skyline, I guess it was. They got the grade up to the bridge. Then they strung this skyline across, with a Tyler system on it. They could pick up a Caterpillar or a shovel. And they'd send it across the canyon. They went on building grade. And they used this big rig for building the bridge, to pick up those 180-foot towers they had there. They were built all standing too, one tower at a time. I think they were about six months on that thing. There was a big crew because you had to have a lot of truss work, you had to have framers, they were all working on the beach. When they got a section made they'd just pick it up with the skyline, and put it in place.

Cliff Hallberg

Above and left: The Haslam Creek bridge under construction for Comox Logging, near Ladysmith, 1940. The bridge was 196 feet high and 546 feet long, and had a five percent grade. The bents were 80 feet wide at the bottom, tapering to 16 feet at the top. Eighty-foot-long stringers were used between the bents. (Mauno Pelto collection)

The Bear Creek bridge, engineered by Kelso Blakeney and Russell Mills in 1939 for Malahat Logging, was 249 feet high and 517 feet long, making it the largest wooden bridge in Canada. The following year, Bill Eastman built the elegant 196-foot high Haslam Creek bridge for Comox Logging at Ladysmith. Most of these bridges did not survive the railway logging era, as it was cheaper to build truck roads around the ravines and canyons. The exception is Canadian Forest Products' Nimpkish operation, which continues to operate a railway and still requires bridges. Many of Canfor's beautifully designed trestles were built by Russell Mills and his son Bob.

Bridges were the most vulnerable part of the whole logging system. If a spark from a train set them on fire, or they burned in a forest fire, the entire operation came to a halt. Consequently, they were usually equipped with water barrels or sprinkler systems to keep them wet in hot weather. In some camps, special tank cars were used to wet them down. When fires did occur, one of the first actions taken was to get all the logging and railway equipment out of the area and across the vulnerable wooden structures.

Inclines were often part of railway shows. These were short sections of track, usually less than a mile long, built to lower loaded rail cars

A steam engine wasn't built to run something like a snubber. It was designed to pull, not hold a load back. On Mount Sicker for about a mile down the hill Joe Kerrone had this incline. On the snubber there was this big brake drum, eight inches wide and twelve feet high. They were lined with pine because there is sort of an oil in pine and it lasts better than anything else we could find. Every week those linings had to be renewed—that was my job—so they would be ready for Monday morning. I had Sunday to do it in.

The blocks were already cut. We bought them from a woodworking shop in Duncan. The holes were drilled, everything was there. All I had to do was put them on, put the bolts in and tighten them up.

But apart from these brakes being such a size, they still had to have a three-quarter-inch pipe running cold water on each brake drum or they'd take fire. There was that much pressure.

Before they got this snubber they had this big twelve-by-fourteen Willamette, a big yarding donkey. They put the line on that. I was there one day when it blew up. There was so much compression built up—the brakes were too small to hold—the steam oil in the cylinder blew up. Just like you'd shoved in half a stick of dynamite. Chunks of cast iron flew in every direction. When we looked there was the piston. No more cylinder. It was gone. That's when Joe decided to get a better machine that would handle this job.

Joe Garner, born on Saltspring Island in 1909. He started logging in 1918 and has never stopped. He has spent the last ten years writing and publishing a series of bestselling books on the industry.

Gustofson Brothers Logging used this enclosed, single-drum steam snubber on its incline at Misery Creek in Jervis Inlet, 1923. (Elphinstone Pioneer Museum 232A)

down steep slopes of up to seventy percent. They came into existence about 1920 and were used until World War Two. In the usual configuration, a steam engine with a large drum was located at the top of the hill and a single car pulled up with a wire rope. The car was loaded and lowered to a landing at the bottom of the hill.

In BC it was most common to use a steam yarder for this purpose, although some of the US equipment companies produced a special "snubbing" machine designed to hold a load back, rather than pull it forward. Also in the US, special bull cars equipped with huge, eight- to ten-foot-diameter blocks were used to give an advantage, but such refinements do not appear to have been used in BC. Inclines were used in situations where a large volume of timber was located close to a main line, but up too steep a grade for a locomotive to climb, at least without a lengthy and expensive series of switchbacks.

One of the best-known inclines was used by Joe Kerrone to bring logs off Mount Sicker in the 1920s. At about the same time, the Ellis Lake

Joe Kerrone's incline engine on the north side of Mount Sicker, near Duncan, late 1920s. This machine replaced a conventional donkey that had been used on the incline until the cylinder blew up from the pressure. (Darryl Muralt collection)

Green Point Logging's 2500-foot incline at Ruby Creek. (W.F. Montgomery photo; Frank Rustad collection)

Railway company on Redonda Island used an incline down 4,000 feet of thirty-five-percent grade. David Jeremiason's Vancouver Bay Logging also used one in Goliath Bay at the same time. In the 1940s, the Englewood logging division had an incline powered with an Empire skidder engine. When lowering heavy loads, the interlock was engaged and the brakes on both the haulback and mainline drums applied. Logs were hauled to this incline with a Waukesha 160-horsepower gas engine, and the cars lowered one at a time to the beach. Bainbridge Lumber used an incline to lower cars to its mainline in the Alberni Valley in the mid-1920s, and the Alberni Pacific company built several before World War Two.

Another well-known incline was the one built by P.B. Anderson's Green Point Logging above Harrison Lake at Ruby Creek in the 1930s. An extensive cold deck and skyline yarding show brought logs to the top of the incline where they were loaded on cars, which were lowered with one of the yarders. The grades on this incline were modest, although a 200-foot trestle was required to cross a small creek.

I built those buildings at Joe Kerrone's camp on Mount Sicker, for the Chinese fallers. The cookhouse would seat about twenty men, I guess. Usually they'd come in for a hot lunch. Joe came to me one day and said, "We've got to have something else up here, two more buildings. You'd better come up and do it." I said, "What are these for?" "Well," he said, "we don't like the Chinamen going into Duncan every weekend, spending all their money. We want these two little houses." I put them up. They had these professionals, hookers. The district was full of hook shops, that's the only way a lot of people had of making a living. I guess there was about six of them in the Cowichan Valley at that time. These girls would come up on payday, ride up the incline, spend the weekend and be gone by the time the rest of the men got there Monday morning.

Joe Garner

The log dump at Rock Bay, 1917. A cable ran from the brow log, under the loaded logs, to the top of the A-frame, and was hooked to the locomotive. It pulled on the cable, dumping up to ten cars at a time. (John Cress photo; author's collection)

By reputation, logging around Harrison Lake was a risky business. Chiefly because the Harrison River, the only access to the area, had always proven impractical. Harrison Mills, who also had a camp at Twenty Mile Point, had failed to overcome the obstacles involved.

Bill Dolmage and I went to look it over. We decided that river navigation was possible, and that it would be feasible to tow logs. Bill started, then, to build river boats.

We had a great deal of trouble in getting our equipment up to the site. The vessels that had brought the equipment from Knox Bay and Greenpoint were useless in the river. We got the *Goblin*, a Gulf of Georgia tug, in to help with the transport. She drew more water than any boat before or since on that river. The draft was just too great, she could not keep the tow line tight. In other words, she could not make time, upstream, against the current, faster than the downstream force of the river. We finally succeeded in moving her by attaching about a dozen auxiliary motorboats along the sides and back of the scow.

It was just the start of the Depression, and we had a few very rugged years. But we had plenty of timber, and all the very best labour we could use at twenty-five cents an hour. We sold our logs, delivered in Vancouver and paid harbour dues, for $7.83 a thousand average during the next two years. We didn't profit, but our financial position improved in other ways. We had credit, where we had started with none, and our depreciation was almost nothing.

Our start was difficult, not only because no one was buying, but also because no one would risk the Harrison logs. In desperation I offered J.D. McCormick of Canadian Western Mills two booms delivered, free and without obligation. If he liked them, he could buy them. He refused this offer, feeling certain I would go broke. But we brought him the logs and he bought them for about fifty cents off list price. They bought more from us after this, when we received orders from Alberni and the Mohawk.

But we were just getting by, and cheaper stumpage was obviously our only chance for survival. We were paying $2.50 to Harrison Lake Timber. They were a Wisconsin outfit, and held a timber licence. Percy Foster, our accountant, and I went back and had the stumpage reduced to $1.50. Later we had it further reduced to seventy-five cents.

In those days logs had to be properly "packaged" or they wouldn't sell. Norman Perry was our grader. We installed an electric saw, and he examined every single log. He took out all those with the slightest stain, and they went to the wood log boom. Only the perfectly clear logs went into the number one boom.

Dewey Anderson

Another version of the parbuckle system was used at Englewood in 1926. (H.W. Roozeboom photo; VPL 1651)

The bottom end of almost all BC logging railways was at the water, where logs were dumped for towing to mill or market. A variety of devices were used for unloading logs. Typical rail cars had log bunks with a set of cheeseblocks (triangular pieces of steel) to prevent logs from rolling off. They could be removed, allowing the logs to be rolled off whichever side of the car was convenient. Most dumps were constructed on the beach, parallel to the water line, or on wharfs built out into deep water. Normally, one of the rails at the dump site was elevated about a foot so the car would tilt. Peaveys or hand jacks were used on small operations to roll the logs off the cars. A more advanced method consisted of a gin pole erected over the dumpsite, from which a block was hung. A cable was hooked onto the locie or a special winch, through the block, under the logs and attached to a brow log on the water side of the dump. When the cable was tightened it hoisted, or parbuckled, the logs off and into the water. At Rock Bay a series of A-frames employed in a similar manner unloaded several cars at a time.

Some of the bigger railway camps used more sophisticated devices. One popular method was a rotating jill poke. It consisted of four sturdy timbers attached to a hub located behind the dump. The outer end of each timber was tipped with a sharpened piece of metal. As the cars were backed up, with the cheeseblocks released, the metal-tipped jill poke was engaged on the bottom log of the first car. As the car moved back the jill poke turned on its hub and the tipped end swung around and forced the logs off the car. The next timber of the rotating jill poke engaged the logs on the next car, pushing them off, and so on. A thirty-car train could be unloaded in less than half an hour with this device.

Comox Logging, at its Ladysmith and Royston dumps, used a self-propelled winch with a boom that travelled on a separate set of tracks behind the loaded cars at the dump, unloading them one by one.

Cathels & Sorenson used a boom mounted on tracks to lower logs into a Davis raft at Port Renfrew, 1926. (VPL 5615)

As they progressed, railway logging operations acquired an array of related equipment, the most obvious and necessary being rail cars. The first cars were known as disconnects. They consisted of two sets of trucks, each with two axles and four wheels, equipped with brakes applied by turning a wheel that tightened a chain and forced steel brake pads onto the wheels. A log bunk rested on top. To load, two sets of trucks were positioned in relation to the length of the logs being hauled and the logs lowered onto them. The weight of the logs held the trucks in place, although on steep grades the logs were often cinched to the trucks to prevent the train from being pulled apart. The advantage of this type of car was that they could haul any length of log. Their drawback was that they could not utilize air brake systems.

Disconnects were eventually replaced by skeleton cars, which soon became mandatory. A skeleton car consisted of two trucks of two or four wheels each, permanently connected with a sturdy timber, usually with a total length of forty-one feet. Longer skeleton cars were used for poles and pilings or longer logs. Air lines from the locomotive ran down the timbers. These cars came into use in the early 1920s and almost completely replaced disconnects within a few years. Regulations allowed for a percentage of disconnects, usually ten or fifteen, to be used in a train for the hauling of long logs.

Flatcars with wooden decks were used by a few companies and by all public lines, particularly for hauling small logs. They too came with air brakes, but were heavier and harder to haul. Comox Logging persisted in the use of flatcars,

Neil McDonald (McDonald Murphy) was a very fine man, and was ready to listen to new ideas, but he was an old-time logger, and like many of this type was much concerned with production. It was a railroad show, as most were in 1929, and the operation was some *distance back from the* beach. He prided himself on having for use as a crummy an old tourist coach, where we could sit back in luxury going to and from work. It used to amuse me to see the fellows regard this extravagance much as you would a commuter train, and Neil made a point of having the daily newspapers on board when available.

Archibald Kerr

partly because they carried larger loads than skeleton cars, and also because at its Ladysmith operation the company made use of the E&N line where flatcars were mandatory. In 1936, Comox owned 200 flatcars and 30 sets of trucks, whereas a small company with four or five miles of line and a single locie might have only half a dozen skeleton cars. In the mid-1920s the seventy-odd railway logging companies in BC were using about 1,500 cars and 500 sets of disconnects. Several Northwest manufacturers produced rail cars, including the Vancouver Engineering Works.

A variety of other vehicles were devised for use on the railways. Small, self-propelled speeders using gas engines performed a number of

In 1926 Merrill & Ring Lumber's loggers at Theodosia Arm were occasionally hauled to work on flatcars equipped with benches. (H.W. Roozeboom photo; VPL 1541)

This Brooks, Scanlon & O'Brien speeder, shown on the Pasha Lake line in the 1940s, provided some protection from the weather. (PRM)

This type of gas speeder became the standard railway crummy, used to haul supplies and do light maintenance work on tracks at Salmon River Logging, 1940s. (BCFM 3-2)

Two Barnard Timber loggers at Orford use a speeder on fire patrol, 1926. (H.W. Roozeboom photo; VPL 1420)

Barnard Timber used this well-equipped water car, pulled by an ex-Capilano Timber locie, for firefighting, 1926. (H.W. Roozeboom photo; VPL 1413)

functions. Some were used as crummies, while others hauled ties, rails and track maintenance equipment. Although most of these light machines were manufactured in company shops, a few were manufactured by machine companies throughout the Northwest. The crummies used to transport workers indicated the company's attitude toward its employees. Some were simply flatcars, with loggers going to and from work in whatever the weather had to offer. Others provided some shelter with the addition of a roof, and some were boxcars with no or few windows. It was not until late in the game that heated speeders, with comfortable seating and protection from the weather, were introduced.

In-camp repair and maintenance facilities were sparse by today's standards. Most were simple blacksmith shops with a forge and a few hand tools. The train crew did most of the maintenance work. Bigger camps had shops large enough to bring a locie inside, but at most camps work was done out in the weather. The VL&M Chemainus shop, Comox's Ladysmith shop and the shop at Elk River Timber's Camp 8 were examples of large, well-equipped, well-staffed facilities capable of rebuilding locies or yarders, but they were the exception. Severe damage caused by derailments or wrecks was usually repaired by barging the engine—or what was left of it—to one of the repair shops, such as Tyee Equipment, in Vancouver.

One of the major changes in the life of loggers brought about by the advent of railways was the creation of the mobile camp. Bunkhouses, cookhouses, offices and other camp buildings gradually assumed a common modular design. They were fourteen feet wide and forty feet long to fit on a skeleton car, and built on log skids so they could be dragged off and on the rail cars easily. Photos of old camps show a collection of almost identical buildings set up on blocks, with the railway running through the middle. Typically, they were located in the middle of a clear-cut, without a standing tree in sight. The camps were usually moved to keep the workers within ten miles of the active logging site. By the time it converted to trucks and abandoned camps after the war, Elk River Timber had gone

This small blacksmith shop served the needs of the Powell River Company's 120-man railway camp at Kingcome Inlet, 1916. (Link & Pin Museum)

through nine camps. BS&W at Menzies Bay had five main camps.

Train crews on a logging railway consisted of the locomotive engineer, a fireman and one or more brakemen. Their primary jobs were, respectively, to drive the train, to keep the steam pressure up, and to control switches and hook up air lines. A steam logging train consisted of a locomotive, weighing between 25 and 110 tons, and up to sixty or seventy cars weighing, loaded, as much as 50 tons each. Loads this size were only hauled on a main line track with a big rod engine. On the steeper, rougher spur lines, fewer cars would be brought down with a geared engine. At the smaller camps, of course, one engine, usually a geared Shay or Climax, was used from the landing to the dump, trailing five to twenty cars, depending on the grades.

Running a steam locomotive in the early 1920s was a formidable task. The locie consisted mostly of a boiler and firebox. The cab was a crude wood or steel box with canvas curtains and equipped with a couple of steam pressure gauges and a few basic controls. Herding this clanking, banging, lurching monster over the crudely built spur lines was a bone-jarring, precarious occupation. Leaving the landing with a loaded train was the beginning of an uncertain adventure. The wise and cautious engineer worked hard to keep the speed to a minimum, particularly on steep grades. Before the early 1920s, his only means of slowing his train was with the engine brakes and the hand brakes, which had to be set separately on the individual cars. Geared engines were rarely allowed above ten miles per hour on any sort of downhill grade, and even on a flat straightaway were not capable of more than double that.

The idea was to ease the load down the hill. Under the best of circumstances this could be a tricky task, and circumstances were rarely at their best. An almost infinite number of hazards could appear—broken rails, spread rails, rockslides, washouts, weakened bridges or trestles, oncoming trains or speeders, livestock on the tracks, mechanical failure. The list is practically endless.

Derailments were common and routinely dealt with. A car running off the track was no great problem; the train crew had various devices for getting it back where it belonged. A derailed locie was a more serious matter, requiring more extensive effort. A wreck with the locie over on its side and damaged, was to be avoided because it could shut down the entire camp. The biggest danger was a runaway, coming downhill

The Headquarters machine shop of Comox Logging, 1920s. (BCARS 90180)

with a loaded train, losing control and not being able to slow down. Depending on the nature of the curves and track, it might be possible to ride it out. More often than not, cars derailed. In the worst case scenario, the locie plunged off the tracks, with the loaded cars piling in on top of it. Standard procedure in a runaway was for the fireman and brakemen to bail out. Sometimes an engineer would stay aboard in an attempt to bring it under control, particularly if there was a possibility of a runaway train roaring into camp or off the dump. But on many occasions the engineer also took to the weeds, and the entire crew would trudge down the tracks to discover what damage had been wrought.

The work of the fireman changed dramatically between 1900 and 1920. Initially he had to maintain steam pressure with a blazing wood fire, pitching chunks of fir through the firebox door. In most camps, wood-cutting crews provided him with stacks of split wood, which he loaded aboard the water tender. His main concern was to keep up a full head of steam in preparation for an uphill haul. On a hard pull, wood was not added because the draft would suck blazing chunks of wood right out the smokestack.

After a day's work you'd bring the locie in. She'd have a full head of steam on her so you'd put the fire out and shut all the niggerheads off, the main stop valve on the boiler in the cab that supplied steam to your injectors, and your atomizers and blower for your fire. You shut all your valves off, the dampers are all closed. Go out and put a lid on top of the stack, to hold as much heat in as you could. And the boiler's all insulated with asbestos, so that helped hold the heat in pretty good. She'd hold steam all night, pretty good. Maybe in the morning she'd have fifty or seventy-five pounds on. You turn on your main stop valve and that gets steam to your atomizer and your blower and you get a piece of rag and pour a little diesel oil on it, light it and toss her in the firebox. Crack your blower a little bit, enough to get a little pull started, turn on your atomizer and open up your oil control valve, and let the oil start to run and the atomizer catch it and throw it out into the fire and she'd ignite and away she goes. Adjust your fire and get a nice fire going in there and give her a little more blower, and pretty soon you had a good fire going and a little steam starting to come up, and in about an hour you had a full head ready to go. I used to get on about six o'clock in the morning and start her up. Used to pull out about seven.

Woodrow Runnells

Two blacksmiths use an improvised shelter at Bloedel, Stewart & Welch's Myrtle Point camp, 1927. (Albert Paull photo; VPL 1950)

They started in with oil when Brooks Scanlon come in here in 1910. They started with oil right off the bat. Some places they burnt wood, on the coast. Bad enough with the old steam donkeys in the bush, setting fires. That's one reason, I guess, they went to oil pretty quick, because wood's pretty dangerous. But even with oil they used spark arresters, because a lot of carbon and stuff would form in the firebox. Chunks of brick would break off and get pulled out with the exhaust. The bricks take one heck of a beating from oil. Terrific heat. We used to have to do a lot of repair work on the brickwork in the firebox. Do all the brickwork in the firebox. Bricks, they'd go a long time, you know, but after a while they'd get a glaze on them, they'd get so darn hot they'd melt right on the surface. Stuff would run down, and a lot of that stuff would drip right into the firebox and sometimes it would get pulled out through the stack and wind up in the dry timber alongside the track and start fires. So that's why we used spark arresters.

When the weather was real dry sometimes, after a long dry spell, and you do a lot of heavy braking, bits of red-hot metal would fly off the brake shoes and land on the track. The brake shoes are made out of what they called pig iron, it's some kind of soft stuff. It would last quite a while, they'd replace them once in a while. Made a good braking surface against those steel wheels. The wheels have what they call a chilled surface on them, almost like case hardened. A real hard surface, cut down on the wear.

Woodrow Runnells

In some camps, coal was available and was used in preference to wood. It made a hotter fire and held it longer, but still involved a lot of hard, hot, dirty work for the fireman. In the 1920s most operations switched to oil in the locies, yarders and loading engines. Firemen thought they had died and gone to heaven. Their task was still the same, to keep up the steam pressure, but the effort now involved the turning of a tap. In fact, in some operations or under certain circumstances, the oil-fired locies ran without firemen, the engineer performing all the work of running the engine.

The most hazardous job was that of brakeman. One of his first tasks was to attach the cars and make up the train. In the early days this was done with a link-and-pin coupler attached to the trucks. This simple, cheap coupler worked—and was perfectly designed to remove a brakeman's fingers when he was holding the pin and the cars were slammed together. Although thousands of patents were taken out on automatic couplers, it was not until the late 1920s that they were widely adopted, and in some camps the link and pin was used until World War Two.

In the days of the disconnect cars, the brakeman's task was to ride on the footboard of the car, applying or releasing the hand-cranked brake with the aid of a steel brake stick. On a steep downhill grade there might be a brakeman

The blacksmith shop at Caycuse. This photograph, taken in 1988, shows the shop much like it was in the 1930s. (Judith Hudson photo; Kaatza Station Museum P989–35–4)

What we used to do there, would be to have all the hand brakes on. The engineer needed his air, he couldn't use that up 'cause they only got two applications, and if you lose them both, well then you're gone. You bail out and just let 'er go. So what we used to do there, we used to run alongside the God-darn train when he was taking off and have these hand brakes on. Then when the shoes warmed up, well, then we'd start releasing these brakes, on the fly, while he was still going, because he couldn't stop. If he stalled, well then we'd be in a heck of a fix. And we'd catch the tail end, and walk the loads all the way back down to the engine. Just jumped from load to load to load till you got back to the engine.

I fell off one time. I gave the engineer the high-ball to go ahead on 'er, and I've got to catch the tail end. We were moving a unit and we had the heel boom and all that on cars, and the heel boom, it's longer than the car and it sticks out, it's got two legs on it. So I was walking up from the tail end. I jumped up on the heel boom and was walking on this one log. We were crossing a bridge, a river underneath, but the bridge was only about eight feet off the water. And as I got in the middle of that darn thing my foot slipped on the wet timber, and I flipped myself under the car.

When I landed, I landed right down on top of the rail, and the front of the truck's coming. So I rolled off the rail all right, but then the trucks caught me right on the shoulder, and they spun me around, kept turning me around. I was trying to grab something, but I knew if I grabbed the wrong thing I'd get my hands cut off. So finally, I rolled off the bridge into the water. When I got into the water, I swam across and caught the tail end again.

George Lutz, retired locomotive engineer. He started logging at Englewood in 1943.

Two Climax locomotives at the shop in Merrill & Ring's Rock Bay woods camp, 1931. (Monty Mosher collection)

for every two cars. They would have to set the brakes on one car, jump off and catch the next car as it came by, jumping aboard or setting the brakes on the run. When it was time to slack off on the brakes, they would run alongside the cars, releasing brakes as they went. One misstep could plunge a brakeman under the wheels, where he would be lucky to lose only an arm or a leg. Until air brakes came along, operated by the engineer in the locie, braking was a job for the nimble footed.

Over time, much of the danger and hard work disappeared from these jobs. From the late 1920s to World War Two, train crews who worked in the better camps held enviable positions in the camp hierarchy. Running a cared-for, oil-burning Shay over a well-maintained line was a much desired job. It still required a lot of skill and knowledge, and the unexpected hazards never disappeared. In large part the improved conditions in BC were due to the efforts of the provincial Department of Railways, which introduced a set of tough safety standards in the early 1920s and resisted the efforts of haywire owners to avoid enforcement by railway inspectors. One of the first chief inspectors was William Rae, whose files are full of letters persuading, imploring, pleading and occasionally threatening reluctant owners and managers to make the necessary changes. In one of a long series of letters in 1926 to A.J. Hendry, superintendent of BC Mills Timber & Trading, concerning the deplorable condition of the equipment used at Rock Bay, Rae wrote: "This Department has no other alternative than to see that companies comply with the regulations respecting safety appliances on railway equipment, i.e. automatic couplers to be of standard height, power brake control on the locomotive and cars so that the brakes on the train can be handled by the engineer in the locomotive hauling such train."

In a letter to his deputy minister, Rae described the situation at Rock Bay: "this is the only railroad operating under the jurisdiction of this Department which has not showed any attitude towards complying with the regulations."

The best-known railway inspector of them all was Bob Swanson, who not only enforced the rules and regulations, but was also known to roll up his sleeves and repair a damaged or unsafe locie on the spot. Swanson, more than any other person in the Pacific Northwest—with the possible exception of his brother, Seattle Red—was responsible for immortalizing the era of steam logging in general, and railway logging in particular. He was a superb logging poet, and in his nineties at the time this is written, he is still the chief custodian of the romance of railway logging. Being an astute observer, he also foresaw its end.

THE TRUCK BUG

by Robert E. Swanson

You've heard of the hemlock looper
And the hungry budworm too.
You've heard of the deforestation
These parasites can do;
But of all the deadly insects
Down to its last descendant
Is the deadly "Truck Bug" when it stings
A logging superintendent.

When once he's stung he'll rant and rave
Of trucks and stinking rubber,
And grades of thirty-six percent
Without a brake or snubber.
He'll suffer grand delusions
Of finely graded roads,
Of rooting railroad down the hill
And hauling ponderous loads.

He dreams of locomotives—
All melted down for junk,
The hogger—thin and scrawny—
A lowly whistle punk.
I don't know yet the antidote
As I pen this little rhyme,
But railroad men please pray to God
They'll waken up in time!

VI The Birth of Truck Logging

Previous page: Through the 1920s, this was the preferred logging truck and trailer combination in the Pacific Northwest: a four- to five-ton truck and an eight- to ten-ton trailer. This unidentified model at a planked landing is getting the second 2000-board-foot log of what will probably be a three-log load weighing about eighteen tons. (BCFM 3–5)

The coastal forest industry that emerged from World War One had a vastly different outlook than a few years earlier. Some significant events had occurred that would have an enormous impact during the 1920s. First, the completion of the Panama Canal late in 1913 opened the markets of eastern North America and Europe to British Columbia lumber. The impact of this development was not immediate due to the unusual circumstances of the war, but with the return of peacetime trade and commerce it was now possible for lumber producers here to ship into these markets at competitive rates. Previously, only high-quality Douglas fir spars and timbers commanded the prices that made shipping them around Cape Horn feasible.

As well, after 1918 the world economy was struggling back onto its feet. Twenty million people had died in Europe during the war, and the traditional sources of supply, including timber, were in a state of disarray. The situation in the Pacific Northwest forest industry, however, was just the opposite. Not only was the devastation of the war not experienced here, but the mobilization of the industry as part of the war effort had created a new energy and capability. All it needed was a direction and an opportunity. That opportunity came during the boom of the 1920s. Lumber markets soared and virtually every stick of timber cut could be sold at a good price.

In BC, at the conclusion of fighting, the established logging industry consisted mostly of a few hundred handloggers and several large, integrated companies like Comox Logging and the Victoria Lumber & Manufacturing Company. The former were in the process of finishing off the timber accessible to Gilchrist jacks, while the latter were comfortably ensconced on the large blocks purchased from the railway companies or on leases of Crown timber made available during the McBride staking frenzy a decade and more earlier. The tracts of timber held by these companies were, for the most part, large enough to support railway logging operations and were of sufficient size to keep them occupied for the next thirty years.

In 1912, the BC government introduced a new form of licence granting cutting rights on Crown timber. These were called Timber Sale Licences and, until 1948, they were the only significant means of acquiring publicly owned timber. The mechanics of issuing Timber Sales were simple and straightforward. Any individual or firm wanting to buy the timber on a defined area of land applied to the Forest Service, which then conducted a public auction. The highest bidder obtained the timber, which he was required to log within a specified period and for which he paid stumpage to the government.

The introduction of Timber Sale Licences added an element of competition to the industry. It was the beginning of a truly unimpeded free enterprise forest economy. The smallest, most insignificant logger was able to compete on an equal footing with the biggest companies. If he was clever, lucky and hard working, he could make a go of it. It was a system that encouraged and rewarded entrepreneurial effort, and it remained in place for several decades.

Another important development affecting the forest industry was the operation of the first automobile assembly line in the world, established by Henry Ford in 1913. One of the many consequences of the assembly line was the phenomenal acceleration of the development of gas and diesel internal combustion engines, particularly to supply the war machines of the industrial world. One measure of the impact of the mass production manufacturing process is that by 1918 there were about 400,000 motor trucks in use in the United States. From the outset, loggers attempted to use these vehicles in their operations: a 1919 meeting of the Pacific Logging Congress was told that 500 trucks were hauling logs in the Northwest. Some of these were at work in BC.

There were many problems to overcome in

Rarely used in BC, this truck and trailer equipped with flanged steel wheels ran on railway tracks in 1920. (Leonard Frank photo; BCARS 39087)

Although thousands of these four-wheel-drive Nash Quads were built for the US military during World War One, very few found their way into BC. This model, at an unidentified location, is equipped with a very early version of a fifth-wheel trailer hitch. (BCFM 1-5)

adapting the first production-model motor trucks to the logging business, but the potential rewards were even greater. In 1920, the typical truck used for logging in the Northwest cost less than $6,000, including a trailer, and could haul 4,000-foot loads. On short, one- to two-mile hauls, these trucks made ten or twelve trips a day, vastly outperforming the cheaper alternative, horses. Their initial cost was about one-quarter that of a locomotive, not to mention the cost of rail cars. Although road costs were high, in some areas public roads could be used.

Trucks were popular in the US because they hauled logs more cheaply than alternative methods. In BC, they appealed to small loggers who could afford neither railway equipment nor a sufficiently large stand of timber to warrant construction of a logging railway.

Manufacturers made significant advances in truck construction during and as a consequence of the war. One reason, of course, was that the military used large numbers of motor vehicles. And, while wartime conditions restricted the production of trucks for non-military use, an edict by the US War Industries Board decreed that manufacturers and owners build and use trucks in such a way that they would stand up to hard use. Finally, truck logging received a real boost near the end of the war when the US government launched a massive airplane spruce logging operation which included the purchase of 150 Standard trucks. In spite of this enormous order, the Vancouver Standard dealer, Sullivan Taylor Motors, promised immediate delivery of BC orders.

By the early 1920s, there were perhaps twenty truck manufacturers selling into the Northwest logging market. Although there was a lot of variety among them, the size range was small—from three to five tons or a bit more. The smaller three- to four-ton models were generally used on public roads, where load limits were usually restricted to less than 3,000 feet. The heavy, off-road trucks were rated at five or five and a half tons, and could haul up to 5,000 feet.

For many years loggers preferred a truck with a wheelbase between 160 and 170 inches to facilitate handling on turns and landings. Also, the frames of trucks with longer wheelbases were subject to buckling under the big loads which were piled on them. They had steel or, preferably, solid rubber treaded tires, about twelve to fourteen inches wide on the rear, less on the front. Until the 1930s they were single-axle vehicles, as were the trailers. Truck and trailer bunks ranged from seven to nine feet and were equipped with chocks or cheeseblocks to hold the logs on. On good roads and favourable grades they were capable of speeds of ten to twelve miles an hour. On dry, pole or plank roads, these trucks handled twelve percent grades, and with cable rip-rap nailed on the road surface or cable-wrapped tires, drivers could navigate these roads in wet weather.

A typical engine in one of these trucks had four cylinders and developed about forty horsepower. The engine in a Duplex, one of the more popular brands in the 1920s, had a four-and-one-quarter-inch bore and a five-and-one-half-inch stroke. Electrical current was supplied by an Eisemann magneto and fuel fed through a float carburetor. Although most of the early trucks had a four-speed transmission, the Duplex boasted eight forward and two reverse speeds. It was one of a few four-wheel drive trucks available, utilizing an internal gear drive on each wheel that was a forerunner of the planetary drive systems used today. A sixteen-tooth spur gear at the end of each axle meshed with a sixty-four-tooth ring gear on each of the wheels, avoiding much of the strain endured by conventional rear-axle drive trucks. Several trucks produced during this era utilized chain drives. Many operators favoured them because the chains were always the first thing to break under a strain and, being uncovered, were easily and cheaply repaired.

Logging trucks faced conditions and terrain not easily withstood by most production model trucks. Beginning in the mid-teens, almost every motor car manufacturer also produced a line of trucks, and dealers throughout the Northwest sought to sell their products to loggers. In this early flush of enthusiasm, such well-known auto builders as General Motors, Ford, Nash, Peerless, Studebaker and Packard all produced trucks that were modified and sold to loggers. Manufacturers beefed up the frames, added heavier springs and made other changes to

Harry Jones, who owned an operation on the Malahat, had a bush mill. He cut planks from logs not wanted. The donkey moved along beside the road and pulled in logs and chunks to block up the stringers. The mill cut road planks out of anything not suitable for ties. We used to pack these things on our shoulders to put them in place. I was running the donkey during this time. When a spur road was finished we would take up the planks and use them again. I was getting forty cents an hour. It was economical to use a lot of labour on these jobs because wages were relatively low.

Frank Norrie

strengthen what were, in retrospect, pathetically underbuilt vehicles for the work they performed. Early models of all these brands were used in BC in the 1920s.

A large number of other truck builders disappeared after a brief period: Standard, FWD (for four-wheel drive), Fageol, Linn, Gotfredson, Day-Elder, Fisher, Little Giant, Thornycroft, Garford, Kelly-Springfield, Riker, Reliance, Knox. Evidence can be found of a few of these models being used in BC. The Royston Lumber Company bought a Garford in 1918, and Tom Plimley in Victoria sold several Garfords and Little Giants on Vancouver Island.

In 1927, Commercial Lumber in Haney was operating a new Gotfredson—reputed to be the first tandem axle logging truck in BC—down a six-mile plank road. From 1920 to 1932, these trucks were built in Walkerville, Ontario, and until 1948 production continued at a Detroit plant. Four years earlier the logging company then known as Webber Lumber had acquired a number of Model T Ford trucks that could haul about 3,000 feet on their trailers. The attraction

I went up to Harrison in 1934 with the M&T Logging and started out chasing and working on the rigging, and getting piling out with Harold Clark and his little Fordson three-drummer. The piling for the booming ground and the skids for the bunkhouses. I went in on a scow with four or five guys. I was second rigger, I guess you'd call it, when we were moving the machines. I always went and helped the rigger, rig up the next tree or a cold decker. I'd do anything they had—hand falling, help the booming guys. So I learned every damn thing there was in the woods.

A contractor was supposed to build a donkey sled and something went wrong, and I ended up helping that guy. Working on the fore-and-aft road, helping build the road itself. They had a little Allis Chalmers gas donkey and a steam pile driver out in front, driving the piles for the road. I could drive piles, or broad-axe—you just learned everything there was to do.

Some places they drove piles for fore-and-aft roads. Harrison Lake was one of the places you were up in the air in these rock bluffs, and that's where the timber was left. Les Trethewey had a railroad engineer in laying out this fore-and-aft road. I pulled chain for him, chopped brush out of the way. I was with him when he laid it out, right from Harrison Lake.

We started out, it was reasonably flat, and then we hit the side of the mountain and he had to start going up grade. It seemed like we were trying to maintain ten or eleven percent or something. To me it seemed like a joke, it was pretty near like a railroad. But it was steep when there was snow or ice, you knew it then with those hard tires.

We were going around rocks and did some blasting, but there were no Caterpillars or shovels to move dirt. They had a small steam pile driver and this Allis Chalmers. They drove piling, put the cap on it and laid the stringers. You'd lap the stringers. There was a bent about every thirteen feet and we used twenty-six-foot culled logs. There was nothing wrong with them, but they were hard to sell—hemlock. On flat ground they even used some cedar, if it was just a short spur. The cedar would go to haywire after a while, it wouldn't take the pounding. But up in the real big bridges they mostly put fir in. Broad-axed it. They were pretty careful with the mainline. A fore-and-aft road you could build right there in the bush. Weaver Lake used planks. We were fore-and-aft up in our camp.

There was a garage out at the end of the wharf. You backed in and there was a gin pole to unload.

You took those roads into where you could get. Then you cold-decked and skylined the timber to the end of the road. They had to keep them fairly straight 'cause you had to make as big a turn as a railroad to get around corners or your trailer wouldn't come around, it would catch on the guard rails and jump over. The broad-axe people would go along and snipe everything nice and smooth so your trailer wouldn't catch on something.

A lot of guys said, "Yeah, you can drive down there when you're asleep." Hah. You wouldn't go five feet if you were asleep. The hind wheels were wide and your front wheels on those hard-tired Lelands were narrow. The trick was to get around corners where they didn't box it right in. When you straightened out, that's when you really had to watch and keep well centred. The older the road was and the guard rails wore, you had to stay in the centre of that stringer. Every now and then the guys would forget or get sloppy, light a cigarette, and they'd drop down in between. Generally they never got hurt or anything because they had big hubs on them and it would bounce and slide along on the hubcaps and instantly you'd try to stop. But you didn't stop those things that quick, when you were sliding on snow, ice or something. They had riprap on it, cable.

But every now and then there would be one jump off, kick over the guard rails and go down ten, fifteen or twenty feet. That was the hospital cases right there. It was my idea, when we had finished with one of the spurs, to make a runaway thing there, where it was flat ground and the road was down low on skids. I said why not give us a run out there and then we can drop off the end into the mud. So they did that three times, and every once in a while we had to use them.

W.W. (Curley) Chittenden, a pioneer truck logger in the Fraser Valley. He grew up on a farm at Bradner and began logging as a boy. He has co-authored two books on valley logging history.

An obscure make, a National, on a fore-and-aft road at Alberni. The protection offered by the cross-timbers behind the seat was largely psychological. (W.L. Montgomery photo; BCARS 73644)

of these small trucks was that they cost about $500 new, compared to the $2,000 to $6,000 price tag on most other models used for logging. In 1932, the Webber brothers joined forces with Art and Virgil Stoltz in Weaver Lake Logging and used three trucks on a plank road west of Harrison Lake. Webber later hauled over plank roads at Rainy River on Howe Sound and back in the Fraser Valley, at Ruby Creek. In the 1930s, Don Mackenzie, who later became a prominent truck logger, chose a Gotfredson as his first truck to contract haul at Cowichan Lake. In the early 1920s, H.L. Chittenden used a Fisher to log his farm at Bradner, and later he obtained a used Kleiber. Westminster Iron Works, the BC dealer for FWD trucks, may have equipped some of them with steel wheels for use on railway tracks. Two FWDs were rebuilt by Westminster in 1923 into thirty-two-passenger commuter trains for use on the Pacific Great Eastern Railway.

Several truck manufacturers remained in business long enough to refine and develop their models. Sterling, in Milwaukee, built a truck popular with loggers from 1913 until White took over the company in 1951. Many of them were used in BC, available from Sabine Trucks in Vancouver. Another well-liked truck was the Chicago-made Diamond T. It was first produced in 1911 but not sold in the West until

A Republic hauling logs on Headquarters Road near Courtenay, 1928. Makes such as this were rare in BC and were probably purchased in the US. (CDM P200–1099b, 988.225.4)

The Commercial Lumber mill yard in Haney, 1929. (Leonard Frank photo; VPL 3975)

In 1927, Commercial Lumber at Haney used the first tandem axle truck in BC, a Gotfredson. The use of dual, pneumatic tires was also unusual at this early date. The driver is Herb Smith, one of the first logging truck drivers in the Fraser Valley. (Leonard Frank photo; VPL 3978)

For almost twenty years the most popular logging truck on the BC coast was probably the White. The basic, four- cylinder, hard-rubber-tired model produced around 1920 was still used on fore-and-aft and plank roads until World War Two. This heavily loaded White at Salmon River Logging's dump was purchased when the company converted to trucks in 1935. (UBC Anderson Collection)

When we first started logging I had a car—I had $160 in the bank and a car, a 1935 Ford sedan. I traded it to the Duncan garage, even trade, for a 1935 Maple Leaf three-ton truck. I had been working for a guy running a bulldozer and he owed me some money. In those days, in gyppo logging, it was hard to collect your wages sometimes. He had a trailer sitting around doing nothing so I said, "I'll tell you George, I'll take that trailer for my back pay." So, then I had a truck and a trailer, and that's what we used.

Frank Norrie

1915. In 1967, the Diamond T company combined with Reo, founded in 1904, and began marketing a new brand called Diamond Reo. Fletcher Logging had four Diamond T's working at Port Coquitlam, but when the company was moved to Fanny Bay during World War Two under the name Tsable Logging, it replaced them with Kenworths. A popular truck in BC that was rarely, if ever, used in the US was the Leyland, built in England and sold at factory outlets in Vancouver and Victoria. Campbell Logging ran a fleet of five Leylands at Forbes Bay in the early 1920s.

The White Motor Company of Cleveland began as a sewing machine manufacturer, went on to produce steam-powered automobiles, the Cletrac crawler tractor and, in 1910, trucks with internal combustion engines. In 1913, White began selling what was probably the first motor truck specifically designed for logging, a steel-wheeled machine. By the early 1920s they were the most popular logging truck in the Northwest, and BC's White dealer, Consolidated Motors at 1230 West Georgia Street in Vancouver, had sold as many as any other brand at work in the coastal woods. In 1967, White opened a plant in Kelowna and began to produce a model called the Western Star. White is one of only a handful of logging truck manufacturers from 1920 that survived beyond World War Two.

Two other survivors from this early period are GMC and Mack. The General Motors Truck Company was formed in 1911 through an amalgamation of several existing small truck builders. It was not until 1916 that the company produced a five-ton model suitable for Northwest logging conditions. The four Mack brothers established the Mack truck company in Brooklyn in 1903. In 1916, the company introduced the AC model, known as the Bulldog Mack, that was produced until 1939. None of these appeared in BC until the late 1930s when Charlie Philp got the dealership and sold four fifteen-ton models to Bill Schnare and his son Stan, who put them to work at Port Douglas for Trethewey Logging.

Almost all the trucks mentioned so far were built in the eastern US. Curiously, it was in the Northwest where most of the early use of trucks in logging took place. As a result, early logging trucks were not designed and built with Northwest conditions in mind. The first steps to remedy that deficiency took place with the establishment of the Kenworth Truck Company in Seattle in 1923. The company had operated since 1915 in Portland as the Gersix Manufacturing Company, which produced a few trucks used in logging. When it was re-established in

Seattle under John Holmstrom, Kenworth adopted a policy of offering customized trucks to fit individual loggers' requirements. The first custom Kenworth came off the line in 1928.

A year before Kenworth set up shop in Seattle, a group of BC businessmen came to similar conclusions concerning the unsuitability of eastern-built trucks for local logging conditions. In 1922, Douglas Hayes and W.E. Anderson, one of Quadra Island's leading businessmen, set up the Hayes–Anderson Motor Company in the 1000 block of Granville Street in Vancouver. Hayes came to Vancouver just after World War One and started the Vancouver Parts Company. He was the driving force in the company, with Anderson, who withdrew in 1928, providing financing – $6,000 in cash and loans. When the company went public in 1932, Hayes withdrew and ownership was dispersed through many shareholders. His son Don started logging in Bute Inlet in the early 1950s and the family group, which includes Pat Carson Bulldozing, is now managed by his grandsons, Harold and Don Hayes.

Starting with logging trucks, Hayes–Anderson eventually branched out into other types of trucks and even into building other logging and industrial equipment. In its early years it acquired the Fisher truck company. In 1928, it relocated on 2nd Avenue, in the industrial area favoured by many logging equipment companies, under the name Hayes Manufacturing. Over the years the Hayes company achieved an

Above: A 1920s-vintage White at the Byles & Groves Port Neville camp, 1938. (Mel Parker photo; CRM 5511)

BC Pulp & Paper used a fleet of Whites at Halfway River in 1936 to supply the Port Alice pulp mill. Ed Eason is front row centre. (Wilmer Gold photo; IWA Local 1–80)

A 1928 White at an unidentified location. The use of pneumatic tires and tandem axles on both truck and trailer, and the practice of loading the trailer on the truck for the back haul, indicate an advanced stage in the evolution of truck logging. (Leonard Frank photo; VPL 11836)

impressive list of "firsts" in logging truck design: dual axles, diesel engines, self-loading trailers. By 1937, Hayes was selling a six-wheel logging truck with a dual-axle trailer, a twelve-speed transmission and air brakes. Like Kenworth, Hayes trucks could be custom ordered. For more than fifty years Hayes shared with Kenworth a pre-eminent position in the production of standard and off-highway trucks.

A third western manufacturer joined the field in 1939 when Peterbilt opened a plant in Oakland, California, and started making a heavy duty, dual-axle truck.

A mid-1920s-model trailer built by Hayes–Anderson, equipped with air brakes and an adjustable reach. (Tom Everett collection)

The modifications pioneered by Kenworth and Hayes were part of a wider process of adapting trucks built for other purposes to the task of logging. The earliest and most significant modification was the development of the trailer. Most of the first production model trucks were capable of carrying no more than five tons, but to be useful in the Northwest they needed to do better than that. The solution was to add a two-wheeled trailer that could carry between five and ten tons, thereby doubling or tripling the load the truck could haul. At first these trailers were steel-wheeled, but soon hard rubber tires were used. A key feature of the trailers was that they had an adjustable reach, the long beam that attaches the trailer frame to the truck. The ability to adjust the reach made it possible to haul logs of different lengths without having to change trailers.

In 1913, Pacific Car and Foundry began selling logging trailers, and during the wartime spruce effort the US government actively encouraged research. It was not until the 1930s, however, that dual-axle trailers began to appear. They were easier on the roads, they permitted heavier loads and, because they provided better weight distribution, they allowed loaded logging trucks to travel on public roads. But they did not track so well around curves on fore-and-aft roads.

Cheeseblock bunks were used well into the 1930s, and in a lot of cases much longer. Cheeseblocks were better than stakes for big one-log loads because they could centre the log on the bunks to balance the load. So long as big logs were being hauled, there was no need for high stakes to hold the logs on, and the higher the stakes the harder it was to load.

Probably the major problem faced by logging truck operators was that of brakes. During the 1920s trucks were equipped with adequate brakes. Some were mechanical, even though hydraulic brakes were available by 1920. The problem arose with the addition of trailers, and was compounded by the heavy loads these trailers carried. Clearly the solution was to provide trailers with good brakes, but this development was slow in coming.

Through the 1920s and well into the next decade, the most widely used trailer brakes were mechanically operated with a separate lever in the truck, connected by cable to the trailer. This arrangement required a constant adjustment as the cable loosened and tightened going into corners and straightaways. The next phase of braking technology was vacuum brakes, which operated off the intake manifold of the engine. They were popular for a while because they worked better than mechanical brakes and were cheaper than air brakes. Their big drawback was that engine vacuum varied, being highest when the throttle was closed and lowest when it was open. The final refinement in brakes was air

Brake shoes on the rear wheels of a 1920s-model White. Before water tanks were added to cool brakes, it was very easy to overheat and seize the brakes on these trucks. (Wilmer Gold photo; IWA Local 1–80)

An early Leyland with a one-log load weighing more than fifteen tons hauling for Slater & Sons along the Koksilah River, mid-1920s. Lionel Hanison is driving. (Fred Gifferd photo; BCFM 1-1)

We first logged with tractors in Ladysmith because we had a big blowdown of timber. We had twelve tractors, D-8 Caterpillars, hauling logs. We had a little A-frame deal to load with. Part of the idea of this operation was to develop pneumatic tires and log with trucks. It had been done a little bit where you only put two or three logs on your truck because they all had hard tires, hard rubber. For several years that was our biggest difficulty, pneumatic tires. They kept blowing out, in spite of everything we tried. The tire companies had men there, engineers, right along riding in the trucks watching them, taking temperatures. They'd get so hot they'd blow out. They increased the sidewalls, you know, and apparently that was worse, because they retained the heat. Then they went back to thinner walls. They tried every kind of fibre they could possibly find, even the steel belted that they rave so much about now. They tried that in 1937, built right into the tires. It looked beautiful. It wasn't a straight piece of steel. It was webbed, about six inches wide. It just looked marvellous, but in no time at all it was sticking out through the rubber in chunks, and all broke up. The Michelin tire people did that. The Goodyear and Firestone were also there, but Goodyear turned out the best tire and spent the most money. Marvellous. I don't think Goodyear made a nickel there for two or three years, but they finally got the answer.

You can't complain now. They've got the right kind of air pressure, and right kind of width of a tire, all that sort of thing. If you blow out tires today, it's pretty well your own fault—overloading or letting the air down too low and you're not looking after them. Of course they never could log these mountains the way they are now without the pneumatic tire. It would cost too much money to put a railroad up there.

Joe Cliffe

brakes. They required the addition of a compressor and pressure tank, so were relatively expensive.

Brake drums presented a further problem. The drums function as heat sinks, converting the kinetic energy of a vehicle in motion to heat. On a long downhill grade, with a load of logs, it did not take long for early brake drums to turn red and even white-hot. Then they either seized to the brake shoes or set the tires on fire. The solution to this problem was to install water tanks to direct a stream of water over the drums on long downhill grades. Initially, these tanks sat under the bunks; later, larger tanks were positioned behind the truck cab, providing added weight over the drive wheels. Additional traction was obtained for return uphill travel by loading the empty trailer onto the truck. This practice began in the 1920s, and early trailer design took it into account.

The evolution of rubber tires was another significant factor in the development of logging trucks. Although pneumatic tires were available in the early 1920s, they were rarely used by loggers. The first inflatable tires were expensive and lacked the deep treads needed in the wet Northwest woods. They also had thin sidewalls that did not stand up to the crudely built logging roads, or to the constant rubbing on the guard rails of the fore-and-aft roads. The first pneumatic tires used cotton cord in the sidewalls, which was not strong enough for truck tires. In the early 1930s, after the Goodyear Tire Company came out with a 12-ply, rayon cord tire with a good tread, pneumatic tires were more widely used. This coincided with the more widespread use of dirt roads on which pneumatic tires could drive empty at speeds of up to forty-five miles per hour. The invention of nylon in the 1940s made the development of even stronger tires possible.

Perhaps the most obvious change in trucks by the 1930s was the addition of enclosed cabs. The first logging trucks had little in the way of a cab. Some models had a windshield, others some kind of canopy or overhead protection from the weather. Judging by old photographs, most of these first cabs did not long survive the hazards imposed by levermen who had learned their skills loading the practically indestructible railway skeleton cars. Given the fact that most of the first trucks were owned by small loggers, it was likely that a lot of them were loaded by economical, Fordson-powered Skagit donkeys which hoisted and, unfortunately, lowered logs at great speed. The drivers of many of these more-or-less cabless trucks were provided with the illusion of protection by the addition of a stack of railway ties u-bolted across the frame behind the seat.

By the mid-1930s, however, new logging trucks came equipped with cabs featuring windshields, doors with retractable side windows, heaters, defrosters, windshield wipers and numerous other luxuries. Many drivers, particularly in hot summer weather or after taking their first trip down the precipitous route to the dump, elected to remove the doors so they could easily exit their vehicle in the event of a runaway.

Until the beginning of World War Two, there was a slow and steady development in truck design. The major effect of the Depression was to eliminate most of the small, obscure manufacturers and to concentrate the building of logging trucks in fewer more specialized companies, such as Mack, White, Kenworth and Hayes.

One significant development that occurred before the war was the introduction of diesel engines. Although they became available during the early 1930s in crawler tractors, it took a few more years for diesels to become commonly used in trucks. This was probably because the early fuel injection systems were cumbersome affairs, taking up a lot of space.

Kenworth made the first US-produced diesel truck in 1932. It was equipped with a six-cylinder, 125-horsepower Cummins. The first diesel-powered logging trucks in BC may have been two new Leyland diesels Fred Clark bought to use at Errington. He later obtained the Cummins Diesel dealership for BC, and his son George operated it for many years.

The advent of World War Two put a freeze on the further development of logging trucks. As long as the war lasted it was practically impossible for loggers to buy new trucks, so they got by with what they had. However, one thing was clear: a significant new sector had arisen in the logging industry during the previous twenty years—the truck logger.

We decided to go truck logging—contracting for Canadian Robert Dollar—about 1920. Previous to that we had worked at different places. And we started truck logging with steam donkeys at Mud Bay in 1920 on a plank road, with solid tired trucks. And, as the business improved and grew and expanded, we went into more modern trucks until eventually we got into the pneumatic tire gas trucks and from there into the pneumatic tire diesel trucks and we finished up logging at Buckley Bay.

We moved to Iron River for Batco and operated there—logging and trucking—until we finished the claim. Comox Logging at that time were considering going into truck logging across here on the Sullivan. They came to see if we were interested in doing their log hauling by truck or had we anything else in mind. So we decided to give up logging—sold our logging equipment and went over and did the trucking for Comox at the Tsolum. We were there for thirteen years and it was a very, very satisfactory connection until they made us a very inviting offer to buy us out. Getting along in years and having been thirty-five years at it, we thought it was a good time to get out from under, and we sold. Since then I've just been wasting my time, I think.

Bob Grant, born in Nanaimo. He and his brother Charles grew up and lived all their lives at Royston. Grant Brothers was one of the first truck logging operations on Vancouver Island.

Precisely where and when the first truck logger appeared on the BC coast is another one of those matters subject to a lot of intense disagreement. There were just as many claimants to the distinction of owning "the first logging truck on the coast" as there were for the first steam donkey. It is a debate that will never be resolved to everyone's satisfaction.

What appears most likely is that trucks of various descriptions began to appear in BC in the years just before World War One. In a few areas, particularly the Fraser Valley, there were, for that time, decent public roads on which a truck could be driven with some confidence, at least in dry weather. It makes sense to suppose that some of these truck owners, probably farmers hoping to make a few dollars from the timber they were clearing from their fields, somehow got logs onto the decks of their vehicles and hauled them to the nearest mill.

In one of the earliest published articles on truck logging, George Challenger confirmed this explanation: "In 1915 a start was made in BC in the use of the motor truck in the woods. At first the truck was found to be of use only in taking out scattered tracts of timber in ranching districts or in valleys served by good highways. Various types of motor trucks were experimented with."

In most cases these early log haulers probably did not think of themselves as loggers, but as farmers trying to make a few dollars in the perpetual struggle to wrestle a stump ranch out of forested land. This was the formative experience of many subsequently successful coastal truck loggers. The teenaged sons of hardworking homesteaders convinced the future lay in agriculture quickly came to realize there was a more prosperous life in logging than farming. Two well-known truck logger authors both started in this manner, Joe Garner on Saltspring Island and Curley Chittenden in the Fraser Valley.

In Terrace, T.R. Davis built a sawmill in 1917 and used a truck of some sort to haul logs. He constructed a trailer out of the front axle and tongue from a horse-drawn wagon and devised a unique method of loading and unloading it. He simply removed one of the trailer wheels, rolled the logs off or on, then jacked up the axle and replaced the wheel.

Most of this sort of logging was done with affordable Model T Fords, and for some years it only occurred during the dry summer months. Roads of that era did not stand up to wet weather use, and few farmers owned the bigger, more powerful trucks just then coming on the market. No doubt there were serious loggers who noticed these occasional loads of logs headed to the mills on trucks, but no one was willing to underwrite the cost of building a road into a decent stand of timber to log with trucks when other means did the job.

It wasn't until about 1920 that the first full-fledged truck logging shows got underway. At this time several companies established logging operations in which trucks were the primary or only means utilized to move logs from the landing to the dump. All of these operations entailed the construction of roads built especially for trucks, although different methods of utilizing both roads and trucks were tried.

Robert Campbell's Campbell Logging Company at Forbes Bay in Desolation Sound was one of these first camps to depend on trucks, based on Campbell's conviction that "there are many tracts of timber from 10 to 100 million that can be logged more economically with trucks than with any other system that has come before the logger so far."

Campbell had a fleet of three Leylands hauling two miles down a fore-and-aft road that cost $10,000 a mile to build. This road was built with hand-hewn timbers laid on crossties, providing an eighteen-inch flat surface on which the flexible rubber-tired trucks travelled. A twelve-inch guard rail was built along the outside of the hewn poles. This was rough country: several large bridges were required. The first mile and a half of this road was a continuous uphill grade of between four and twelve percent, which was rip-rapped with five-eighths-inch cable for traction. This provision did not overcome the problem of overheated brakes, so Campbell installed a Berringer snubbing device, consisting of four sheaves twenty-six inches in diameter to which brakes could be applied. This solved the problem by lowering the trucks down the steep portions of the grade. Numerous turnouts were built into the road to allow trucks to meet, and at each of these, as well as at the landing and dump, telephones were installed. A McLean boom was used to load the single-axle trailers and, according to Campbell, did no damage to the trucks. With average loads of 4,100 feet, the Campbell fleet put about 2 million feet into the water each month.

In 1920, Grant Brothers used a truck on a logging contract for the Robert Dollar company at Mud Bay, south of Courtenay. In 1922, they bought the first truck built by Hayes Anderson. The same year, Northern Pacific Logging started a truck logging show at Roy in Loughborough Inlet, using four White trucks owned by Harold Mann. The superintendent at Roy, Tom Murphy, went on to develop truck logging for the Powell River Company. Northern Pacific also loaded onto trailers and hauled

A four-wheel-drive Duplex skidding logs down a plank road at George Challenger's Beaver Inlet camp, c. 1924. This method was known as the Buck System. (Henry Twidle photo; CRM 5134)

A White headed back up the hill at Loughborough in the early 1920s. This may be at the Northern Pacific Logging show at Roy. (Henry Twidle photo; CRM 5106)

Loading a White with a McLean boom in Loughborough Inlet, probably at Roy, 1920s. (CRM 5132)

down a one-and-one-half-mile fore-and-aft road that cost about $5,000 a mile to build.

A similar operation began a couple of years later, across Loughborough at Beaver Inlet. A few years previously, George Challenger had worked in Loughborough as managing director of Grassy Bay Timber, a short-lived railway logging company owned by Dewey and Clay Anderson. When that camp closed, Challenger started his own logging show a few miles away, using two Duplex trucks. Challenger employed what was called at the time the Buck System. This method involved towing logs, connected in turns like those on skid roads, behind the trucks. The plank roads negotiated very steep grades, and one stretch, built on a trestle, dropped 360 feet over 2,300 feet of twisting, serpentine grade. It was an expensive road to build, averaging about $2 a foot.

Challenger's show was plagued by bad luck. Possibly the first logging truck fatality occurred here. One of the trucks stalled and the driver left it in gear when he got out to crank-start it. When it started the truck ran over and killed him. In 1925, the claim was burned out when the donkey started a fire.

The Buck System of truck logging owed its name to F.S. Buck, owner of two truck logging companies that started about 1920. Buck also used Duplexes to drag logs down plank roads at Deep Cove Logging in Indian Arm and at Cedar Creek Logging in Howe Sound. A longtime logger, he also became Vancouver's Duplex dealer. Buck was one of the real pioneers of truck logging. He first tried to use crawler tractors to skid logs, but came to the conclusion that the ground was too wet and soft in coastal BC for this machine. The road system Buck devised consisted of sawn planks laid end to end on crossties. These roads were eight to ten feet wide, with a thirty-inch-wide track on each side for the truck wheels. In the middle was a thirty-six-inch-wide trough, built two inches lower than the wheel tracks, in which the logs were dragged. To build the Deep Cove road, a portable sawmill was installed near the summit to cut planks. Buck used three trucks at each site, hauling down grades of twelve to fifteen percent.

There were several other intrepid truck logging shows in the early 1920s. The Rivers Logging Company at Sointula used trucks on a road steep enough to require a snubbing engine, and Merrill & Ring's Vancouver Bay Logging, under David Jeremiason, may have added trucks to its railway show in Jervis Inlet about this time. This surge of activity was slowed by the Depression, but resumed again with the recovery in the last half of the 1930s. From then

Unloading trucks with a McLean boom and a choker-and-tong combination at Brown & Kirkland's Brown Bay operation, early 1930s. One of the three donkeys then swung the logs down to a boom in the bay with a skyline. (BCARS 73769)

until the war, large numbers of one-truck gyppo shows, along with several substantial independent truck loggers, got themselves launched in the logging business.

In 1934 Ernie Mann and Les Trethewey's M&T Logging started up near Twenty Mile Creek on Harrison Lake. This was a substantial show using eight early-model Leylands that had formerly seen service in North Vancouver as garbage trucks. Three years after it started, this camp was taken over by Fred Brown and Fred Kirkland's B&K Logging, which by this time was one of the major truck logging companies on the coast. B&K ran three more truck shows—at Brown Bay just north of Campbell River, in Sechelt Inlet and on Powell Lake. These were in addition to a railway operation at Elk Bay on the east coast of Vancouver Island.

Mann & Trethewey's fleet of Leylands at Harrison Lake, c. 1934. Curley Chittenden is second from right. (W.F. Montgomery photo; Frank Rustad collection)

A White truck at the junction of two fore-and-aft roads at a Pioneer Timber camp, Malcolm Island, 1936. (Wilmer Gold photo; IWA Local 1–80)

We contacted Bob Filberg, who was the general manager, and had no trouble getting a parcel of timber on Comox Lake with about 3 million feet in it. The front of the property had been logged back from the lake about 1800 feet, above which it became pretty steep. We moved in what equipment we needed and started to log. We had to buy another gas donkey to use loading. Until this point we had used our yarder to load.

At the first setting we logged there was a gradual slope from the lake to the spar tree, and we used a truck and trailer. On the second setting the road was so steep a truck could not get up to the landing pulling a trailer, and it was hard to keep it from running away on the down slope. We did away with the trailers, placing the front end of the load on the truck and letting the back end drag on the ground.

We hung a big block at the bottom of the spar tree and strung a line down the road to where it levelled off a bit. We tied each end to one of the trucks. When the loaded truck started down the hill it pulled the empty truck up the hill. The trucks passed on a wide spot halfway up. This system worked well, and without any trouble we logged a million feet on that setting.

Harper Baikie, one of the founders of Baikie Brothers Logging of Campbell River. He was born and raised on Denman Island and logged all his working life in the Campbell River–Courtenay area.

Brown was a partner in a second company, Earle and Brown, that logged on Watson Island with two trucks.

In 1934, Pioneer Timber began truck logging on Malcolm Island, with Jack "Step-and-a-half" Phelps as superintendent and Archie McKone as master mechanic. McKone had started with trucks a dozen years earlier at Roy with Northern Pacific. Three years later Pioneer moved across the channel to Port McNeill to contract log for Broughton Straits Timber and the Powell River Company. McKone, whose inventive genius helped make this one of the foremost truck logging camps on the coast, built one of the first truck turntables in BC, permitting a logging truck to turn around in a confined space. This camp mainly used 1920s-vintage Whites, hauling on trailers down a plank road.

BC Pulp and Paper, which used A-frames to log in Quatsino Sound for its pulp mill at Port Alice, set up a large truck logging operation at Spry Camp in the early 1930s which utilized fore-and-aft roads and hard-tired trucks.

Elsewhere, truck logging was flourishing by 1936. P.B. Anderson's Granite Bay Logging was using trucks on Quadra Island. Grant Brothers, near Courtenay, was hauling with four Leylands. Ole Buck and W.B. Turner had two of the earliest diesels on the coast, six-ton Hayes Andersons, at their camps at Mount Benson and Boat Harbour. The Baikie brothers had acquired a timber claim on Hornby Island, two miles from the beach, and got their first truck, a Day Elder, for $500.

These were all full-phase, independent loggers who acquired trucks as part of their operations. It was at about this time that a new type of logger, the trucking contractor, appeared on the scene. Before anything else, these people had a passion for trucks. Most were not particularly interested in any aspect of logging except driving an overloaded truck with a pitiful excuse for brakes down some of the steepest hills in the country. Logging company owners, for the most part, were quite content to contract such work to this new breed of log jockeys. One of the first trucking contractors was Pacific Coast Truck Loggers, a division of Smith & Osberg Logging, which worked at Powell River and several other locations in the 1930s.

Meanwhile, at a cluttered little garage in Abbotsford, one of the most famous trucking contractors of all time was about to enter the business. Bill Schnare was a mechanic who had a fascination with trucks. He couldn't have picked a better place to set up a garage. In the 1930s trucks were taking over the freight business between Vancouver and the Interior and they all went right past his door. As well, Fraser Valley farmers were buying trucks, and there was Schnare, right in the middle of it all. Given

Above: Winter logging at BC Pulp & Paper's Spry Camp, January 1937. A snow-covered fore-and-aft road presented some of the most hazardous conditions in the entire business of logging. (Wilmer Gold photo; IWA Local 1–80)

Dumping logs at Pioneer Timber's Port McNeill camp. Archie McKone's turntable is in the foreground. (Link & Pin Museum)

Pacific Coast Truck Loggers, one of the first trucking contractors on the coast, used this White on a fore-and-aft road at Rock Bay, mid-1930s. (W.L. Montgomery photo, BCARS 73662)

the realities of the Depression, it was only a matter of time before he ended up with a few trucks he'd acquired as payment on bad debts, most of them Fords. Schnare was not a logger, and had no interest in or knowledge of logging beyond its potential for his great passion—trucks. He wasn't even interested in driving them, just fixing them and modifying them to do jobs they were not designed to do, like hauling a big load of logs down a steep, poorly built mountain road.

Eventually Schnare's son Stan got hold of one of these repossessed trucks and took a contract hauling for a little logger in the valley. When that worked out, he got some more trucks from Bill, hired some drivers and got more contracts. It was a natural fit. Most loggers at that time had little or no experience with gas engines. They had been brought up on steam equipment. To them, the internal combustion engine, with its electrical and carburetion systems, was a mystery. When one broke down, the tools in the blacksmith shop were useless. This is where Bill Schnare came in. He anticipated whole generations of seat-of-the-pants mechanics whose happiest moments were spent staying up all night, deep in the bowels of a hard-used logging truck some contractor needed to have on the job in the morning. Stan Schnare eventually got a contract with Vedder Logging, a sister company to B&K Logging, owned by the same Brown and Kirkland partnership. They had been truck logging for several years now, with their own trucks, and were beginning to see the advantages of contract hauling.

According to Fraser Valley logging lore, a lot of the kinks of truck logging got straightened out on the Vedder Logging show on Vedder Mountain. As fast as Schnare's drivers encountered problems, he modified the trucks to overcome them. Breakdowns were no problem with a garage nearby at Abbotsford. Schnare began adding second axles to single-axle trucks to improve traction for climbing back up the hills and to bring down bigger loads. He fitted water tanks under the bunks with lines to run water onto the brake drums. And he began to tinker with the engines, adding turbo-chargers and making other improvements so his trucks could haul bigger loads faster.

In 1938, loggers working the steep valleys running into Harrison Lake convinced Schnare that what he really needed was bigger trucks—specifically, Macks. He went to Charlie Philp,

Schnare & Schnare's fleet of Macks at the Iron River Logging camp, 1945. Some of these trucks were also used at nearby Elk River Timber. (CRM 13308)

Truck logging contractor Harold Burritt, right, with one of his first trucks, a Hayes Anderson, at Hale Creek on Harrison Lake, c. 1930. Burritt was hauling for H&R Logging. (BCARS 73742)

One of Buck & Turner Logging's trucks, possibly a White, fuelling up in Nanaimo, 1932. (NCMA P1–14)

who had just acquired the Mack dealership after a stint at Hayes Anderson, and bought four new fifteen-ton monsters. The trailers on these trucks had twelve-foot-wide bunks and, in place of cheeseblocks, three-foot stakes to hold the logs on. Thus equipped, they carried bigger loads than anyone had ever seen on a truck. Schnare put them to work on a contract for Trethewey Logging at Port Douglas. Within two years any lingering doubts about the future of truck logging were dispelled, and there was no stopping the Schnares and their S&S Trucking. They began taking contracts outside the Fraser Valley, for B&K at Nanoose and, eventually, for H.R. MacMillan when he began his relentless acquisition drive on Vancouver Island in the early 1940s. Before they sold out in the 1950s, great fleets of S&S Macks were working at various camps from Elk River Timber's at Campbell River to BS&W's Sarita River operation.

During the era that Schnare dominated the contract trucking business, a lot of other independent-minded truckers were following the same path, particularly in the Fraser Valley where it was easy to get hold of an old truck, tear it down, put it back together, add a trailer and some bunks and go logging. Two of the most successful and well-known trucking contractors on the coast, Ed Eason and Harold Burritt, both got their start in the Fraser Valley, with one eye on what Schnare was doing and the other looking around for work. There were all sorts of small loggers around with gas donkeys who knew how to get the timber down and out to the road but were not very adept at keeping the underbuilt trucks of the period on the road. And there were a few larger companies, such as B&K or Pioneer Timber, demonstrating that these new rubber-tired machines could successfully operate on the same scale as the railway camps.

What was occurring in the late 1930s and during World War Two was the beginning of an industry-wide retooling that was facilitated by, and at the same time created a need for, other types of equipment. The first and greatest need was for roads.

Laying the hewn stringers on H&R Logging's fore-and-aft road, Harrison Lake, 1934. Charlie Trethewey is at right. (W.L. Montgomery photo; Frank Rustad collection)

When the first trucks came into use in about 1920, the lack of good pneumatic tires, combined with the unavailability of machines suited to dirt road construction, led to the use of the only other material at hand, timber. Smaller trees—by 1920 standards—and unmarketable species were readily available for construction of two kinds of timber roads. Both were nothing more than lightweight versions of the timber trestles constructed by railway loggers.

The fore-and-aft road was made with hewn stringers, fourteen to sixteen inches on the upper, flattened face, laid on cross skids resting on the ground or cap timbers on trestle bents. Guard rails laid either inside or outside the stringers were spiked into place to keep the trucks on the narrow tracks. Cable, usually worn out strawline, was zigzagged down the face of the stringer and nailed into place. The same crews and equipment used in railway trestle building were employed—hand axes, adzes and a pile driver. Depending on the terrain and the availability of good building material, one of these roads cost anywhere from $1,000 to $10,000 a mile.

We built our own roads but there was lots of independents that didn't. Some of the independents didn't have the equipment to do it. They'd just yard and load and get somebody to build their roads. But we found out that didn't pay. You had to build your own roads to make it pay.

Johnny Drenka was a contract yard and load man. That was the big thing with him. The fact that he couldn't get trucks was what put him into trucking. Earl Watt was the same. Earl and I split, I don't know what year that was, and I went on my own and Earl went on his own, and he stayed with Empire. I wanted to get my own timber, so I left and Earl stayed with Empire right 'til the end, until he sold out.

Harold Burritt hauled lots of logs for us. He used to call it Hendrickson's ski jump. Over thirty percent grade. He'd take one truck and go up to the top and come down with it, and let his kid drive the flat. The young guy, who owns the outfit now. And the kid would drive the truck down here and dump it and come back onto the flat, and then Harold would take the truck up and come down. Hendrickson's ski jump, he called it. Hell of a thing to call it, I thought it was a hell of a good road. Steep mind you. Yup, we had over thirty percent on some of those hills, and there was no way you could stop them. They'd just slide. Just like roller bearings underneath the tires.

Al Hendrickson

Pioneer Timber's plank road at Port McNeill curved around the camp and onto the log dump in 1937. (CRM 17354)

The second type was the plank road, utilizing a similar superstructure of skids or bents with stringers, and with three- or four-by-twelve-inch planks laid across the stringers. This road was quite a bit more expensive, requiring up to 200,000 board feet of lumber—usually unsaleable hemlock—in every mile. However, cross planks offered better traction and it was easier to manoeuvre a truck on a planked surface than on fore-and-aft stringer, particularly around corners. At many truck logging shows it was common to use fore-and-aft roads, with short stretches of plank road at the landings, turnouts and dumps.

There was also a low-cost version, sometimes called the puncheon road, in which small logs

By 1937, Comox Logging had acquired several Hayes gravel trucks to build roads. (Wilmer Gold photo; IWA Local 1–80)

A shop-built, steel-wheeled crane, towed along the plank road, loads logs at BC Pulp & Paper's Halfway River camp, 1936. (Wilmer Gold photo; IWA Local 1–80)

were laid crossways on the ground, close together in the manner of a corduroy road. Either hewn stringers or planks were laid fore and aft for a running surface. This type was used on easy, level ground or for a short spur off an otherwise more carefully constructed road.

Not until the late 1930s, when bulldozer blades for crawler tractors and improved pneumatic tires became available, did truck loggers begin to abandon wooden roads. Even then there was a major obstacle to easy road construction—the task of loading gravel in a truck to haul it to where it was needed. Until the development of front-end loaders in the 1950s, loggers too small to afford track-mounted shovels had to build ramps to load trucks with a bulldozer. As better road building equipment became available at a reasonable price, the construction cost of logging roads fell, and it became economically feasible to build a decent all-weather road.

Truck loggers also needed better loading systems. The various systems used by railway loggers for stacking logs on cars were rejected by the truckers, whose lightweight trailers could not stand up to the rough treatment a duplex or crotch-line loading system handed out. The first, most favoured loading systems were the McLean and hayrack booms, both of which lifted both ends of the log, swung it over the trailer and lowered it into place. When heel-boom loading systems, like the one used on the Lidgerwood steel spar, were reinvented in the 1930s, they began to replace the two-tong McLean and hayrack loaders because they were faster.

Dumping logs also required modified systems. The jill poke devices so popular at railway dumps were no good on lightweight truck trailers. Tilted roadbeds were part of the solution, particularly when the early trucks were hauling one-, two- or three-log loads. The next development was to use a spar or an A-frame to parbuckle the logs off the trailer. But with smaller logs and larger loads it became advisable to lift

Loading with a McLean boom on a loading spur at H&R Logging's Harrison Lake show, mid-1930s. Cable rip-rap can be seen on the main haul road at right. (W.W. Chittenden collection)

Loading a 1940-vintage Mack at Port Neville Logging's Fulmore Lake load-out for the short haul to salt water, 1947. (CRM 11146)

They were dynamite, some levermen, boy, they were something. Then of course, with the trucks, they found out that wouldn't do. You had to be so careful with the trucks and that evolved to tong lines and heel booms. The old MacLean Boom was the best, with two lines coming down. Two blocks out from the boom, instead of one like a heel boom. And that really was the most successful of the whole works. The only thing was if you were really a high-ball outfit, lots of logs coming in, then you had to have two tong men out there. They also did the chasing. In my day lots of outfits had two tong men and a chaser as well. And one head loader of course. But that meant you were loading all the time, one truck went out and, bang, the other one was right back in.

And of course you hauled with cheeseblocks, which were on chains that you adjusted on both sides. But they would roll over if you had a squeeze, one log prying down, it would turn over, you could lose your whole load. The cheeseblock would turn over. Then some smart guy came up with a Seattle bunk, but the trouble with a Seattle bunk was that the stakes were stationary right on the outside of your bunks. The good thing about it was if you had a big log you could smash the log down on top of it, then it would stay there.

Al Hendrickson

the logs off the trailers to avoid damaging the bunks. One of the most common methods, adopted in the 1940s, was an A-frame that could lift the entire load and drop it into the water.

Right from the beginning some attempts at primitive pre-loading systems were made. At Joe Kerrone's Mount Sicker incline, logs were stacked in load-sized piles on a brow log. When the trucks backed the trailers in, the donkey lifted the front of the whole load, the truck backed under it and the logs were lowered onto the bunks. This was referred to as the leaning-tree system because the spar tree was erected with a tilt so the main block hung over the road. In such a setup, the yarder often served as the loading machine.

For those who actually got behind the wheel of a 1920s-vintage logging truck and herded it

Above: Comox Logging used a pre-loading system at Ladysmith in 1937. The tail end of this load rested on a brow log behind the trailer. The steam donkey behind the A-frame hoisted the front of the load while the trailer was backed under, then lowered the logs onto the bunks. (Wilmer Gold photo; IWA Local 1–80)

The first truck I worked around was at Caycuse. Don Mackenzie came into Wardroper. They had this godawful canyon, no way to get a railroad in there. Don had some trucks working somewhere else. They were the old Diamond T, single-axle. He come up and looked at it and they figured out. They had it well planned out on paper, grades and one thing and another. He brought in two of them Diamond Ts. It was a mile-and-a-quarter haul to the beach. I run a loading donkey there, I was kind of the camp handyman there for a while. I was loading them trucks when they first came there, that's the first logging trucks I seen.

We used a hayrack with two tongs, I just picked the logs up. I remember when one got away from me. It wasn't really my fault. The head loader, when he got a big log, too big for the tongs, he used to put a strap around it, hook it to both eyes with the points of the tongs. They got this one, it was big, but it was kind of rotten. But he finally got the tongs to hold. I had to pick this up pretty high, there was another one standing up on end kind of close, I had to pick it up pretty high to get over that, and I just got over the truck when one of those tongs broke loose and it landed right in the middle on the trailer, it just went Whoosh, right to the ground. I wasn't very popular with Mackenzie at that time.

Cliff Hallberg

You couldn't order parts for those trucks. Some of those camps didn't even have decent welding equipment. At old Port Neville Log, they had no arc welder there, just acetylene, and a cutting torch. Everything, what welding they had, was done with acetylene, which is much slower than the arc welder. But they seemed to get by.

A lot depended on the mechanics. They were usually cranky, too. They figured they had the weight of the world on their shoulders, which I guess they did. And they had to work all hours. Some of them worked all night so somebody'd be going the next day.

They'd have a big shop. Just a big shed, usually, open on one side. You could roll the trucks in and out. Wasn't very warm in the winter. In fact, nothing in those days was insulated or very warm, anyways. The sheds were pretty big in most places because they had to have the height so a truck could come in with the trailer loaded if they had to. They'd have a pit so's they could get under it. A lot of it was pretty primitive, compared to nowadays.

Mel Parker, born in 1920 to an Oregon logging family. His mother was a Palmer, and the family company, Parker & Palmer Logging, operated at Astoria and at Duncan as Scottish–Palmer Logging. Parker started punking whistles at age eleven and falling at eighteen. He was involved in Hanson & Parker Logging at Port Neville, Parker Logging with his brother, and Wyatt Logging. He retired in 1980 at Campbell River.

down a fore-and-aft road to the beach, the glory days of steam logging were a dim and distant memory. Running a locomotive and driving a truck had little in common, as many an engineer who had his locie sold from under him found out. Locie crews made up a kind of social unit: an engineer, a fireman and one or more brakemen. They were out there together, helping each other along. This was not so with a truck driver. He was a single man, sitting out in the open with no protection from anything, especially the ten tons or so of logs balanced precariously on the trailer right behind him. If a train took off downhill out of control, there was a good chance of derailing the cars or losing their loads before the whole thing piled up in the ditch. With a truck, the most likely place for the logs to end up was on top of the truck and driver.

A twelve percent downhill grade on a fore-and-aft road on a rainy day struck terror into the heart of a young guy who thought he'd learned everything there was to know about driving on

Above: A 1935-model Ford taking on a huge load at an unidentified location, while a second, similar model waits behind. (VPL 4100)

Byles and Groves hard-rubber-tired crummy on a fore-and-aft road at Port Neville, 1938. The crew rode this crude machine in rain and snow, usually falling to the ground when they jumped off because their legs were asleep. (CRM 3883)

Comox Logging's crummy at Ladysmith in 1937 offered some protection from the weather, but not a comfortable ride. (Wilmer Gold photo; IWA Local 1–80)

Old Floyd Byles, when he'd need another truck he'd get a load of scrap iron, dump it at the shop. An old motor, generally wore out or three-quarters wore out; tell the mechanic there: "Make another logging truck."

They had every type of motor and radiator—you couldn't go by the name on the radiator, 'cause the motor was apt to be a different motor. And the steering wheels was made out of iron—you can imagine on a frosty morning, when there was no cab . . .

They had a windshield in front and a tarpaper roof with two posts to hold it up in back. When they took the load off he'd just lift the cab up and it would fold on hinges and lay on the radiator. At least he had a windshield in front of him. And that's all he had, and a bit of a roof. No sides.

The brake system they built, it was just a big hand lever and a ratchet, wound up a little drum and a piece of cable, small cable, to the back, attached to the trailer brakes. You started cranking that up before you ever came to a grade, made sure you got all the slack out of the line 'cause you didn't have time after you started downhill. The brakes weren't water cooled so they couldn't haul them on too steep of grades. Of course if it was very steep those things wouldn't go back up anyways. Not much traction on those timbers when it was wet, or a little frost. It was bad.

Mel Parker

his old man's farm or herding milk trucks around the Fraser Valley. The only thing worse than a wet pole road was one covered with snow or ice. Once a truck started skidding down the poles, it took a steady hand on the wheel, nerves of steel and a lot of luck to keep the tires between the guard rails. In dry weather some of those roads were just about as bad because the brakes on the trucks were minimal devices at best. Trying to keep the cable on the trailer brakes tight or loose enough while negotiating curves on a fore-and-aft road, and throwing in a change of gears here and there, required a driver with four, preferably five, hands.

It was exhausting work driving one of those trucks, and it only got harder when they moved off the planks and onto the dirt roads. Then the driver had to fight mud, rocks and ruts, all of which could grab the front wheels, tear the steering wheel out of his hands and aim his vehicle for the ditch. All the while the driver was sitting a couple of feet behind a screaming gas engine in an age of primitive muffler technology, and just above a gearbox that by itself made

Above: Wellburn Timber's Diamond T, purchased directly from the factory shortly after World War Two. (Royal BC Museum)

A postwar model Hayes at Englewood, 1947. Canadian Forest Products acquired its first trucks in late 1942, shortly after taking over Englewood, and used them to supplement its railway. (BC Jewish Historical Society; O.F. Landauer Collection)

more noise than all the steam engines on the south coast combined. When you talk to old-time truckers you usually have to yell, or remind them to turn up their hearing aids. That was the one uncontested advantage of steam machines: compared to a four-cylinder White or a brand new 1938 Mack humping a full load up a six-percent adverse, a steam engine gave off only the slightest whisper.

After a day of this sort of brutality, early truckers, and certainly the owner-operators, could look forward to spending most of their evening and maybe all night monkey wrenching. The first logging trucks were not foolproof pieces of machinery under the best of conditions, and logging rarely provided those. Most of the trucks operated far from garages and parts depots. Anyone working up the coast could count on a few days' wait for parts to come in on the Union boat. The result was that truck loggers learned to improvise. By the time most trucks had been away from civilized surroundings for a year or two, they wouldn't have been recognized by their makers.

There was, however, an entirely new form of logging evolving in this entrepreneurial soup. The skills and knowledge acquired before and during World War Two were creating a system of logging that was technically better, more efficient and a lot easier on the forests than the railway logging systems that had dominated the coastal landscape for the previous twenty to thirty years.

Given an equal and fair advantage on an open, competitive playing field, the small and some not-so-small truck loggers would likely have gained control of the timber supply, driving the big timber companies out of the woods and back to their mills. But this was not to be. Even as this new sector in the industry was getting on its feet, steps were being taken to deprive it of direct access to timber.

VII Final Mechanization

Previous page: Green Point Logging used Caterpillar RD8s in the mid-1930s to complement their railway show on short hauls to Harrison Lake. (W.J. Moore photo; UBC Anderson Collection)

During the interwar period, there were a number of other significant technical developments that came to fruition, or were launched on their way to general acceptance in the logging industry. Perhaps the most important of these was the crawler or track-laying tractor, known to the world ever since as the Caterpillar—the Cat—no matter what the brand name might be.

In logging, the development of the Cat paralleled the evolution of the truck. They were highly compatible pieces of equipment and, both being powered by gas or diesel engines, they allowed a full and final abandonment of steam power in woods operations.

The origins of the Cat are found in the early steam tractors, or road steamers, of the type used unsuccessfully at Jerry Roger's logging camps in English Bay. Steam-powered tractors were not used very much in logging, especially outside dry areas such as California. Tens of thousands of them were produced, however, for the North American agricultural, construction and mining markets. It was in large part out of the attempts to adapt these steamers to logging that the crawler tractor emerged. To a great extent this development took place in California and led the shift in industrial power from eastern North America to the West.

During the last century there were a number of separate small innovations which, combined, culminated in the creation of a successful crawler tractor. In 1858 the first North American crawler tractor, known as the W.P. Miller Traction Locomotive Steam Car, was built in California. A small steam engine powered a single-drive axle connected to wheels that ran inside two endless tracks. The machine was steered with a pilot wheel at the front. It was not a great success and never went into production.

For almost fifty years manufacturers, including those in the West trying to serve the logging industry, pursued the development of steam-powered, wheeled tractors. Although several of these models were used with some success and with a variety of log hauling accessories, they were all fairly useless on rough or wet ground. As a result they never had any success in the Pacific Northwest rain forests. One of these manufacturers, De LaFayette Remington of Oregon, set up a factory to produce a three-wheeled steam tractor he hoped would meet the needs of loggers. Although he sold a few machines, Northwest loggers did not beat a path to his door, and when his factory burned down in 1887, Remington took one of his models, called the Rough and Ready, to Daniel Best's California tractor factory. Best bought Remington's patents and began incorporating some of his designs into the Best Steam Traction engine, probably the most popular steam tractor used by dry-land loggers. Best was also an early pioneer of gas engines, and in 1892 he sold 200 of them for use on irrigation pumps.

During the same period, a New England family, the Holts, established a wheel factory in Stockton, California. One of its main products was the Big Wheel, used with horses and steam tractors in the dry-land forests to skid logs. The company also entered the farm equipment business and designed a combine that used an innovative chain-drive system instead of gears. In 1890, Holt produced its first steam tractor, which also used chain drives. Some of these were used by loggers, and Holt also made a ten-ton log wagon to pull behind the tractor.

In 1904, Holt came out with its first track-laying tractor, the first practical crawler tractor in the world. This crude, forty-horse steam-powered monster was capable of outpulling the company's most powerful sixty-horse steam tractor. From the beginning, because of the way it moved, it was called a Caterpillar, a name the company registered in 1910. In 1905, Holt also got into the gas engine business, setting up a subsidiary to build them. It soon began installing these engines in the Caterpillar and by 1910 about 100 gas-powered Cats were operating.

A Holt five- or ten-ton model built in the early 1920s, working at Brooks, Scanlon & O'Brien's Stillwater camp, 1926. The donkey is loading an Athey bummer, a skidding device that preceded the logging arch. (H.W. Roozeboom photo; VPL 1564)

Working in the same business and selling into the same markets, it was no surprise that the Holt and Best companies became fierce competitors, ending up in court in a bitter patent dispute in 1908. The upshot of this fight was that Best sold its patents to Holt. But a few years later C.H. Best was back in business, producing the first crawler tractor without a pilot wheel, the legendary Model 60, in 1919. To a great extent the Best machines were better made copies of the Holts. Best even went so far as to use Holt's copyrighted name, Caterpillar, in its advertising. Naturally, more court battles ensued.

Meanwhile, a number of other manufacturers began producing crawler tractors. In 1916, the White brothers, builders of one of the first logging trucks, started the Cleveland Tractor Company, which two years later came out with a small gas-powered crawler called the Cletrac. It had differential steering, was light in weight and had good clearance—all features that were

The tractor of 1920 was a very inefficient tool to operate in the woods. The tractor might have done in short-log country in the Interior on high, dry ridges, but certainly those that I saw were pretty ineffective for logging on the coast. They had some use in the woods for construction purposes, but not for logging. Anyone who tried them on the coast in the early years just had one hell of a time with them. However, they kept at tractors and improved them so that they were not only a fine tool for construction and building and grading roads, but also became very effective for logging. It was not as revolutionary as the truck. You didn't need the tractor in order to log with trucks, and when we went into Ladysmith we didn't go in as a tractor operation but as a donkey operation. When we went in there first we had some tractors, but most of the logs were produced with donkeys, not tractors.

Robert J. (Bob) Filberg

Shaughnessy–Lidstone Logging's Caterpillar Sixty pulling a bummer at Cowichan Lake, 1928. This was one of the models made when the Holt and Best companies merged in 1925. It sold for about $4000 in 1929. (BCARS 73709)

part of an advertising campaign aimed at loggers. Many Cletracs were sold, but they were never particularly successful in the logging industry, largely because they did not stand up to the work. The company produced about forty different Cletrac models before the Oliver Tractor Company bought it in 1945.

During the 1920s, at least twenty-five other companies entered the track-laying tractor business. About a third of these were on the US west coast and made particular efforts to develop machines for logging. The Blewett company of Tacoma produced a tractor called the Webfoot in 1920. Union Tractor in San Francisco came out with one called the Bulldog a few years earlier. At least two companies sold modified Fordsons, equipped with track-laying devices—the Belle City Trackpull in 1926 and the Trackford in 1928. Virtually all these manufacturers disappeared during the Depression or World War Two; the major survivors were Allis-Chalmers and International Harvester, both of which appeared in the 1920s.

Until the late 1920s the primary use of crawlers in the logging industry was to skid or haul logs. Initially, they were also used to pull log-loaded wagons on dirt roads or log cars that ran on rails. As well, tractors occasionally pulled a string of logs along a chute.

Although records are hazy on the subject, the first crawler tractors used in logging on the BC coast probably appeared in the Cowichan Valley in 1923. Local lore has it that the first Cat on Vancouver Island was owned by the Evans brothers, who owned a small steam-powered sawmill at Duncan and logged in the area. This machine was likely a Best 60, and came accompanied by Albert McNeil, an operator sent from the Best factory, who remained in the Cowichan area. Lake Logging also started up that year using a Cat at Craig's Landing. During the last half of the 1920s, several other small contract loggers used Cats at various sites around the lake—the Tractor Logging Company at the south end, Kerrone–Morin Logging at Nixon Creek, S&K Logging at the west end and Sam

A Caterpillar with a Carco steel-wheel arch skidding logs for Virotsko Brothers at Extension, south of Nanaimo. The frame over the tractor held the winch line when it was used to control a bulldozer blade. (NCMA P1–41)

Handra at the foot of the lake. Gerry Wellburn, at Duncan, replaced his horses with a Cat 60 in 1928.

Elsewhere, tractors were slow to appear. Harold Clark at Harrison Lake was one of the first in the Fraser Valley to try Cat logging in 1927. Probably the first Cat on the north island was brought into Quatsino Sound by Albert Moore in 1931. Moore, a former logging manager for Whalen Pulp and Paper at Swanson Bay, came to Quatsino Sound in 1928 to contract A-frame for the Port Alice pulp mill. He bought his tractor from Morrison Tractor & Equipment. Morrison and a former partner named Kaufman had been the BC dealers for Best tractors and when the Caterpillar company was formed they took over that dealership. Kaufman pulled out in 1926, and Morrison continued until 1933 when Earl Finning, a California Cat dealer, took over and launched what became the largest and most successful equipment company in BC.

In 1925, the Holt and Best companies settled their legal differences by merging and adopting the name Caterpillar. They retained two highly successful logging tractors, Models 30 and 60, that for many years were the mainstay of coastal Cat logging. However, the real explosion in crawler tractor usage lay in the future, following the development of auxiliary equipment.

In the US coastal dry belt, Cats commonly used Big Wheels to assist in towing logs. In BC, however, the Douglas fir logs were too heavy for the wheels and the ground was generally too wet. One partial solution was the development, in Washington, of the pan, a broad piece of sheet metal with an upturned front end, on which the forward end of the logs were laid. It prevented the logs from digging into the ground as they were towed. A further development was the bummer, a refinement of the Big Wheel, in which the logs rested on a wheeled frame that was towed.

In the early 1920s Willamette Iron & Steel of Portland, makers of steam logging equipment, marketed the Erstad winch, which was adapted

---

Merrill & Ring were Cat logging before they trucked, up around Cat Lake, that was all Cat logged. That was the first Cat show in Squamish. They had a Caterpillar D-8 and an arch. That was roughly... they had a shutdown in '30 for a couple years... in '32, something like that. It was one of the first Cat shows, I guess, in BC. That's how Cat Lake got the name.

To move that D-8 Cat they had a bull car, it was just a big girder-framed flatcar actually. The deck wasn't wide enough to accommodate a D-8 Cat, so they bolted a couple of big timbers along the outside of it so the tracks could sit on them.

And that was all dragged down in tree lengths. Brought it down with the arch in tree lengths. He used to bring that down and dump it into Cat Lake. There was a man, Joe Ozanich, who used to buck in the water there, just hand buck them into log lengths.

Ed Aldridge

Beginning in 1937, Comox Logging used several Caterpillars with arches at Ladysmith to salvage blown-down timber. (Wilmer Gold photo; IWA Local 1–80)

Below: This stripped-down model was used to ground skid logs by Bell–Campbell Logging at Sointula, 1936. (Wilmer Gold photo; IWA Local 1–80)

In 1942, Comox Logging used a D4 Caterpillar and a wheeled arch to salvage pulpwood from logged-over areas at Ladysmith. (Finning Tractor collection)

to run off the rear of Caterpillar tractors. The winch was quickly used in combination with a new type of logging wheel produced by both Pacific Car & Foundry in Seattle (Carco), and ESCO, a Willamette offshoot in Portland. This rig had two relatively small wheels on an axle to which a steel A-frame with a fairlead at the top was attached. It hooked on to the tractor's drawbar. A line from the winch ran through the fairlead and attached to the logs. When the line was reeled in, the front ends of the logs were lifted off the ground and easily skidded. A further advantage of this setup was that scattered logs could be bunched with the winch line, without moving the tractor.

The final addition to the fairlead logging wheel was the Athey track-laying wheel, developed in 1920. It was attached to the axle at only one point so the tracks pivoted when they encountered an obstacle. They travelled over very rough ground without tearing themselves or the tractor apart, and provided good flotation on wet ground. In 1929, ESCO, Willamette and Erstad formed a new company, Hyster, to produce this new device, which was called a logging arch. It bunched and skidded several large-diameter logs in one turn, quickly and efficiently. Over the next twenty-five years or so this was to become the primary piece of Cat logging equipment, and one of the major methods of skidding logs on the coast. The welded Hyster arch and a cast model produced by Carco dominated the field. After World War Two, with improvements in rubber tires, balloon tire arches also became popular. They were followed by the development of an integral arch that was attached to the rear of the tractor.

But initially there was a serious problem with the arch. The weight of the logs transferred to the drawbar of the tractor, combined with the drag on the winch line, raised the front of the tractor off the ground, reducing traction and making it more difficult to steer. Solving this problem was one of the reasons for developing the bulldozer blade—as a counterweight.

Hyster also took over production of winches designed for attachment to Caterpillar machines, including a large two-drum winch that was first produced by Ersted in 1925. This rig was used for high-lead yarding, with one winch

Following the 1938 Campbell River fire, Comox Logging used Caterpillars to salvage timber north of Courtenay. (W.J. Moore photo; Finning Tractor collection)

Three-drum Hyster winches, like this one mounted on a Caterpillar D7 near Sumner, Washington in 1943, were rarely used in BC. After World War Two these units, called tractor donkeys, were combined with portable steel spars, with very limited success. (Finning Tractor collection)

A Comox Logging Caterpillar skidding logs to a pre-load A-frame, 1937—possibly the same pre-loader shown in Chapter 6. (W.J. Moore photo; Finning Tractor collection)

on the mainline and the other on the haulback, and for loading. By removing the yarding or loading lines, it could be used with an arch. This was a notable attempt to make the Cat an all-purpose logging machine. Although BC loggers purchased several of these units, it was not a very successful accessory. Compared to other gas donkeys then on the market, it was not a particularly good yarding engine. There were other, better uses to which the Cat could be put on a small logging show, and loggers needing a cable-yarding system found it was better to use a separate donkey.

The evolution of bulldozers—blades mounted on the front of tractors—began with devices used with horses to plow snow. As the various makers of crawler tractors developed their machines, they and independent accessory manufacturers attempted to develop suitable blades. One of the best designs appeared in 1927, produced by construction contractor Robert LeTourneau for use on Caterpillars. The blade was controlled by a cable running through a block hung from a frame on the front of the tractor, back to a winch on the rear. When it was combined with the Hyster double-drum

Viv Williams, left, operated this Caterpillar Diesel 75 at Fred Clark's camp at Errington in 1936. (Viv Williams collection)

Pioneer Logging at Sooke used a Caterpillar Diesel RD8 and Hyster arch on a one-and-one-quarter-mile haul in 1937. (W.J. Moore photo; Finning Tractor collection)

Harold Bendickson operates an RD8, purchased new from Finning in 1937, at Hardwicke Island near Wellbore Channel. (CRM 12446)

winch, the result was the basic Cat logging machine that would be used for decades. One winch controlled the blade; the other was used for skidding logs, with or without an arch.

Other companies were developing hydraulic-controlled blades, two of the most popular being LaPlant–Choate and Isaacson of Seattle. In 1933 Garner Brothers bought one of the first International tractors on the coast, a TD-14 narrow gauge built for logging and equipped with an Isaacson blade. It was sold and serviced by BC Equipment, and is still being used on Galiano Island.

The development of the bulldozer blade turned the Cat into one of the most versatile pieces of equipment in the woods. In addition to skidding logs efficiently, it built roads on which trucks, which had been pretty well confined to fore-and-aft or plank roads, could travel. Road building costs were slashed. With one machine, a small operator could get his logs to the beach or to a road where a reasonably priced truck could haul them.

In 1932 Thomsen & Clark Timber used one of the first bulldozers employed in BC for road building at its Chehalis River camp. This machine, a Caterpillar 75 diesel equipped with a double-drum winch, Hyster arch and a Carco hydraulic blade, was put to work skidding logs to clear the right-of-way and to construct the railway grade along the Chehalis River.

There was one additional improvement to tractors that was of major significance to the logging industry. It was known in the early years of the century that diesel engines had many advantages over gasoline engines. Fuel was cheaper, and diesel engines were more efficient and had more lugging power. But the first diesels were incredibly heavy, due primarily to the bulk and weight of the fuel injection devices then available.

In the 1920s Caterpillar installed Atlas diesel engines in a few machines, but they were too heavy and did not fit well. Best began development of a diesel engine in 1915, and Dow, the pump manufacturer that would later introduce one of the first power saws, produced a diesel engine for several years until 1923. Caterpillar introduced its first diesel in 1931, followed by the RD series of diesel engines in 1935. Cletrac came out with a Model 80 diesel in 1933, and Allis–Chalmers with a spark-ignition diesel in 1934.

The adaptation of the diesel engine for use in tractors had wide repercussions. Tractor design had evolved to accommodate relatively compact gasoline engines and the need to develop a diesel engine that would fit into a tractor led to the creation of diesel engines small enough for much wider application. In 1932, the first Kenworth truck diesel came out, a beneficiary of the intensive research and development that had gone into production of diesel engines for tractors.

By the mid-1930s, the various improvements described above had made the tractor a popular piece of logging equipment. One of the first of the major companies to acquire Cats on a significant scale was Comox Logging, when it opened its Ladysmith operation in 1937. Three years earlier a heavy wind had blown down trees over a huge area and, for the most part, Cats were used to salvage this timber. The Ladysmith operation started with small RD-4 and RD-6 Caterpillars, all equipped with blades.

In 1942, Comox Logging and the Powell River Company launched an experiment to prelog and relog stands in both the Ladysmith and Courtenay operations, shipping the salvaged timber normally left in the slash to the Powell River pulp mill. The success of this venture was a major consideration in the subsequent building of the Elk Falls pulp mill at Campbell River. A large portion of the salvage operation was done with Cats, using larger RD-8s with Hyster arches.

Another company making extensive use of Cats in the mid-1930s was P.B. Anderson's Green Point Logging at Harrison Lake, where new RD-8 Cats with arches skidded logs from donkey landings into the lake. In 1936, several other substantial loggers had incorporated tractors into their camps, including Brown & Kirkland, Buck and Turner, and the Cameron Lumber Company at Cowichan. On Lake Cowichan, Joe Kerrone was using a gas and a diesel Cat, and Royston Lumber up the island had a Caterpillar and two Fordson crawlers. A typical gyppo operation of that era, Larson, McLauchlin & Carney, was running a seven-man camp at Fanny Bay, using a Hayes Anderson truck, a fifty-horsepower Caterpillar diesel and a gravel truck to build roads.

This was merely the beginning of the use of Cats in the woods. Shortly after the end of World War Two, there was hardly a camp of any size on the coast that did not have a Cat working in some phase of its operations. Some of the enthusiasm for these new machines was

I had started in the woods in 1934 as a young fella. In those days, most people who lived in that part of the world fished, logged and trapped in season. During the winter months you trapped. During the summer months you fished, then back logging again. The area I was born and raised in, at the mouth of Knight Inlet, was where the Powell River Company had started logging, at Kingcome, I guess about 1910. By the time of the First World War, just before I was born, the Powell River Company had put in some camps on Turner Island, next to Village Island where I grew up. By 1924 they were winding those operations up. They turned around and had a whole bunch of independent loggers A-framing mostly, and a few with small railroads, gas locomotives. Though they called them contractors, Powell River Company actually paid them a market price that was established in Vancouver. You sent the boom south and they deducted the cost of towing to Powell River. The company got a good seventy-five percent of the requirements of the mill for twenty-odd years from what is referred to as the Jungles—the mouth of Knight Inlet in those islands. They had thousands of contract loggers. Some of them would pick up timber sales, others would log Powell River's timber licences and pulp licences. But they still got paid the market price. There were some pretty big loggers there, with three or four skylines strung out back of an A-frame. Well over a mile, a mile and a half. One machine would transfer it to another.

The development of the A-frames allowed those fellows to log areas that were quite steep and rugged. About the early fifties the Cat came out. Then a lot of independents were able to go into isolated areas. Punch in a road and go back a mile with just a Cat and an arch. Hardy Trucking on Minstrel Island were the first to put in a crushed rock road. By the mid-fifties there were quite a few trucking operations.

When we first started truck logging we shunned rock. Today we look for it because it's cheaper to build out of rock than soil. When we first started we had to haul gravel in and the roads were never that great on account of that. Today we turn the roads against the rock bluffs because we can drill it and make the road out of it. It's cheaper in the long run. To haul rock any distance is prohibitive. Go into the bank and spread it out.

In the forties, road building was pretty primitive. It was all done with jackhammers. And when the jackleg was invented, this was the thing you could use a bit of air to push it, attached to the handle of the jackhammer. It saved a lot of teeth. We built a lot of roads with those jackhammers. Then the wagon drills and the airtracks came in. Rock isn't a problem today unless it's real hard rock. Most of Vancouver Island is relatively soft rock. There's very little real granite on Vancouver Island. A lot of it is semi-decomposed rock, quite soft.

Dan Hanuse, who founded Cheslake Logging with his brothers in 1949. He logged in the Blackfish Sound area for fifty years and was president of the Truck Loggers' Association 1986–87.

Above: A Caterpillar 75 diesel with a Hyster arch, used to skid logs to a landing at Fred Clark's Errington operation, 1936. The truck being loaded is the first Kenworth diesel used in BC. (Finning Tractor collection)

H&R Logging was one of the first companies to use a bulldozer building roads at Harrison Lake. (BCARS 73740)

By the 1960s, Caterpillars such as this D6 were equipped with integral arches for skidding logs, and roll-over protection for the drivers. This is the Harper Logging site near Kelsey Bay, 1965. (Finning Tractor collection)

due to the possibilities they offered for a less destructive type of logging than that being done by the dominant high-lead and railway shows. This was before the days of plantation forestry, when some far-seeing operators were concerned about the future timber supply. Cats were seen as a means of harvesting timber without damaging the new growth coming up under the old.

But there were other, more immediate reasons for their popularity. When they were not building roads, they could be used to skid logs, drag the first primitive steel spars to a new setting and perform dozens of other tasks that made loggers wonder how they ever got along without them.

Cats also made possible the survival of small, often one-man logging operations of the kind that began with handlogging. At any given time for almost twenty-five years after the end of the war, until the development of rubber-tire skidders, there were scores of independent Cat loggers scattered between Vancouver and Prince Rupert.

Although they were eventually squeezed out by bigger companies, these small operators contributed a substantial volume of timber to the open log market, enabling numerous small mills to buy logs. They competed for their timber at public auctions, providing the provincial treasury with revenues far higher than those the big companies were required to pay.

A typical independent operation of this era consisted of a family living in a float house, tied up in a protected bay. Although there might be an employee or two, it was more likely that the logger worked alone, with the part-time assistance of his wife. With this single machine, a Cat, he could gradually push a rough road back two or three miles through the timber. With one of the new one-man power saws then available, he fell and bucked his timber and, in the late afternoon, skidded out the day's production. One of the most satisfying feelings in the world, even in the pouring rain, had to be the long two- or three-mile-an-hour haul to the beach with a few thousand feet of number one fir sawlogs hanging from the arch, the throaty roar of a D-8 and the aroma of diesel exhaust filling the air. After dumping them in the bull pen, the small logger could shut off his Cat—unlike the horses he once had to care for—and get in an hour's fishing before dinner.

When there were enough logs in the water, they were boomed and a radio call put in for a tug to tow them to a mill or, more likely, Vancouver where one of the dozens of log brokers shopped around for the best price. While this kind of life was not everyone's cup of tea, it provided a good living for hundreds of loggers for many years.

Opposite: Bucking at a Lidgerwood steel-spar landing with a beavertail saw. Franklin River, 1947. (UBC BC1930/188 J1358)

A second important piece of logging equipment adopted between the wars was the power saw. During the boom period of the 1920s, it became apparent that a better method of falling and bucking trees was needed. All other phases of logging had been mechanized. The primary phase, falling, was still performed by hand, using essentially the same methods employed in the late nineteenth century.

Until the 1920s, large numbers of European immigrants, primarily Scandinavians, had come to the BC coast and taken most of the falling and bucking jobs. Few others were willing to take on the arduous tasks of chopping undercuts with an axe, or falling and bucking with hand-pulled crosscut saws. But now it was becoming clear that the supply of husky Swedes was not inexhaustible, and various individuals and organizations in the West Coast logging industry were starting to look around for mechanized falling equipment. Furthermore, as logging progressed up the valleys and mountainsides into stands of smaller diameter trees, the average production of fallers and buckers steadily declined. Clearly, what was needed was a mechanized saw.

In the 1920s it was common practice for builders of early power saws to demonstrate prototype models in Vancouver, often before large crowds. In 1927, for instance, the BC Loggers' Association staged a demonstration on the front lawn of the courthouse by an American inventor named Arseneau. He showed the assembled crowd that his saw, a monster weighing several hundred pounds and mounted on two wheels, could easily buck thirty-inch logs. But the machine was deemed too cumbersome for use in BC, and he failed to sell any saws. Nor did the Arseneau saw ever go into commercial production.

The search for a power saw began long before this demonstration. The first attempts focussed on devising machines utilizing more or less conventional crosscut saw blades, which ultimately

An original model of the Wolf Electric Drive Link Saw, first built in 1920 at the Peninsula Iron Works in Portland, was demonstrated the same year at a loggers' convention in downtown Vancouver. The seventy-pound saw, powered by a 1.5-horsepower, 110- or 220-volt electric motor, had a three-foot beavertail bar. There is no record of this saw being used in BC. (VPL 4097)

led to the production early in the century of gas-powered drag saws used to buck firewood for steam boilers. In the 1870s, Ransome, a US company, placed on the market a saw called a steam treefeller. It was a drag saw equipped with a device that clamped onto the butt of a tree. A steam-driven piston moved the blade through the tree. At least one of these treefellers was used in BC. A Saltspring Island farmer, John Brown of Burgoyne Bay, imported one in 1879 for cutting firewood. It was not a successful industrial saw: it required a boiler to generate steam, as well as pipes that were laid through the trees and needed constant repositioning.

During the early years of this century, several different design approaches were attempted, mostly in the US. One used a special wire rope equipped with cutting teeth that was rapidly pulled back and forth between two powered drums. Another utilized an electrically heated red-hot wire to burn through the tree. A machine using an auger that drilled a series of holes through the tree was tried, and another rasp-like device mounted on the end of a rotating shaft was used to chew through the tree. At one stage even the Lidgerwood Company was working on a power saw design.

One of the most interesting designs was produced by James Shand, a Manitoba farmer and millwright. In 1918 Shand patented his "endless chain" saw and built two working models using a bicycle chain with embedded teeth for a cutting chain and a one-cylinder gas engine for power. Hoping to find a market for this saw, Shand moved to BC in 1919, but aroused no interest among loggers. The bar on his saws was only twenty-four inches long, too small to be of much use on the coast, and although he was on the right track, the gas engines available at the time were too heavy and unreliable to be practical.

In 1916 we came over to Port Alberni and went and seen the foreman here. Oh yeah, he's got a job down the canal, go down there, falling. A tug towing boomsticks back, he took us down. We showed up at the office. "What are you doing down here?" he says. "I didn't hire any men, I don't need any men, you'll just have to go back, there's no room in camp for you."

That made me mad, get down there and get in a pickle like that. So we thought we could get a ride back on the tug, he was going back to town that afternoon with a boom of logs. He says, "Nope, I can't take any passengers when I'm towing logs." Anyhow, I took a walk out in the woods, the camp was not far from there, right on the beach, but they were logging only a little ways out. I heard the tree falling and I went in there after it fell, kept my eyes open, two guys there, two old Swedes. "Where'd you come from?" I told them. They said, "Oh don't worry, we're quitting tonight, there'll be a job for you." They quit and they picked us up.

There was one old fella, his partner quit, and the bull bucker wanted to know if I could go with him. So they sent me with him. And you know, he wouldn't even talk. He'd grab that falling saw and the axe, and he'd go over there, you didn't know where he was going to fall the tree and he wouldn't tell you. All he had was a mouthful of snoose. So when he picked up his axe and started chopping, well I knew he was going to fall a tree that way. The first thing he says, "What hand do you work?" So I says, "I work the right hand." "That's what I work, and I'm not going to change now, you'll have to do the other." I said, "That's okay, I can work both hands."

And he worked the devil out of me, and we was only working day work, three dollars a day. I never worked so hard. No sooner he'd fall a tree than he'd run to the next one. I worked down there for about a month.

One day I went back falling with my dad. There was a slide on the mountain, steep, where no trees grew or anything, it was rock. You can see these slides, sometime on a sidehill. Right up on the top end of it was a great big dead tree. It had the limbs on it, but it was dead. It was raining, snowing. We had to fall this and the bull bucker says to fall this straight down the hill. I said, if we do it will go for miles.

So my dad could stand on the ground and chop. I got up two springboards. It was steep. I just got the two holes in, and got up on the board, and I thought, gee, there's something moving, what the heck's going on, I felt funny. The whole darn bark broke off up the tree and slid down, but when it hit the ground it broke up and spread out. Knocked my springboards out and me out and I jumped twenty feet from the springboard level down to the ground ahead of the bark. And I landed on my hands and knees and went sliding down the hill a hundred feet on my hands and knees over this rocky mountain rock, not pebbles, broken rock. Out of the corner of my eye I could see the bark coming after me. Then I was wondering if I was going to stop because I was going a pretty good speed. Somehow or another I stopped all of a sudden. And the bark stopped behind me. I went a hundred feet.

I was walking back up the hill; my dad was pulling the bark out like a madman, he thought I was underneath it. He didn't see me jump. We left the tools right there and walked into camp and quit.

Alvin Brown, born and raised at Coombs. He and his brother Gunny worked together falling and bucking, including forty-three years with MacMillan Bloedel and its predecessor companies.

Falling snags after the 1938 Campbell River fire. This is probably one of the first imported Stihl saws, although an almost identical model was produced in Vancouver when the war began in the fall of 1939. (CRM 16852)

During the 1920s, at least four designers in scattered locations were working hard on developing an acceptable power saw. One, George Dow, a California pump manufacturer, came out with one of the first production model saws, the Dow Low Stump Portable Power Saw. It weighed close to 500 pounds and was mounted on two motorcycle wheels, between which was slung a four-cylinder Indian motorcycle engine. A small hand winch was used to raise and lower the saw. The bar was equipped with a swivel so it could cut on either a horizontal or a vertical plane.

In 1920, Charles Wolf of Spokane began to manufacture the first portable chain saw, an electric model that required a generator to power it. This saw was equipped with a new type of chain, a scratch chain with rakers that cut in either direction. A faller could turn it around halfway through the day after one side of the teeth was dulled. Wolf also made a gas saw. His company was taken over by the Reed Prentice Corporation of Massachusetts, which worked through the 1920s in an unsuccessful attempt to develop a four-cycle gas engine for the saw.

Meanwhile, two German companies were forging ahead of their American competitors. Emil Lerp refined a fifteen-cubic-inch, two-cycle gas engine to power a saw using a cutting chain, and in 1930 he set up a 200-man factory in Hamburg under the Dolmar name. Also in Germany, Andreas Stihl produced an electric chain bucking saw in 1926 and four years later developed a gas-powered version. It was a 120-pound

The Dow Low Stump Portable Power Saw brought to Great Central Lake by Bloedel, Stewart & Welch in 1936. It was a cumbersome failure, weighing almost 500 pounds, and was raised with the crank at the top. (Leonard Frank photo; UBC BC1930/391 16609)

In the first part of 1937 this young Jack Challenger was a bull bucker and he was keen on getting power saws, any kind. He said it would beat hand work, and be cheaper. Jack used to come to me and talk for hours. It cost me a lot of money, we were working contract and he'd be there two or three hours sometime, and we didn't get anything for it.

In 1937 there was a loggers' congress in California. The company sent him down there. The company that made the Dow Low Stump power saw, took him up in the sugar pine country in the hills of California and demonstrated this saw to him. He had a lot of pictures he'd taken down there. In that country, dry, and the trees were about twenty to thirty feet apart, didn't grow close together like this, not the ones that he showed me. But they were big. And growing right out in the open. And sheep could be feeding out in the grass—all grass instead of brush. Well, they had this power saw there, and it was on wheels, like two motorcycle wheels. And a frame built in there, and a little winch on top. The power saw slipped up and down between these guides, like a pile driver hammer. You could move it right down to the ground, but you had to wind it by hand, with a little winch. The head would swivel around.

It had a four-cylinder, sixteen-horsepower, air-cooled Wisconsin engine that weighed about 200 pounds, the engine itself. He showed me these pictures for months. He says, would I run it if they got one? I says, yes, I'd run it. He couldn't get anyone to listen to him. It was six months before they got one.

In the fall of 1937, one Sunday morning, he come by my float on Great Central Lake. I see the tug come by with a big crate on it, it went to camp. I was about a quarter of a mile from camp. So I went up to camp to see what they had in this box, and sure enough it was the power saw. So Jack says, "Get some tools and unpack it." I says, "It's Sunday." "Well that don't make any difference," he says, "get it out of the box." And I took it out of the box. It was about eleven o'clock then. He said, "Go home and get some lunch, and I'll send it right up the lake. I want to see it run today." He couldn't wait until dinnertime.

I went home to have a bite to eat. They took off, but I had a speed boat that made about thirty miles per hour, so I overtook them, they were only going about ten miles up the lake from camp. There was a nice flat spot there, flat right out to the water's edge, about a half mile along the lake. A bank about a foot high. So we got there, it was past dinnertime, but he said he was hungry. They had another camp on a float down the lake about a mile and a half from where he was going to work on the claim. So him and the boat man decided to go for dinner. He give me strict orders not to fall a tree 'till he got back. We monkeyed around there. You could freeze to death with cold. I said to hell with Jack, I'm going to fall a tree.

There was a pine tree about that big growing right out of the water's edge. It took two of us to handle it, one fellow on the head and one fellow on the machine. I whipped the undercut into it and down it went into the water—it made a splash like a rifle shot. It didn't have any brush on it, and it hit the water with a bang.

He was about a mile and a half away down in camp, by the time the sound got down there. You could hear his boat starting. He came right up there. But he didn't get in on the first tree. I just laughed at him, I didn't give a damn.

We went to work falling, the three of us—Jack was helping us—and we took every tree that we could cut. I guess in about three hours there we felled about a hundred thousand feet of timber, just felled it out in the lake. He thought it was great. I said wait until we get back into the bush.

Lifting that damn 500-pound thing over logs—you couldn't lift it. After three or four days we got some two-by-ten planks and one of them had a little hook on it, to hook on the log so it wouldn't slip back. You had to holler for a bucker to come and help you get it over the log.

It'd really cut, that thing. It was different than these power saws today. The chain had a raker on there, and you had a machine, a swedge to swedge the raker out—squeeze it out wider, cause it was such a wide kerf. When I come to file this thing I said, "Have you got a swedge?" "No, I ain't going to buy one," he says, "they want $300 for one." "Well," I says, "you ain't going to get by without filing the saw and if you don't swedge it, it won't pull the shavings out."

Anyhow, I worked it for two weeks. Timber all over the place. Five buckers behind us and you didn't know where they were, some were good buckers, some were lazy buckers and it was getting in rocky ground. The big difficulty was, being on wheels, even when you start off level, the roots flare up. When the wheel came up on the tree or rock or whatever it was, the bar would bind. God darned groove in the saw would get red hot. Red! It'd stall the machine, because it was crooked, see, it was binding, one way it was going back in the hole, and the other was rising up on the root. You couldn't keep it level. Otherwise it would have been a success.

I quit it, and I didn't bother with a power saw until 1950. I went back to hand falling and made more money.

Alvin Brown

A power saw crew at Wellburn Timber, 1940, using a Timberhog to make the falling cut after the undercut has been made with axes. This saw, a copy of the Stihl, was made by Reed Prentice. (Wilmer Gold photo; IWA Local 1–80)

machine with a single cylinder, two-cycle, high-speed engine driving a scratch-type chain on a sixty-inch bar.

By 1936, the shortage of fallers was becoming critical, and Sid Smith, Bloedel Stewart & Welch's manager, resolved to do something about it. "I believe the most important problem which loggers have to solve is the question of securing a practical mechanical falling and bucking saw," he said before departing to a Pacific Logging Congress meeting in Spokane. "It would be money well spent if each camp on the coast would assess itself, say one-quarter of a cent per thousand feet, which could be placed in a fund in charge of a number of engineers who could decide when an invention relating to a mechanical saw was submitted, if it would justify financial assistance to perfect it." That year, as a result of Smith's proposals, the Pacific Logging Congress made the development of a workable power saw a high priority.

Late in 1936, with the backing of some of its member companies, the BC Loggers' Association took the initiative by ordering five Stihl saws and inviting Andreas Stihl to BC. BS&W undertook field testing and development of the new saws. When they arrived in December, two of them were sent to Port Alberni and turned over to BS&W's falling and bucking supervisor, Jack Challenger—the son of pioneer truck logger George Challenger. During the first half of 1937, Challenger put the saws to work at Franklin River and launched a program to recruit and train fallers.

These first machines were plagued with breakdowns, and replacement parts had to be made by hand at the Franklin River shop. However, a major problem was that Stihl had employed cast magnesium parts to decrease weight. At that time no one knew how to weld magnesium, and until Bill Collister, one of the BS&W fallers, taught himself how to do this, replacement parts had to be ordered from Germany. By the time two Stihl engineers, and then Stihl himself, arrived later in the year, Challenger's crew had devised several modifications which were taken back to Germany and incorporated into improved models. While in BC, Stihl appointed his first North American dealer, Don Smith of Vancouver, who set up BC's first power saw shop in the Alcazar Hotel.

Just before Stihl's visit, BS&W acquired one of the Dow Low Stump saws and took it to the camp on Great Central Lake. Challenger gave it to Alvin and Gunney Brown to try out. After a few weeks, everyone came to the conclusion it was not a practical saw for the BC coast. Eventually the company turned it into a fire pump at Franklin River.

Some of the first Stihls found their way into the hands of other forward-looking loggers. Wellburn Timber got two in 1936 and started training fallers in their use. Two of the original Wellburn fallers, John Evans and Dick Reid, worked falling at MacMillan Bloedel's Shawnigan Division until they retired. BS&W obtained the remaining Stihls from the BC Loggers' Association and concentrated on developing this saw, in co-operation with the dealer, D.J. Smith Equipment, and the Stihl factory. It was clear to Challenger and his people at Franklin River that these were the best saws available, and since Stihl held a number of important patents he was likely to remain in front of the field. Over the next two years, BS&W acquired another thirty Stihls. The company became one of the manufacturer's primary proving grounds, with one of the camp's machinists, Bob Shade, dispatched to the Stihl plant in Germany.

When war with Germany broke out, the power saw program was in trouble. The first big shipment of saws was seized on the New York docks. But the war ended Stihl's patent protection and almost immediately D.J. Smith began producing a saw that was identical to Stihl's. These saws, along with the few real Stihls that had reached BC, were known throughout the industry as "Hitler" saws. Several other companies in Vancouver jumped into the void, as well as many more in the US. Soon after the war started, Reed Prentice bought the D.J Smith operation and began selling the saw under the name Timberhog. Smith moved to Ontario and started the Hornet power saw company, which he later sold to the Mall Tool company in the United States.

When Bob Shade returned from Germany, he established Shade Engineering on Granville Island, with financial backing from BS&W. Prentice Bloedel and Jack Challenger were directors of the company. Shade made a saw called the Forest King, powered by a 125-cc single-cylinder two-cycle Villiers motorcycle engine sold by Fred Deeley. When BS&W had a dispute with Shade, Bloedel and Challenger sold their shares to Deeley, who in turn sold out to Dick Burnett in 1944. Three months later Burnett fired Shade, who was called into the army. Burnett began working with Challenger and the BS&W crew at Franklin River and by the fall of 1945 had five saws in the field. By the end of that year the Vancouver plant was turning out six saws a

---

The thing people said about the power saws at first was that they couldn't stand the noise, the noise was so terrible. I believe there was some truth to that. Hand falling was so quiet. All of a sudden this terrible monster came along and the noise bothered them to start with. I guess something we should have learned earlier was to get ear muffs. A lot of people lost their hearing because of that noise.

The danger part of it? They were afraid of them because of the noise, I think primarily. At that time the mufflers didn't cut the noise down, they were very loud. It took some time to get the new people used to it, to that terrible noise. But it was lighter work. It wasn't the heavy slogging. The younger people took to it quite readily. But the old timers fought it tooth and nail. They didn't think it would ever take the place of hand saws. But productivity in those days, the late thirties, was important. Hand fallers would fall about 7000 board feet per day. With the advent of the first power saws, the Stihls from Germany, production was about 10,000 board feet per day, which is quite an improvement. Considering the saws at that time were 130 or 140 pounds.

There were people that quit the woods because they wouldn't pack that monster around. It was a new life. If we wouldn't have trained a lot of new fellows it would have taken a lot longer to get it going.

When you were hand falling and you hit a pitchy old fir you were pouring oil and water on it to pull your saw through. When you got a chain saw that pitch didn't make any difference. When a faller saw that a few times . . . Or when you were long-butting hemlocks—we had to long-butt all the hemlocks in those days and they did it by hand saw. That's one thing the falling saw could do, it could take the long-butts off. When the fellows could see that chain saw take the long-butt off . . . And in those days you didn't get scale for the long-butt where you cut it, you got it up at the top of the next log. So, it was very poor. You did a lot of work for nothing. With the chain saws it was so easy. That did encourage some people to go for them.

It was getting more difficult to get hand falling crews. Most people didn't want to hand fall. It was very difficult work. There were an awful lot of cuts with the hand saw. And with the sharp axes. I believe chain saws are not as dangerous as the old hand saws. For one thing a guy with a chain saw does the work of three or four people, so that cuts your accidents down right there. There's only a quarter as many people out there to get hurt.

Olaf Fedje, who began falling in the 1930s. He was one of the founders of Fedje & Gunderson, one of BC's largest and oldest falling contractors.

Martin Fossum, a foreman at Bloedel, Stewart & Welch's Menzies Bay division, with a Burnett saw, 1945. (CRM 18512)

During the hand falling days mostly Scandinavians and central Europeans were involved in the falling—very few Canadians. They were used to hard work back there. I hate to say it but I know when I was in charge we tried out Canadians. We had a class up in camp and I had an instructor for each group. In spite of the fact that those boys were young fellows and very husky, only a very small percentage turned out. I'd say about five percent of the whole lot carried on after graduation. They just gave it up. Too much work, too cumbersome. But the Scandinavians, they're used to pulling the Swede fiddles. After the power saws came in a lot of those Scandinavians and central Europeans carried on as head fallers on the power saws. They didn't like to run a saw. They'd rather be on the head end, and guide the saw into the tree. But very few of them turned into machine men. And they were hard workers too, you know. The old hand fallers gradually petered out. Most of 'em ended up as snag fallers. After the timber was felled and logged off there'd be residual snag areas and the snag fallers would knock the stumps down. We had to cut any snag ten feet high or over. A lot of those guys served the purpose of doing that.

Martin Fossum, born in Norway in 1907. He came to BC in 1926 and began working as a faller at Merrill & Ring Lumber in Squamish. He spent his working life falling, bucking, scaling and bull bucking, and was involved in the development of early power saws. He is retired at Campbell River.

month, which were snapped up by various camps on Vancouver Island.

In 1943, Reed Prentice had dropped out of the power saw business to concentrate on machine tools for the war. The business was sold to a co-operatively-owned employee group headed by Ray Pitre and renamed Industrial Engineering Ltd., commonly known as IEL. From the outset and for many years, IEL was the pre-eminent power saw manufacturer in North America, largely due to its design engineers. The IEL invention of a direct drive system, replacing gear drives, placed the company well ahead of its competitors.

The availability of locally manufactured power saws, combined with a shortage of fallers in a growing wartime timber market, resulted in an explosive growth in their use. By 1939, BS&W's annual falling production with power saws was 100 million feet, compared to 6 million two years earlier. In a June 1942 analysis of one logging site at Lot 110, BS&W power saw operators felled more than a million feet at a cost of $1.48 a thousand, while hand fallers accounted for 93,000 feet at $1.60 a thousand. Power saw fallers averaged 6,200 feet a day, while hand fallers averaged 4,400 feet. The fallers, working on contract, were paid $1.00 a thousand for power saw falling and $1.60 a thousand for hand sawing.

Through the war years and for a short period after, the power saws used in the industry were all two-man falling saws, and their use did not substantially alter the organization of falling and bucking crews. They were big, heavy machines—the Timberhog weighed 117 pounds—that were awkward to use. Hand falling crews at this time usually consisted of a head faller and a second faller who dropped the trees, and a bucker to cut them into logs. In some camps there might be a separate crew of two axemen to put in the undercut for the fallers. The introduction of power saws caused a small change in this organization. The new saws needed a strong man on the engine end of the machine, to carry it and run the engine, and another on the tail end, holding the handle and guiding the bar as it cut. It was this last, less strenuous task that fell to the head fallers. Many second fallers off hand crews were older workers, who were unable to take on the job of lugging a 120-pound saw through the woods. Companies such as BS&W sought out husky young men, at least five feet ten inches tall and weighing 175 pounds.

The great appeal of the first saws was that once they had been wrestled into position, they cut at incredible speeds with relatively little effort. However, with the cutting chains then available, they could only cut across the grain. These chains, called scratch-tooth chains, would not cut along the grain or at an angle, so they were useless for making conventional wedge undercuts.

Also, the first bars were usually no more than sixty inches long, with a helper-handle at the tail end. This meant that to fall trees with a diameter larger than the clear length of the bar, side notches had to be cut into the tree before the undercuts and falling cuts were made. Also, instead of a V-shaped undercut, a series of horizontal undercuts were made, and the undercuts chopped out. George Burns of Franklin River designed a picaroon-type axe to chop out undercuts; it was eventually produced by the Walters Axe Company in Quebec.

Although these saws were crosscutting saws, for several reasons they were unsuited for bucking. The carburetors used in the first years were a float type, which meant the saws only ran when they were in an upright position. Some falling saws had a fixed horizontal blade that could not be used to buck a fallen tree. But even a saw with a vertical-cutting bar was of limited use if the bar was not longer than the diameter of the tree. Also, a good proportion of the trees fell in positions that required undercutting so the saw would not get pinched if bucked from the top. When the bar did get pinched, it had to be disassembled to be removed. It was practically impossible to cut from the bottom upwards with

Above: Removing a parallel undercut at Industrial Timber Mills Camp 3, 1943, with a picaroon-type faller's axe designed at Franklin River. (Wilmer Gold photo; IWA Local 1–80)

Dick Burnett (centre) with Martin Fossum (right) and mechanic Art Backlund at Menzies Bay working on the development of the Burnett saw, 1945. (CRM 18510)

In those days when they started with the power saws they had a school down here at Nanaimo. A guy by the name of Ole Buck started it. He was a logger. And they'd train these guys on the maintenance of the saw and how to run it and look after it. They called them "machine men." They'd go out to camp and they'd all start buying power saws. He'd run the saw and they'd put a faller with him; he's probably been a hand faller. He's on the light end, the faller. He took care of where the tree's going, the falling of the tree, and the young guy just run the machine. He was the second faller, or the machine man. Yeah, he was the muscle. That's the way they started out. Then they got down to one-man saws. Now it's every man for himself.

Mel Parker

They come out with a little one-man saw made in Seattle called a Titan. It was pretty good. Everyone got one but my brother and I. So I says to the scaler, "Fetch me out a power saw," and he gave a big laugh. He says, "Oh, so you're going to go with a power saw." So he fetched me out a power saw. I used it for bucking. I used to fall by hand and buck with a power saw. My brother wouldn't look at the power saw. I used to buck and when it come time to buck, I could buck three times as much as he could and he was using a crosscut saw. But when I took the power saw, wages went up five or six dollars a day. You could cut more timber, cut thirty-five, forty thousand feet. We cut like that, falling by hand and bucking with a power saw, for a couple of years. One day, he was a long time bucking, he had a tough piece of bucking and I chopped undercuts in a half a dozen trees. We worked on the right-of-way, I chopped him about a half a dozen undercuts in these trees we were to fall and he still hadn't finished so I said, "God dang it, I don't know why I can't fall with that saw."

I started it up and cut a tree down with it. Oh, I cut another one, I cut another one, three times faster than he could cut them by hand. So that finished it, no more falling by hand, forget it. So for a long time, I done all the falling, all the bucking. All he done was measured them.

I got the company to get me one with a longer bar. The bars were too short, they were only thirty inches, and it was big timber, so I got a forty-inch bar. I could get all around the tree and cut a pretty good size tree.

Alvin Brown

these saws. And finally, the only place for the man holding the tailstock to stand was on the downhill side of the tree, an impossibly dangerous position to be in.

Because of the increased production of power saw fallers, the first crews consisted of the two fallers and three buckers to keep up with them. When a shortage of buckers developed, fallers were forced to buck as much as possible with power saws, a hazardous and inefficient exercise requiring fallers to haul their big machine into the felled timber. Clearly, there was a need for a suitable bucking saw. It had to be a saw light enough for one man to operate, without a handle on the tail end of a bar long enough to cut at least halfway through big Douglas fir.

Most manufacturers began producing one-man saws at some point in the 1940s. IEL came out with the first successful one-man saw, the Beaver, in 1948. It weighed twenty-nine pounds and took a twenty-two-inch bar, known as a beavertail bar. The chain ran around its large curved end in a groove, as opposed to an end sprocket in earlier bars on two-man saws. It had a manual clutch, float carburetor and a small, 1.25-horsepower engine. This saw was aimed primarily at the eastern Canadian pulpwood industry; with its short bar it was of limited use on the West Coast. An improved model introduced in 1949, called the Pioneer, could use a longer bar. It was so successful that in 1951 the company adopted the name for all its saws. Also in 1949, a model called the Twin, which used two Pioneer cylinders, was developed. It would handle even longer bars and was lighter than earlier two-man models. In 1959, John Westin and Jack Miller used a 1951 model Pioneer Super Twin to fall the famed Caycuse Fir, a twelve-foot diameter Douglas fir growing in the Cowichan Valley that was 1,266 years old, the third largest tree found to that time in BC. It took them less than one hour to put it on the ground.

Probably the most successful of the early one-man saws was the Titan, produced by a Seattle company, Mill & Mine Supply, set up in 1937 to distribute Stihls in the US. When the war began, it too made a Stihl replica. Titan also produced a good two-*man* saw, the Blue Streak, and the one-man Junior, probably the first saw used for both falling and bucking. Previously there was no concept of a dual-purpose saw that would eliminate team falling entirely and turn the whole falling and bucking phase into a one-man operation.

In 1949, the McCulloch Motor Corporation, which had built two-cycle engines for target drones during the war and then components for chain saws built by Reed Prentice, brought out a saw—the Model 12-25A, with an all-position carburetor—that swept the market. The next year the twenty-five-pound Model 3-25 followed, with the same kind of success. By this time there were dozens of chain saw makers in a booming market. By the end of World War Two, half a dozen manufacturers or assemblers were in BC, as well as a large body of fallers skilled in their use. During the war years, BS&W alone trained between 250 and 300 power saw fallers. They learned not only how to operate these saws efficiently, but also how to maintain and repair them. The ability to keep a power saw functioning all day became one of the requirements of the job. Only major repair jobs were performed in the camp shops. Fallers who had grown up on prairie farms turned out to be particularly adept at keeping power saws in good working order.

The adoption of the power saw, perhaps more than any other logging tool, symbolized the end of one era and the beginning of another. The tools and techniques of falling had remained the same since logging had begun in the Northwest, while other phases of the job evolved.

Olaf Fedje with one of the first one-man, beavertail bucking saws. (UBC BC1930/391)

McCulloch started in Los Angeles, I think right at the end of the war. The oldest McCulloch I've got is a model 1225 and that would have been made in 1947. He was very innovative. He came in with die-cast castings when everyone else was using sand-cast castings. That makes a big difference in weight. Beavertail bars was another one of his ideas. He made his own carburetors. Everybody had been using Tillotsons prior to that. Quite a few little things. They were the king of the heap when I started in '51; there were more McCullochs and IELs around than anything else.

They were hard to work on. Jesus, they were hard to work on. The starter housings, the bolts going in the starter housing, they had captive nuts behind them. Instead of having threaded castings they put a nut in a little socket and if you ever dropped that out of there, of course, you couldn't tighten the goddamn screw. You'd have to take everything off and try again. It was a nightmare.

The carburetors were not as good as they should have been and the ignitions got damp. They were always a problem. But they were really good engineers, those McCulloch guys, because they developed things other guys just hadn't done at all.

Dave Challenger, sales manager, Pacific Equipment Company; grandson of pioneer truck logger George Challenger; nephew of Jack Challenger. He has one of the world's largest collections of power saws and other falling tools.

It took a certain kind of person to hand fall the big timbers of the Pacific rain forest. Steady nerves were needed to balance on a narrow springboard, sometimes far above the ground of a steep sidehill, and to chip away calmly and methodically at an eight-foot-diameter Douglas fir, waiting until the last moment when it began to fall before walking to safety with dignity. It required a certain artistry to select and hone an axe or saw, in reality a delicate and trivial instrument to use in bringing down some of the largest living things on the face of the earth. And it required an enormous strength and endurance to swing an axe and pull a hand saw all day, hour after hour in the hot sun or pouring rain.

From the beginning, fallers were a special breed, workers who stood apart in a logging crew. There was a kind of arrogance about these old hand fallers. They demanded and received respect from the rest of the crew and bosses alike. In an occupation where a supervisor's ultimate authority derived from his ability to

The head faller watches a big fir as it begins to topple. The machine man has already moved to a safer position. Industrial Timber Mills, Camp 3, 1941. (Wilmer Gold photo; IWA Local 1–80)

replace any man on the crew, fallers were usually left on their own by foremen who did not know how to fall. They had their own bunkhouses, their own preferred tables in the dining room where no one else dared to sit. They were clannish, working together in small crews apart from the rest of the operation. No one watched over their shoulders. And they were the only members of a logging crew who spent their time in a standing forest, the only ones to really know what it was they were destroying.

Power saws changed all that. They shattered the quiet rhythm of axes chopping and saws cutting wood, broken only by the occasional crash of a falling tree. They were noisy, smelly, cantankerous machines, demanding entirely different skills and personalities. They shook and vibrated, turning old fallers' fingers into gangrenous stumps. They were a horrible intrusion on the last remnant of the good old days.

But they could cut; God, could they cut. As they improved and the skills required to master them were painfully acquired, a new elite emerged with its own rituals and protocols. By the 1960s, the one-man falling and bucking saw was a refined instrument, an artist's tool, almost. And after that they only got better.

To fall a six- or eight-foot fir or cedar without damaging any of the premium-priced wood it contains, is a delicate task requiring enormous skill and judgement. It still demands most of the abilities of the old hand fallers, plus new ones to operate and maintain one of the most refined industrial tools ever produced. And the job still has to be done in the hot sun or pouring rain. The springboards are gone now, and power saw fallers do not take their tools to bed as old axemen sometimes did. Within the rigid hierarchy of a logging camp, they still retain their status and the air of detached aloofness that comes from knowing how to cut cleanly through a tree until the last split second, before calmly putting down the saw and moving to a safe place while a 100-ton tree comes crashing down a few feet away.

Industrial Timber Mills' power saw gang #3 on a lunch break at Camp 3, 1943. (Wilmer Gold photo; IWA Local 1–80)

The advent of the power saw offered new opportunities to fallers that had not existed earlier. With the end of the war and resumption of normal working conditions, several of the BS&W-trained fallers went into business as power saw distributors, falling contractors or both. The new division of labour had created a class of autonomous, independent-minded workers who did not fit well into the rigid hierarchy of a conventional logging camp. Quite simply, supervisors did not want to deal with individual fallers so there was a general trend to contract out falling and bucking to someone who understood the new saws and could keep a crew in line.

An initial problem faced by the contractors and the supervisors was that of poor falling practices. It was relatively easy and quick to fall a tree if you didn't worry whether it hit a high stump and broke. It was no great loss to the faller. He could easily drop another one and make a good day's wages by falling fast and carelessly. Contractors who understood saws and the business of falling were far more successful at ending careless falling than general logging

Right: Power saw falling at BC Forest Products' Nitinat camp, 1948. (Wilmer Gold photo; IWA Local 1–80)

In the old hand falling days, when it took a lot of time and work to fall a tree, the fallers were careful. I can remember packing limbs to cover stumps. It took a lot of work to fall a tree, and if you had to buck out broken pieces with a hand saw it was a lot of work. To get more production he would try to save each tree. With a chain saw, though, a guy could just cut more trees to make more volume. Chain saws cut so quickly. Management then had to take another position and bring in parallel falling, and proper falling.

With the first chains, the scratch chain, you had to cut cross-grain. You couldn't cut at an angle. You put an undercut in and then another cut and chip it out with a picaroon. But you could still direct the fall as well with a chain saw, maybe better. With hand falling it took an expert axeman to make a perfect undercut.

We could do with chain saws, right from day one, as good a job falling as with hand saws, if you handled the tools properly. Now, if they put a narrow undercut in so it would be easy to chip out, then the tree wouldn't go where you wanted it because the lips would meet and it would twist and turn, maybe barber chair the tree.

Right from the start, power saws did improve productivity. You may get some greenhorns that did poor work. But if you gave a power saw to a fellow who had the ability to fall and buck timber properly he would do as good a job with a power saw as with a hand saw, and he would do more. But when you are training a lot of people you are going to get some problems with quality of work.

If you were hand falling you had a three-man crew, at least. The head faller decided where to fall it. With chain saws you did the same thing. You took your best man and he was the head faller. He was the boss of the crew, said where the tree was going to go. After a few years every man out there is a head faller because he falls his tree and he bucks it. In the old days there were guys that bucked timber all their lives, maybe thirty years or more. They never tried to head fall, wouldn't take a job head falling. They didn't think they were capable of it because they were always the bucker. With some second fallers it was the same thing. Then all of a sudden they are all head fallers, all responsible for the falling of each tree. Not one man out of a crew, but each man. Then it took a lot of training to get these fellows to do the quality of work needed. It was a big problem.

Olaf Fedje

Two Baikie Brothers fallers, Jack and Les Coe, using an Atkins electric chain saw at Brown Bay, 1943. (CRM 18345)

On the Queen Charlottes they brought in during the war the Atkins electric saw. I was camp mechanic for Pacific Mills during the war, when they brought these saws to do the falling of the big spruce. They had generator units that looked like a big welder. They ran electric cable about one inch in diameter out into the woods. You left your unit sitting wherever you could get it to on the beach. We were logging off the water at that time, there were no roads at that time—all Cat roads or A-frame logging. Wherever they could get a road in they'd put a power unit in and run the line out. They'd run a Y off it and have two different sets of fallers working. They went out 800 feet with the cable.

Of course, there were all kinds of problems. Getting the crew trained to respect the cable, because if they fell anything across the cable . . . The connections weren't very good in those days, they were using old aircraft connections. They'd take the load for so long and there'd be a big puff of smoke and you didn't have any connection. The rain would short the electricity. You spent most of your time fixing cables.

The motor on the saw was maybe eight inches in diameter, twice as long, a very powerful little motor for its size. The electronics of the whole thing meant you had to really understand electricity.

They had up to a ten-foot bar. You side notched to get the big trees down. You had a terrible time with those saws because of the weight of the bars. The drive head wasn't real heavy. There were so many problems with the long bars because the minute you got a bit of a bow in the bar your chain would jump off. There was a lot of chain running around at a pretty high speed, tearing up anything it touched. They were double ended, they weren't a beaver saw in those. The electric bucking saws were beaver tails.

The chain was like a chisel chain and you used a round file to sharpen it. After, they had what was called a planer chain and you used a three-cornered file. The blade across the top was flat, the shavings would pile up just like from a planer. It was pretty tough cutting so they had to take off a thin cut. To get a cut started they worked out many innovations to keep the bow out of the bar. They would take an axe and cut a bit of a groove for the bar to sit on.

Most of the crew had been hand fallers. Half of them wouldn't go anywhere near it. But there were always some that wanted to try something new. That was the start. When they wanted to change over from hand saws to chain saws, Tom Murphy and Panicky Bell talked me into taking a contract to look after all the chain saws. Panicky Bell said to take any man on the crew who was interested and wanted to go falling. They were desperate, they had to get more timber down. They couldn't get hand fallers and they needed more. They took rigging men and made them into fallers. They were the best ones to train because they knew the woods.

Viv Williams, born in Saskatchewan in 1915. He started logging in the Fraser Valley in 1935, and after acquiring his own logging company in 1946, he logged on the Queen Charlotte Islands, South Bentinck Arm and the Fraser Canyon. He was president of the Truck Loggers' Association for two years, was the first BC logger to buy and pilot a helicopter, and for many years has been actively involved with the Junior Forest Wardens of Canada.

supervisors who most likely began their careers in the days of steam, rail and hand saws.

One of the first to go into the distribution-contracting business was Don Challenger, Jack Challenger's nephew, when he set up Power Saw Sales and Service in Vancouver. Ron Eliason was the first power saw contractor at Elk River Timber, using IELs. Sells Brothers Contracting, started by three Westsells in 1945, started out with McKenzie and Flavelle at Halfmoon Bay and went on from there with contracts to convert from hand falling at several of the big coastal camps. One of the most successful members of Jack Challenger's original Franklin River falling crew was Olaf Fedje, who started a contracting company and a chain store in Nanaimo, both of which are still in business almost fifty years later.

The wartime spruce logging program made use of a different type of power saw. In the Queen Charlottes, Viv Williams was assigned the task of keeping Aero Timber's Atkins electric chain saws running. The Atkins company was an old established hand saw company with factories in the US and Ontario. Its chain saw had a chain with a curved tooth that was the prototype of the modern saw chain. Atkins saws were sold in BC by Finning Tractor, primarily because the generators used to power them were designed for installation on a Caterpillar tractor. After the war, when the Powell River Company took over Aero timber, Williams contracted the falling for Northern Pulpwood at Miller Creek, starting with twelve-horsepower two-man saws made by the old Disston Saw Company. These saws used Mercury motors made by a company famous for its outboard motors. At this stage Williams' bucking was done by hand. In 1947, Disston–Mercury came out with a nine-horsepower one-man saw that took a sixty-inch bar; Williams switched to these, with fallers doing the bucking as well.

Power saws improved steadily for about thirty years following World War Two. As many of the manufacturers disappeared or were swallowed by competitors, the survivors brought out new models almost yearly, in a widening range of sizes. But in the mid-1940s there were still many problems to solve.

The first and most obvious difficulty was weight. In 1945, a saw capable of handling a sixty-inch bar weighed 115 or 120 pounds. Today a saw with the same capability weighs 20 pounds or less. The weight reduction was achieved in dozens of ways—improved casting methods, the use of lightweight alloys and plastics, improved bar and chain materials and design, and so on. The constant reduction in weight, combined with faster engine speeds,

Using a twelve-horsepower, two-man Disston to buck at Viv Williams's Miller Creek camp, 1947. (Viv Williams collection)

The unreliability of the motors in the early days was a big problem. We had lots of down time. Due to vibration the ignition wires would break off. There were no air filters on the breathers. Sometimes, if the operator wasn't careful, he'd get it in the pile of sawdust and it would suck in the sawdust and seize the motor up. You'd take your tools and take the cylinder head off, chip out the sawdust in the cylinder and go to work. Today they're so highly machined a bit of dust will stop them. In those days they were pretty crude and they could be monkey wrenched on the job quite easily.

The early fallers weren't mechanically sophisticated at all, to put it mildly. Myself, raised on a farm, I knew a little bit about machinery For most of them, to be put on a machine, it was baffling to them. In those days we had some spares with us. But if you broke a chain you'd have to fix it in the field yourself. Repairing a broken chain on a stump, when the no-see-ums and mosquitoes were killing you, took a long time. Now, we'd never ask anyone to repair a chain on the job. Repair it in the shop. Now the cost of chain, in the whole cost of keeping a falling and bucking crew going, is nothing.

We did most of the repair work in the shops. Of course, if you were working on contract and the saw broke down you didn't get paid. You had to try to improvise. In the start there were still axes out in the woods, and maybe a hand saw as a spare. You could finish off the day with a hand saw.

Olaf Fedje

Falling with a Disston one-man saw, DA2-11, at Freda Creek. O'Brien Logging, 1954. (PRM 1677A)

Below: Wedging loose a pinched bar while bucking with a Disston DA2-11 at O'Brien Logging, Freda Creek, 1954. (PRM 1673B)

created another problem—vibration. Fallers discovered that a few years' use of a power saw impaired the blood circulation in their fingers, leading in severe cases to gangrene and amputation. During the late 1960s, anti-vibration handles and mounts were added.

For many years power saw users experienced difficulty in starting their saws, particularly in wet climates like the Pacific Northwest, because of poorly developed electrical systems. Improved insulation, followed by sealed electronic components, eventually eliminated this problem. A similar problem had been encountered earlier with carburetors. Until 1953, when Tillotson introduced a diaphragm carburetor, float-type carburetors that functioned only in an upright position were all that were available. The first Tillotson was a three-inch cube weighing three quarters of a pound, presenting a problem for makers trying to reduce saw weights. As the size and weight of this carburetor was reduced, most manufacturers began to use it.

Improved mufflers and engine design were the answers to one of the first challenges power saw manufacturers had to confront: the noise problem. Many hand fallers objected to the loud sound of the early saws, claiming they were unable to hear branches and tops breaking loose above them as they worked. Others said that if a faller heard a branch fall, it was too late to move anyway. In the early years numerous buckers were hurt or killed when trees were felled on them because the engine noise prevented communication between workers. Power saw fallers were the first loggers required to wear hard hats in the early 1940s; others did not have to use them until about 1950. But the real damage caused by noise was to hearing, and a whole generation of fallers ended their days deaf or with hearing aids.

One major improvement was in cutting-chain design. Saw chain first came into existence in the seventeenth century and was used in a variety of minor applications until the rapid development of power saws in the 1940s. The first chains used with the new power saws were scratch chains, similar in design to crosscut hand saws. Some scratch chains had only two rows of cutting teeth, while improved versions had rakers of the same sort that were added to hand saws in the late nineteenth century. In 1940, Hassler devised a curved or hooked tooth. Several improvements were made on this design during the decade, including the development of the chipper chain by Joe Cox. His chain led to the founding in 1947 of the Oregon Chain Company, which has dominated the world market ever since. The great advantage of the chipper-type chains is that they cut at any angle to the

A faller with an IEL saw at Hillcrest Lumber, 1956. This is one of the last saws made under the IEL name. The new Ontario owners changed it to Pioneer and closed the Vancouver plant the same year. (Wilmer Gold photo; IWA Local 1–80)

P.B. Anderson (left), examining a McCulloch 5-49 at an equipment show in Victoria, 1950. (UBC Anderson Collection)

The power saw shop at Canadian Forest Products' Woss Camp, 1953. (BCARS 81760)

grain, allowing them to be used for putting in angled undercuts. The new chains also cut faster and require less power to drive them.

Once the power saw business got rolling, there was a proliferation of saws and companies that made them. By 1957, for example, there were about thirty power saw manufacturers in the US. In 1962, McCulloch made its millionth saw. Part of the reason for this expansion was the growth of the casual market. In 1960, about 7,500 chain saws were sold in BC, seventy percent of them to loggers. In 1968, annual sales were 9,500 in BC and 100,000 in Canada.

Through the 1940s and 1950s, Vancouver was a major centre of chain saw development and production, with about ten companies operating at different times. The most significant of these companies was IEL. By the early 1950s, the company was turning out 4,000 saws a month and exporting them in boxcar lots to the US, where it was one of the largest selling saws along with McCulloch and Homelite. In 1955, Ray Pitre sold the controlling shares of IEL to an Ontario company, Outboard Marine Corporation, which set up a new company, Pioneer Saw Ltd. Board members included a member of the Vancouver Bell-Irving family and Briggs, of Briggs & Stratton engines. In 1956, the Vancouver plant was closed, throwing 130 of the most skilled power saw makers in North America out of work. The Pioneer company slowly declined, although it maintained a major market share selling saws based on the original IEL designs. It was acquired by a group of employees in 1977, but this group lacked the capital to revive the company. In the early 1980s they sold the enterprise to the gigantic Swedish Electrolux company, which already controlled several major saw companies. Since that time some of the basic IEL saws have been sold as Pioneer–Partners.

Meanwhile, Burnett Engineering enjoyed a brief period of success. By 1946 at least 100 Burnett saws were in use on Vancouver Island. A thirty-man plant built the saws in Vancouver and Bill Collister ran a repair shop on the island. But lack of capital and a decision to develop the two-man Powermatic saw, at a time when everyone else was producing one-man saws, doomed the company by 1950. Burnett became the BC distributor for Homelite and was involved with McCulloch.

Another Vancouver company, Power Machinery, was set up in 1946 by H.D. McDonald, one of the IEL founders. It came out with the lightweight Canadien saw in 1957, and by 1960 was building seventy-five saws a day at its Commercial Drive plant, a large portion of which were sold in the US. It was taken over by Bristol

Some claimed the noise of the old saws made them more dangerous because you couldn't hear anything coming. I guess it's a little more dangerous that way, from a limb or something coming down. But you just watch a little closer. And you can watch a little closer when you're not working on that saw, you're just holding it . . . you can look around a bit. So I don't know actually if it was any more dangerous. A lot of men got hurt, of course, trying to save their saws in a bad situation, instead of throwing it down and running. They'd think, "Well there's a lot of money tied up there." But I always had it in my head that you can fix a saw or buy another one or something, or someone can. Throw 'em as far as you could and take off. Of course, when they started getting the one-man saws that was fine. They didn't weigh anything anyway; you could fire one of them out of the way. Them big ones, you didn't throw them very far, you just dropped them behind the stump and go. Sometimes I was hoping that the tree would fall on it before the end of the day. Big, heavy, clumsy things.

Mel Parker

Power saws were useful for many tasks in addition to falling and bucking. Butler & Wagner of Nanaimo, contract sled builders, used a two-man saw to build a donkey sled at Stoltz Logging near Duncan in 1953. (BCFM 4–3)

Aero-Industries in the early 1960s, and eventually was absorbed into the Electrolux group.

The last BC saw producer was Quadra Manufacturing, set up in 1971 by former Power Machinery people. Quadra built the Frontier saw at a new plant in Trail headed by Jim Hutchinson. It was a small, very successful consumer saw. The company was eventually purchased by Ontario interests and the plant moved east, where it was acquired by Electrolux, producer of Husqvarna, Jonsered and Partner saws. The Frontier saw was sold, with different paint jobs and labels around the world as Husqvarna, Jonsered, Jobu, Pioneer–Partner and Danarm.

Ernie Cannon, who started making power saws at Shade Engineering in 1943, started a power saw component company in 1954, Cannon Machine Works of Vancouver, to make cutting bars. Later, when he moved the company to Burnaby, Cannon concentrated on building high-quality replacement bars and also produced drilling attachments for power saws. As a sideline, Cannon took over the machining work on Bob Swanson's air horns.

Another BC company with a major role in the industry was Windsor, a Burnaby-based maker of bars and chains. Bill Hodges established it in

Before power saws, most falling was done on contract. It made the faller independent; he was out there on his own. At many times fallers were difficult to handle. They were independent. Each tree is different. They had a different spirit. Sometimes management found them difficult to deal with. Fallers probably had more to do with introducing the union than anyone else. Lord knows, they weren't going to be the bull bucker. If you were the hooktender or chokerman you'd say, "I want to be a rigging slinger." But a faller was independent so they were fairly difficult to work with at times. Many logging managers would say, "I've got a hell of a good job—if I didn't have to deal with those goddamn fallers."

In the old days the bull bucker had lots to say, he was the head of falling and bucking. Today the general foreman is more the guy in charge. In some of the bigger camps the bull bucker is still there. There were a lot more bull buckers in the old days because it took a lot of people. Even a small camp would have to have lots of fallers, because individual productivity was so low. Now productivity is high and you might only need half a dozen men. You don't need a foreman for half a dozen men. Some of the bigger camps might have a bull bucker, but he might also look after grade. A well-trained, small crew doesn't take the supervision it did in the old days. It makes quite a difference in the number of bull buckers needed.

Olaf Fedje

Fallers at Hillcrest Lumber, 1956, wedging a tree that leans back onto the cutting bar. (Wilmer Gold photo; IWA Local 1–80)

1944 as a custom machine shop. In 1948, Windsor began making cutting bars for Power Machinery, and within a few years was supplying bars to several saw manufacturers. The company grew along with the power saw business and in 1960 started making sprockets. By 1963, it was supplying bars and sprockets to seventy percent of North American and European chain saw manufacturers. A cutting-chain division was added in the late 1960s when Windsor took over Atlas, a short-lived Victoria chain company. Then, in the 1980s, Windsor got involved in a major lawsuit with the Sandvik company of Sweden. When the suit was lost, Sandvik took over the Windsor operations and closed the Burnaby plant.

The rise and subsequent demise of BC's chain saw manufacturing sector is symptomatic of a much wider condition in the logging and logging equipment manufacturing industries. In response to the local industry's urgent need, resident machinists, mechanics and businessmen produced power saws that not only met local needs but also were saleable in a much wider market. Loggers enjoyed the benefits of a near-at-hand supplier that responded quickly to meet their changing needs. As the manufacturers succeeded and became larger businesses, they were bought up and either folded or moved elsewhere, leaving loggers dependent on distant suppliers not particularly concerned with their needs.

There are many reasons why a distant manufacturer will buy a regional producer of equipment, such as chain saws, and either close the plant or modify its product. A common reason is simply to eliminate competition in a bid to dominate the market. Alternatively, some of the power saw companies acquired had their production lines modified to serve the consumer saw market, a much larger and more lucrative business than the industrial models used by loggers. And, in at least one case, acquisition of the BC manufacturer was done to obtain rights to technical innovations in small engine design so they could be used to build outboard motors, a much larger market than that created by loggers.

Today there is only one power saw manufacturer in Canada, the Electrolux-owned Poulan in Ontario. Indeed, there are few makers of professional-class saws left in North America. Now the field is dominated by European brands, primarily Stihl and Husqvarna. In fact, in Canada, one of the most heavily forested nations in the world, there has been no axe manufacturer since 1974 when the Walters company closed.

This phenomenon is not limited to the power saw business. Of the dozens of equipment manufacturers founded in BC to serve the logging sector since the turn of the century, hardly any remain, and almost none that are still owned by Pacific Northwest residents. Practically all the builders of yarders, donkeys, rigging gear, trucks, engines and hundreds of other items have disappeared. Manufacturers of logging equipment throughout the rest of the Northwest region have met a similar fate, and today an increasing portion of the equipment used in logging in the Northwest comes from the East, Europe or Japan.

Quite apart from the loss of direct economic benefits from the manufacturing sector in the Northwest regions, loggers do not have easy access to the builders of the equipment they use, and they have to take what is designed for use elsewhere. This does not always impose great hardships, but it limits the loggers' ability to respond to new and changing conditions as easily as they once could.

The forces that brought about this change began to affect the industry after the war. Some of those forces led to a period of unsurpassed growth and expansion. But other forces were responsible for an unprecedented phase of corporate concentration, combined with mergers and takeovers by distantly owned financial groups. This occurred in the logging industry as well as in the equipment manufacturing sector. The postwar period of change had a number of phases, the first being a wholesale conversion of the technical basis of logging. The evolution of falling and bucking methods was typical of this conversion, but only a small part of it. Other changes, even more momentous, were about to take place through the 1950s and 1960s.

Falling old-growth Douglas fir at MacMillan Bloedel's Menzies Bay Division, 1968. (CRM 10717)

VIII Postwar Conversions

Previous page: A Caterpillar wheeled loader loading a Kenworth off-highway truck at a small landing with a Madill steel spar, c. 1960. (Finning Tractor collection)

The end of World War Two kicked off an expansion phase in the coastal forest industry that continued with only minor hesitations for the next thirty-five years. There were several reasons for the increase in timber demand—a booming domestic housing market, postwar reconstruction in Europe, the growth of export markets around the world. But probably the most significant factor was the explosive growth in demand for wood pulp.

For almost three decades leading up to the end of the war, coastal pulp production had stabilized more or less at the output level of the four existing mills—Powell River, Port Alice, Ocean Falls and Howe Sound. Until the postwar period, the primary use for wood pulp was in the production of paper. Beginning in the 1930s and accelerating during the war, a vast new industry developed that produced thousands of synthetic products—fabrics, plastics, wallboards, chemicals—derived from cellulose. This created an enormous demand for pulp wood that quickly translated into sweeping changes in the coastal logging industry.

In the twenty-five years following the war, another ten pulp mills began operating on the coast. They were designed to utilize the residues from both sawmills and plywood plants, and

Below: Smallwood salvaged at Crown Zellerbach's Ladysmith operations for use in the Elk Falls pulp mill, early 1950s. (Mauno Pelto collection)

The utilization of smaller logs after World War Two required new equipment for handling them. This small-log accumulator—called Joe's Jaws, after Joe Cliffe, its designer—was used to bunch logs for loading at Comox Lake by Comox Logging. (Mauno Pelto collection)

whole logs not wanted by the solid wood mills. The logging sector profited in several ways from this increased demand. Because sawmills now sold waste wood to the pulp mills, more value was extracted from saw logs, and a wider spectrum of logs was utilized as saw logs. Large volumes of logs previously left in the slash because they were too small or of a species undesirable for lumber, were now run through whole log chippers at the pulp mills. Timber stands passed by because of their low ratio of saw logs could now be harvested economically. The definition of merchantable timber changed dramatically.

In a free market economy, changes in demand of this nature would have led to automatic increases in timber utilization levels, a rapid diversification of the solid wood processing sector, a pulp sector based on solid wood mill residues and a logging industry dominated by the most innovative and efficient operators. But, because of the BC government's almost exclusive control of the timber supply, this did not occur. Instead, the postwar period was one of rapid technological change occurring against a background of legal, regulatory, political and corporate upheaval. By about 1970, the coastal logging industry had undergone an enormous transformation from what it had been at the end of World War Two.

The timber market was always up and down, in prices. Lots of times you couldn't sell hemlock, you couldn't sell cedar. Fir was the mainstay. We had a time here we smashed up the most beautiful cedar you've ever seen in your life. We had to smash it up. It cost us ten dollars for every thousand feet we put on the water, if you could get somebody to buy it. You'd have it down here for months. A small logger like me, I couldn't afford that. So we'd just smash it, we'd fell the cedar first and fell the fir on top of it, smash it all to hell. The Forestry, they didn't condone it, but they turned a blind eye to the fact that we were smashing the cedar up, because they could see there was no way we could log if we had to take the cedar out. I often think about it now, and when cedar went up higher than fir, I said, my God, think of that.

Al Hendrickson

In the latter half of the 1940s, as one of the first consequences of changes in the industry, strains developed in the relationship between workers and employers, as well as within the ranks of organized labour. As described in an earlier chapter, the International Woodworkers of America had established itself as the BC loggers' union by 1944, when it successfully negotiated its first master agreement with the BC Loggers' Association (BCLA). In 1946 the stage was set for a confrontation. The tacit agreement to avoid strikes during the war was no longer in effect and the growing strength of the union was cause for alarm among the major companies in the BCLA. The IWA was feeling its strength and spoiling for a fight, and on May 15 its members went on strike. The strike spread throughout the province very quickly. Even unorganized loggers walked off the job in some camps. When a settlement was reached thirty-eight days later, the union gained major concessions and 10,000 new members.

However, distant events were looming. The US was experiencing the anti-communist hysteria leading up to the McCarthy excesses. The 1947 Taft–Hartley Act, requiring union members to declare Communist Party membership, led to the expulsion of many American IWA leaders and the deportation of Harold Pritchett, a BC member of the IWA executive and a Party member. On both sides of the border the union was divided into two factions: a militant left wing faction composed of a significant proportion of longtime union organizers and executives, and an equally militant anti-communist White Block. In the US, the leftists were expelled and the White Block took over. In BC, the leftists remained in power. A delegation of BC delegates to a 1948 IWA convention in Portland were denied entry to the US when they refused to declare at the border whether or not they were members of the Communist Party. In the uproar that ensued, the BC district council of the IWA voted seventy-four to fifteen in favour of disaffiliating from the IWA and reconstituting itself under a new name, the Woodworkers Industrial Union of Canada (WIU). Members of the White Block in BC regrouped, elected a new IWA BC council and, with the support of employers, the provincial and federal governments, the media, the Canadian Congress of Labour and the Co-operative Commonwealth Federation (CCF) party, laid claim to union assets and certification rights in the logging camps.

The area around Campbell River and Courtenay had been a union stronghold from the days of the Wobblies. Several of the largest camps on the coast were in this area, some of them run by aggressively anti-union managers, so it was the most heavily organized area in the province. When the IWA was organized, the North Island was designated as Local 1-363, with a head office in Courtenay and another in Campbell River. One of the first camps organized was at Oyster Bay, on the Island Highway between Campbell River and Courtenay. This camp was originally built by the federal government as a Depression-era relief camp. It became a logging camp in the late 1930s when Ole Buck acquired it for the British American Timber Company (Batco). It was planned as and soon became one of the biggest truck logging camps on the island. During the early part of the war, Al Simpson took over and changed the name to Iron River Logging.

Simpson soon gained a reputation as a maverick operator: he signed the first industry agreement with the IWA in 1943. It was a trend-setting contract. Though its wage provisions were modest, it provided for a grievance procedure, a fixed schedule of work hours, seniority, statutory holidays and recognition of a safety committee. Simpson attracted some of the best loggers on the coast and was highly successful at logging the low-grade timber in the area. In 1944, he sold out to the H.R. MacMillan

Around that time it was pretty hard to get good loggers. You had to take what you got. I can remember going over to Zeballos after the war and by God, loggers were hard to get. I was running camp in one place there. That was after Gordon Gibson was in there. We never had no road built. I rigged a tree and we got going with a little bit of a high lead. Then we got a bridge crew in from Vancouver to build a bridge up the road about three or four miles out of town, across the river into this timber that we had.

I'd go down and rig on the flats there in the morning. I'd go up to see how the bridge crew was doing in the afternoon. One day, not a goddamn soul around the bridge. Where the hell are these guys? They had a halfway house there, with girls, just down maybe a thousand feet from where the bridge was. Oh Jesus. Here the whole crew is in the cathouse there. What the hell was going on? No work going on the bridge. You would have a hell of a time, what are you going to do, you can't get a new crew at that time. That went on for a couple of weeks and finally they got them all on the boat and sent them to town. Thought we'd get another crew in. Black used to hire out, at the hiring hall. Jesus, the second boat that come in there after that, they had the same bunch on again. It's all you got!

Sam Telosky

Mechanics at Crown Zellerbach's Kitimat division repair a track-mounted Madill steel spar equipped with a Tyee yarder, 1965. (Jack Cash photo; Mauno Pelto collection)

Company, which put in a superintendent to oversee several phase contractors, including Schnare's S&S Trucking.

When the WIU was formed, the North Island local became Local 363 by a two to one vote. But a group of White Block supporters from Courtenay re-established themselves as Local 1-363 of the IWA and, backed by police, took over the Courtenay union hall. When they attempted to seize the Campbell River hall, they were forcibly repelled, leaving the area divided into a WIU-dominated organization in Campbell River and a White Block faction in Courtenay. The Iron River camp was one of the WIU's strongholds, although it too had a minority IWA faction.

In November 1948, the WIU was in the midst of a certification drive at seven North Island camps when three fallers were fired at Iron River. The MacMillan head office refused to recognize the WIU grievance committee and insisted on dealing with an IWA committee. With its back to the wall, the WIU voted to strike, and the camp was shut down. The Campbell River local soon ran out of money, as the strike fund was controlled by the IWA-run Courtenay office. Eventually the factional dispute erupted into a riot between IWA and WIU supporters at the Iron River camp, and two IWA leaders ended up in hospital. The following day a mob of 200 IWA supporters stormed the Iron River camp. They were held off by police, who arrested several WIU members. Soon after, the camp closed for Christmas; in the new year it opened with a new crew. The fight was taken out of the WIU. It declined slowly, and in 1950 its leaders urged the remaining members to join the IWA. The influence of the militant wing of the woods union effectively ended. In the late fifties the IWA reorganized into regions, with BC's 30,000 members forming Region 1. In the years that followed, the coast-based IWA established a foothold in the Interior and gained members in sawmills throughout the province. However, it failed to organize the rapidly expanding pulp and paper mill work force, which came under the jurisdiction of the Canadian Paperworkers Union and the Pulp, Paper and Woodworkers of Canada.

Purged of its radical faction, the IWA became a major force within the industry during the safety campaigns of the 1950s and 1960s. From its earliest days, logging was a dangerous occupation, and coastal logging was particularly so. Since 1916, when workers' compensation began in BC and accurate records were first maintained, the accident and fatality rates in logging have always been at least three or four times

Setting chokers on a steel-spar, high-lead show at Kitimat, 1965. (Jack Cash photo; Mauno Pelto collection)

those of other, comparable industries. The very nature of the work and the setting in which it is conducted ensure that. Some periods, however, have been worse than others.

British Columbia Logging Fatalities
(Loggers killed per decade)

Decade	Killed
1910–19	123
1920–29	628
1930–39	449
1940–49	591
1950–59	644
1960–69	496
1970–79	414

The accident rate during the 1920s is a reflection of speeded-up work practices adopted with the advent of high-lead and railway logging. A similar phenomenon occurred during the technical transformation of the logging industry during the 1950s. It did not go unnoticed, and various industry organizations, including the IWA and the Truck Loggers' Association, devoted much of their resources to instilling an awareness of safe work habits and the importance of developing them.

The safety campaigns of the 1960s confronted a deeply ingrained attitude that had grown with the industry. Crudely stated, this attitude supported the idea that logging is a rough, tough and dangerous business and that anyone working as a logger has to accept the risks without complaint. It was this kind of macho thinking that enabled woods managers in the railway camps to initiate production contests and other unsafe practices. It took many years and a great deal of propaganda to convince managers and workers that the lives and health of loggers are more important than putting an extra load of logs a day over the dump. Looking back on a decade of safety campaigns, TLA president Bill Moore deplored what he termed "the mediocre way of logging" that was permitted previously. "The war brought a change to the art of logging," he told that year's TLA convention.

> Men were scarce after the war due to the competition of other industries, and management groped with the immense problems of stronger unionism, increased costs, radical changes of equipment, and a head-long rush into bigger and more complicated machinery to reduce those costs

The chaser unhooks the turn at the landing. Kitimat, 1965. (Jack Cash photo; Mauno Pelto collection)

> and keep efficient. The professionalism of logging was lost . . . I think most of us remember the late forties and early fifties as a time when anybody and his donkey could go out and make a buck, if he could find enough men to work for him. With the advent of diesel power over steam, it meant that anyone could jump on a donkey and get some logs—no ticket was needed and mediocrity was accepted.

Logging is still a dangerous occupation. Several loggers are killed each year and a few are injured every day. But the old attitude, that real loggers are not concerned about personal risk, has pretty well gone the way of the two-man crosscut saw.

The Truck Loggers' Association emerged during the war to represent the interests of a new sector in the industry, the independent small loggers who had gained a secure position in the industry with the use of trucks and other equipment then appearing on the market. On the eve of war, the coastal logging industry was dominated by a number of relatively large companies using heavy steam yarders and loaders and railways to log big stands of the best coastal timber. They were represented by the BC Loggers' Association, an industry group formed before World War One and now dominated by a few of the largest companies, particularly Bloedel, Stewart & Welch and Comox Logging. By the late 1930s, some of the independent loggers on Vancouver Island realized that to survive they were going to have to fight for forest policies that took their interests into account. New regulations requiring snag falling and slash burning, introduced

Grapple loader prepares to load another log while the truck driver stamps the logs. Kitimat, 1965. (Jack Cash photo; Mauno Pelto collection)

by the Forest Service after the 1938 Campbell River fire, placed an onerous, and in retrospect unnecessary, burden on truck loggers. Opposing the new rules was another reason to organize. In 1939, Bert Welch, of Olympic Logging at Qualicum Beach, invited about thirty small loggers working between Nanaimo and Campbell River to a meeting to establish a loggers' organization. The only one to show up was Wallace Baikie, who at the time was contracting with his two brothers for Comox Logging. After a further two and a half years of informal meetings and discussions, they were joined by several other small truck loggers from the island, and in 1943 they held the TLA's first general meeting in Nanaimo. Some of the big companies with truck camps, or divisions of them, joined early but withdrew later over policy differences.

The TLA quickly became a powerful force within the industry. By the end of the war most of the established truck loggers on the coast had joined, as had several of the larger independent owner-operated companies, such as A.P. Allison and P.B. Anderson. By 1945, the association's annual convention had become the major forum for all sectors of the industry, including representatives of the bigger non-member companies, cabinet ministers and government officials. From the outset the organization included another key sector of the industry, the companies supplying the equipment and materials used in logging. Earl Finning, of Finning Tractor & Equipment, was one of the suppliers who played a major role in getting the TLA off the ground.

By the 1948 TLA convention, the provincial government had introduced legislation based on a secret memo to the government from Chief Forester C.D. Orchard, establishing Forest Management Licences and a new program of sustained yield forest policy. Crown-owned timber had been sold at public auction until this time. Now, under the new plan, large areas of forest land would be leased to private companies and the timber sold at a price established by the government. Orchard's memo had been circulated among the owners of the big companies—who, with a few notable exceptions including H.R. MacMillan and Prentice Bloedel, supported it.

Orchard appeared on a panel at the convention to defend his concept, which was vigorously criticized by Bill Keate, a Port Neville logger and timber broker, who predicted that the free market in logs would be destroyed and independent loggers eliminated. It was quite clear after this session that Orchard's plan was going to be of benefit to the big established companies, which supported it, and detrimental to independent market loggers such as those who had formed the TLA. From this point, the TLA became the major opponent of the tenure system that became law later that year.

The new legislation did two things: it limited the amount of timber that could be logged, and it defined two kinds of sustained yield units, or areas, within which the logging limits applied. One type of unit, known as Public Working Circles or Public Sustained Yield Units, would be managed by the Forest Service and the timber sold by public auction—essentially, a continuation of the existing Timber Sale system, with limits on how much could be logged each year. The other type of sustained yield unit was known as a Forest Management Licence, which would be awarded without competition to companies willing to manage the area over the long term, as well as log the timber.

Independent loggers' first objection to this system was a question of principle: the lack of competition in awarding the FMLs was contrary to the working of a free market economy. This argument was ignored, and when the government began awarding FMLs in late 1948, the true nature of the scheme became clear. FML #1 was granted to the Celanese Corporation, a US company that had never done business in BC. Orchard later claimed his intention was to award several hundred FMLs of modest size to existing logging operators. The first one, which happened to be in Forest Minister E.T. Kenney's riding, covered an area of more than 6 million acres.

The second FML, issued a few months later, revealed another facet of the new order. It was granted to Canadian Western Lumber with the financial backing of the American-owned Crown Zellerbach Corporation, after intense lobbying on the part of Bob Filberg, the superintendent of Canadian Western's subsidiary, Comox Logging. The disturbing aspect of this FML was that it covered an area along Johnstone Strait previously designated as a Public Working Circle where dozens of independent loggers operated. The government's defence of these first allocations was that it was necessary to grant cutting rights of this magnitude, in order to attract investment in the pulp mills being built by the new licence holders.

In a few years, as more FMLs were awarded, yet another facet of the tenure system became apparent. Some companies—and it will never be known for sure how many or which—began making under-the-table payments to provincial political parties and, in some cases, directly to politicians. In the mid-1950s the entire system of timber allocation was thrown into disrepute when it was revealed that the minister of forests, Robert Sommers, had accepted a bribe to award a FML to another new player in the timber industry, BC Forest Products. Sommers was convicted and sent to jail, but charges against the company that issued the bribe were dropped and it kept the FML, also in an area formerly designated as a Public Working Circle.

Throughout this entire period, which lasted until the late 1950s, there were only two voices raised in protest. One belonged to the TLA, and it was discounted because the public saw it as speaking for the self-interest of independent loggers. The other voice did attract public attention. Gordon Gibson, who had been logging in the Tahsis area with his brothers and father since the 1920s, won election to the provincial legislature in 1952. He soon began receiving a flood of information from independent loggers on the various ways the new system was being corrupted. Gibson mounted a sustained attack on the FML system in general, and its administration by the W.A.C. Bennett government in particular, eventually forcing a second royal commission under Chief Justice Sloan to re-examine the issue. It was during the course of this commission that the bribery charges surfaced that led to Sommers's conviction.

In the midst of the Sloan Commission's deliberations, a second public enquiry, the Lord Commission, was set up to consider Gibson's charges in the legislature that "money talks" in the awarding of FMLs. After meeting for three days, Judge Arthur Lord adjourned and reported he had found no evidence of bribery.

The exoneration of the FML system —now called Tree Farm Licences (TFL)—by the Sloan and Lord commissions, combined with the trial and conviction of the forests minister, more or less ended public objections. A widespread sentiment became firmly entrenched in and around

I recall waiting expectantly for the *Malahat*'s first trip into Port Alice with a load of logs from Gibson Brothers at Nootka Sound. She was days late, but advised on our own radio that she was in Quatsino Sound and would be into Port Alice that night. I made frequent trips down to the dock, and finally could see her mast and running lights and could hear the familiar bong, bong, bong of the old diesel.

As she approached the dock I could see the glimmer of a two-cell flashlight moving about on the log deck, and then Gordon hollering, "How are we doing, getting close?" Then God-damning the flashlight not having good batteries. The *Malahat* moved into the dock like a stately princess, and we berthed her. Gordon came ashore exclaiming that he had used his flashlight quite a bit clambering over that deckload of logs on his journey up. I said to him, "Why not get a searchlight so you would be able to land a ship like this properly?" He remarked, "Did you ever see a better landing? If we had a searchlight I would have been so brave I might have knocked your goddamn dock down, besides spending money like it was out of style."

With Gibsons the unusual was an everyday occurrence, and I wish I had a picture of the *Malahat* on this, her first voyage to Port Alice pulp mill with a log cargo. The bridge, being aft, had only been partly dismantled to make room for logs. Just ahead of the bridge was an old two-drum, open-faced donkey engine with a mainline and haulback, and that is how they loaded the cargo, commencing at the bow and working backwards. In the unloading, the reverse of this took place. In fact, there was a whole rigging crew aboard—engineer, chaser, chokerman and signal man; and in addition a cook (Clark Gibson's wife) and a flunky. Our crew unhooked the slings in the water and provided the boomsticks to secure the logs. Gad! What a system, and everyone happy into the bargain.

Archibald Kerr

A loaded off-highway truck crosses a laminated beam bridge. Kitimat, 1965. (Jack Cash photo; Mauno Pelto collection)

the industry: the system of administering public forest lands was corrupt, and the big companies were too powerful to oppose.

Almost forty years after he was asked for a bribe, Joe Garner published a detailed account of his experiences in attempting to obtain a FML, and the price he paid for refusing to go along with the proposed under-the-table deal. "Sometimes I'm sorry I didn't blow the whistle on it all back in the fall of 1953," he wrote in his latest book. "But I was small potatoes and business activities were pressing—and the government was in total control." This account, along with the memoirs of others active in the industry at the time—among them Gordon Gibson's *Bull of the Woods* and Ian Mahood's *Three Men and a Forester*—provide compelling evidence of underhanded allocation of timber rights and the administration of forest policy in a manner favouring the large, integrated companies.

By the early 1960s, from their bases in the TFLs, the big companies began to move into the Public Working Circles. The adoption of sustained yield policies in 1948 brought about the need to limit annual harvests in each of these administrative units. In most of the coastal PWCs the annual cut level was already being logged by existing operators. In an attempt to deal fairly with timber allocation, the Forest Service introduced a quota system which allocated established loggers an annual harvesting quota based on an average of their previous three years' cut. Although the quota system was never defined in law, it was accepted by most as a working solution, largely because it conferred a right to log that in itself was worth something.

Almost immediately, quota became a saleable item. Through one means or another the big, integrated companies began to amass it. Between 1954 and 1974, the ten largest companies in the province increased their control of the timber supply from thirty-seven percent of a 5.6-billion-board-foot harvest, to almost fifty-five percent of a 12.7-billion-foot harvest. By the end of this period only a few truck loggers were able to hang on to their quota. Most succumbed and sold their cutting rights to the integrated companies. In some cases—A.P. Allison Logging and P.B. Anderson, for example—the owners or their descendants simply sold their rights and retired. In others, logging contracts were included in the sales agreements.

A fair assessment of these events recognizes that there were many problems with the timber auction system as it once functioned. Collusion in bidding, spite bidding and extracting cash or other forms of payment in exchange for refraining from bidding were common practices at various times. In some cases, the TFL system resulted in a superior level of forest management after logging than was practised on land managed by the Forest Service.

One reason for the growth of the contract logging sector was the addition in 1959 of a clause in TFL agreements requiring the licensees to have a portion of their logging done by contractors. This was a belated attempt to mollify the independent loggers in the wake of the corruption revealed by the Sommers affair. Some of these contracts were for all phases of the logging process, others were partial phase contracts—falling and bucking, yarding, trucking or booming—where a number of contractors, often in conjunction with company-run crews, worked together.

Through the postwar decades, as it became more and more difficult to obtain timber sales, newcomers to the industry tended toward a single phase of logging. Olaf Fedje, for example, became one of the coast's largest falling contractors. Following the success of trucking contractors like Bill Schnare, scores of trucking contractors appeared, either from the ranks of formerly independent loggers, or as individuals establishing their own contracting business. Road building or some parts of it were often contracted out in later years. On the coast, however, in contrast to the Interior logging business, most contractors have been full-phase, stump-to-dump operators.

The few loggers who remained fully independent of the large, integrated companies did so in a number of ways. Some, like Cliff Coulson in Port Alberni, invested every cent they could scrape together in buying quota. Other more established operators, such as Baikie Brothers Logging, maintained their supply by purchasing private land and logging the timber on it.

As large corporations succeeded independent loggers as the dominant force in the industry, several other regulatory changes affected the nature of logging. With the elimination of competitive timber sales, the means of establishing the price paid to the government for timber had to be changed. An appraisal system was devised that was supposed to approximate a free market value. Logging practices no longer necessarily reflected the best and most efficient method of getting timber from the stump to the dump. Instead, they were adjusted to take advantage of various quirks and loopholes in the appraisal system. Loggers began to lose control of their operations to accountants and lawyers working on behalf of the corporations holding the cutting rights.

Over time the appraisal system, along with other regulations resulting from the conversion to a non-market logging economy, led to some bizarre practices, the costs of which were borne by the public treasury and the contract loggers themselves. One example grew out of the regulations concerning both the stumpage appraisal system and utilization standards. At the time it became mandatory to log and process all coastal timber down to specified diameters and lengths, the normal practice was to dump the logs into the salt water, boom them and tow them to a sorting ground (usually Howe Sound) where they were measured or scaled. The scale reports were the basis of payment to both the government and the contractor. It was to the advantage of the big companies, whose mills were not designed to utilize these smaller or lower-grade logs, to lose as many of them as possible along the way. Neither the contractors nor the government were paid for any of the logs that disappeared. Eventually, under pressure from contract loggers, the government required that sorting and scaling be performed at dry-land sorting facilities before the logs were put into the water.

A boom boat takes away a truckload bundle, unloaded intact into the water. Kitimat, 1965. (Jack Cash photo; Mauno Pelto collection)

The policy of the pulp mills in the 1930s seemed to be to accumulate logs until they had a big inventory and then quit buying until the market fell off. As you know, hemlock lying in a boom begins to sink after it is there a little while.

In those days I'd say seventy percent of the loggers were open market loggers. Today I wouldn't hazard a guess. About four years ago [1955] I heard that a mill was advancing Tom Lamb some money on some sort of a deal. One day on the street he asked me how many of our members were open market loggers. I told him I knew of none except him, and that I wasn't too sure that he was.

As I said, in the thirties most logs were put on the open market, and there was always a fight between the mill and the logger. Loggers tried to uphold or push the prices up and the mills tried to beat them down. Those were rough times.

John Burke

A boom man pulls a swifter line across a bundle boom, preparing it for towing to Ocean Falls. Kitimat, 1965. (Jack Cash photo; Mauno Pelto collection)

The tenure provisions in the 1948 revisions to the Forest Act, of which the TFL was the centrepiece, facilitated an intense period of mergers and takeovers in the industry. Essentially, the process involved the acquisition of a few of the biggest logging companies that existed at the end of the war by out-of-province forest companies or financial conglomerates. The new coastal forest companies then took over and merged with most of the other companies owning timber land or holding cutting rights on public timber in the area, creating local timber monopolies.

As soon as the war ended, Bloedel, Stewart & Welch implemented plans to build a pulp mill in Port Alberni to utilize waste from the Somass and Great Central mills, as well as small logs from its operations in the valley and down the inlet at Franklin and Sarita divisions. During and after the war years, Sid Smith, the company's manager, had been acquiring timber and cutting rights at a hectic pace around existing BS&W operations at Alberni and Menzies Bay.

During the same period, the H.R. MacMillan Company implemented a similar strategy in several locations. By the outbreak of the war, MacMillan, in his ongoing market battle with Seaboard Lumber, had acquired the Canadian White Pine mill on the Fraser River, the Alberni Pacific company and the Rockefeller timber in the nearby Ash River valley, extensive timber stands at Malahat and along the Nanaimo River, and interests in Chehalis Logging near Harrison Lake and the BC Plywood mill on the Fraser. He was the biggest lumber producer in the province. In 1939 he also bought the bankrupt Campbell River Timber, but sold it to BS&W two years later.

MacMillan went on an aggressive expansion campaign during the war, taking over Thomsen & Clark, the Robert Dollar timber interests at Northwest Bay, Shawnigan Lake Lumber, Iron River Logging, Wellburn Timber and, in 1944, the biggest prize of all, Victoria Lumber & Manufacturing. He built the Alberni plywood mill in 1942 and the Harmac pulp mill in 1948.

In 1951, the MacMillan and BS&W companies merged, creating MacMillan and Bloedel, a gigantic operation controlling three quarters of a million acres of the most valuable timber on the coast, and mills producing twenty-five percent of BC's lumber and thirty-eight percent of coastal pulp. The *Vancouver Sun*, deploring the new concentration of industry power, observed that H.R. MacMillan "deserves a better fate than to become known as a great monopolist."

Across the Gulf of Georgia, the Powell River Company came out of the war with 100 square miles of recently acquired timber land, having taken over Kelly Logging and Aero Timber in the Queen Charlottes, Alice Lake Logging, Bell & Campbell, Broughton Timber, Knight Inlet Logging and O'Brien Logging. George O'Brien joined Powell River as logging manager. After the war the expansion continued with the addition of sawmills and the acquisition in 1951 of the Anderson family's Salmon River Logging, which had been sold to Westminster Shook Mill in 1946. In 1960, Powell River merged with MacMillan and Bloedel, creating the largest forest products company in Canada.

Elsewhere on the coast, a similar process was occurring. Canadian Western Lumber, owner of Comox Logging, set up the Elk Falls Company in 1950 on a fifty-fifty basis with Crown Zellerbach, a San Francisco company. While it was named Crown Willamette, Crown Zellerbach had acquired the Ocean Falls mill in 1915 and operated it under the name Pacific Mills. The new Elk Falls entity was set up to build a pulp mill at Campbell River designed to utilize pulp logs from Canadian Western's TFL #2, and wood residues from sawmills in the Lower Mainland. In 1953, Pacific Mills, Elk Falls and Canadian Western with its various subsidiaries, including Comox Logging, all combined in a new company, Crown Zellerbach Canada.

A somewhat different process created another major coastal corporation that appeared after the war, BC Forest Products. During the war, when H.R. MacMillan was serving as head of the Wartime Shipping Corporation, he worked closely with E.P. Taylor, a Toronto promoter. They became good friends, and out of their conversations Taylor developed an interest in the BC forest industry. When MacMillan learned that the Humbird family wanted to sell VL&MC, but was unwilling to sell it to him, he got Taylor to make an offer. Eventually this offer was accepted, and within months Taylor sold the company to MacMillan.

This exercise in deception sparked Taylor's interest in the forest industry, and with MacMillan's assistance he acquired several medium-sized firms with mills and timber on the Lower Mainland and southern Vancouver Island. In 1946, Taylor bought Hammond Cedar on the Fraser, Sitka Spruce Lumber in False Creek, Cameron Lumber in Victoria, a logging and timber holding company owned by Matt Hemmingsen and Cameron Lumber, and Industrial Timber Mills at Youbou. These were combined into a new company, Vancouver Cedar & Spruce, which soon after was renamed BC Forest Products. MacMillan loaned the new enterprise some of his senior executives to with management and handled sales until 1953. BCFP aggressively acquired more timber by purchasing several other companies, including Malahat Logging and San Juan Lumber at Port Renfrew, Oscar Niemi's interests in Sechelt and Jervis inlets, and Blackstock Logging at Pitt Lake. Several TFLs and extensive cutting rights in Public Working Circles were amassed, and by the early 1960s, BCFP was the second largest forest company in the province.

Several other substantial operations came together in the 1950s and 1960s. Koerner's Alaska Pine added to its lumber milling base, taking over the Woodfibre and Port Alice pulp mills in 1952, then selling control of the corporation to Rayonier, a New York firm, in 1954. The Koerner family arrived in BC from Czechoslovakia in 1939 and successfully devised methods to dry and season hemlock lumber so it could be shipped abroad without staining. The company began with a mill in New Westminster and in 1946 bought the Canadian Puget Sound Lumber Company. Rayonier later took control of logging operations at Jordan River, the Queen Charlottes, Port McNeill and at several other locations on the west coast of Vancouver Island, in the Fraser Valley and up the mainland coast to the Gardner Canal.

The Gibson brothers entered into a partnership with the Danish-owned East Asiatic Company in 1949, and it developed into the Tahsis Company. A few years later, when East Asiatic wanted to build a pulp mill and obtain a TFL, the Gibsons sold their interest to their Danish partners and retired from the business, leaving Gordon Gibson free to devote his time and energies to fighting the TFL system in the legislature.

Canadian Forest Products was created in the late 1940s when the Bentley and Prentice families expanded out of a veneer plant set up just before the war in New Westminster. The company bought several small lumber and shingle mills on the Fraser, then acquired the Englewood holdings on the Nimpkish River, which were expanded with a TFL in 1961. Canfor is noteworthy for being the only coastal company to retain its railway equipment. One of the biggest forest companies in Canada, it is one of the few still partially owned by its founding family.

Weldwood, a US-dominated plywood and lumber producer, established a presence on the BC coast in 1964 when it bought Canadian Colliers, a company set up in the early 1950s to manage the Dunsmuir coal and timber interests on Vancouver Island. Before this sale, Canadian Colliers took over the Thurston–Flavelle Lumber Company and the Timberland mill in Surrey. In 1966, Eurocan, a Finnish company, obtained a TFL in the Kitimat area and built a pulp mill to supply its European paper mills. Most of this TFL lay east of the Coast Mountains and entailed an extensive road system over which a fleet of twenty-six Hayes off-highway trucks hauled logs.

These were the major integrated corporate groups formed during the fifties and sixties. There were a number of other firms pursuing a similar course, but with reduced ambitions or success. Scott Paper, a US company producing specialized paper products, became a major owner of private forest land on Vancouver Island at this time. Eventually it sold its two-thirds share of Elk River Timber at Campbell River to BC Forest Products. Other companies, such as Whonnock Industries along the Fraser and Doman at Duncan, which started out as very small operations in the same period, later grew to prominence.

In 1960, the increased concentration of control and the integrated character of the new corporate entities were reflected in the formation of the Council of Forest Industries. In effect, COFI was a merger of the BC Loggers' Association and several forest products producer associations. While it claimed to represent the entire forest industry, in fact it represented only the interests of the large integrated companies.

In spite of the trend toward corporate concentration in the postwar period, coastal logging at the woods level was still dominated by the independents. They did a large portion of the actual logging, they were among the most innovative loggers, they helped develop much of the new equipment that appeared and they became stable businesses in the coastal communities in which they worked. For these reasons the small owner-operated companies were able to survive and to reassert themselves during the crisis of the 1980s.

The massive structural changes the industry underwent during the twenty-five years following the war coincided with equally significant changes in logging technology. When the war began, most coastal logging, particularly in the bigger camps, was done with steam-powered technology developed twenty years earlier. The use of internal combustion powered machinery, particularly trucks, was just becoming widespread when the war began. Wartime restrictions made it difficult if not impossible to obtain new equipment; even obtaining parts was often hard. Consequently, the industry came out of the war with a pent-up need for new and better logging machines.

The increased demand for new equipment, especially trucks, came from several directions. Small independent logging and contracting operations were springing up all over the coast. Returning veterans and young loggers who had picked up some experience and knowledge working for wages, saw the postwar period, with its healthy markets and availability of low-priced logging machinery, as an opportunity to establish their own businesses. The rapid increase in pulp mill construction led to the opening of large areas to logging, all of which were established as truck shows. In addition, most of the railway companies had worked back into areas too steep for rail transport and were looking at trucks to either supplement or replace their rail systems. By 1950, only fifteen percent of the logs moved on the coast were handled exclusively by rail. Another nine percent were moved by a combination of rail and truck, with five percent yarded directly to the water by donkey or tractor. The remaining sixty-nine percent were hauled exclusively by truck.

For a decade or so after the war, the easy availability of surplus military equipment met some of the huge demand for trucks. A surplus Mack truck could be picked up for a very low price and either used with its existing motor, or repowered with one of the hundreds of surplus diesel engines flooding the market. One of the most successful trucking contractors on the coast, Ed Eason, came out of the army near the end of the war and got his hands on a partially dismantled army truck, which he rebuilt. From this modest beginning rose one of the biggest contract trucking operations on Vancouver Island. Others, like the Norrie Brothers, built full-phase, independent truck logging operations with modified surplus equipment.

At one point, in 1951, one Vancouver dealer obtained 200 surplus US-made heavy duty trucks, mostly Macks, from England. They were reconditioned and sold with a new truck warranty at prices well below that of current

After the war, many independent loggers launched themselves into business by piecing together surplus machinery. The truck at right, belonging to Norrie Brothers Logging, was a Mack used on construction of the St. Lawrence Seaway. It is equipped with sixteen-foot bunks. The truck at rear is a war surplus Mack with an engine from a four-motor Quad salvaged from the US Navy. (Frank Norrie collection)

models. For a time a minor industry developed around the conversion of six-wheel-drive military vehicles into logging machines. One adaptation, used widely in the Interior, was to mount a winch and arch on the rear and use the truck as a prototype rubber tire skidder.

Although industry had not been able to get its hands on new equipment during the war, the industrial mobilization effort that occurred brought about many new developments and improvements that had wide application in logging equipment. Two of the most significant areas of technological advance were in engine design, particularly diesel engines, and tire manufacturing. The impact of these changes was most evident in trucks that became available during the late 1940s.

By 1950, a trend in logging truck design was obvious—more power and bigger trucks. This development was more pronounced in BC where a much higher proportion of truck logging took place on private industrial roads. In Washington and Oregon, truck loggers made much greater use of public roads and were forced to comply with load limits and restricted

We were always gyppo loggers. We bought three military surplus Mack trucks that were just borderline highway-size trucks. They had old-fashioned Buda diesel engines in them that were real dogs. We took out those and put in our 671 Jimmys and they'd run rings around any of the trucks that were here in that era. Why I have a hearing problem now is because of those things. They screamed, made a terrible noise. In those days people weren't concerned about ear difficulties. We just had a straight pipe up the back of the cab. They drummed in the cab, and just about deafened anybody at the side of the road when you went by.

Later we got hold of some of those monster Mack trucks that had been used on the St. Lawrence Seaway. They were real big ones. We bought two of those and converted them to logging trucks. They had a V-16 Cummins diesel engine in them. They packed a lot of wood. They were pretty heavy on maintenance. There were a lot of parts in the rear-end assembly that required a lot of maintenance. We eventually sidetracked those and got into Pacific trucks.

Frank Norrie

bunk widths, and were unable to make extensive use of larger off-highway trucks. In the late 1940s, trucks with engines in the 150- to 250-horsepower range were popular, but by the early 1950s loggers were demanding 300 horsepower and more. Adverse grades were getting steeper, trailers were available to carry bigger loads, and roads were getting better, making higher speeds possible.

It was at this time that truck manufacturers began to specialize their designs, and the number of builders of trucks suitable for use by Pacific Northwest loggers decreased dramatically. Before the war there were dozens of manufacturers producing the three- to five-ton rated trucks adapted to hauling logs over fore-and-aft or plank roads. During the Depression and the war, many of these makers disappeared. Those who remained opted to concentrate on larger, more lucrative markets. This left relatively few builders of on- and off-highway trucks useful to loggers, and it was these companies that began to produce bigger and better logging trucks. International had its diesel-powered Emeryville series and a smaller Loadstar series using a six-cylinder gas engine. White came out with a Cummins diesel-powered model in 1949, and Diamond T's 1951 model, aimed at loggers, had a 300-horsepower Cummins diesel.

The demand for big off-highway trucks in BC made the province a world centre of heavy-duty truck manufacturing for a time. The dominant company was Hayes, which produced its first truck in 1922. With a skilled engineering staff and plant in Vancouver, Hayes was able to respond quickly to requests and suggestions from loggers. In 1952, it introduced its HDX series, one of the most popular logging trucks ever built. Twenty years later every HDX ever built was still in active use.

The Hayes success story continued for many years. In the late 1940s, its staff of sixty was turning out fifty trucks a year. By the late 1960s there were six hundred employees in three Vancouver plants producing six hundred trucks a

There were nine other independents at that time [about 1950 in Squamish], maybe more. I'm talking about independents, and they were small, real small. The independent loggers would contract the trucking out. Not many of them could afford to have trucks. The independents then agreed with the existing truckers that they would not have trucks themselves, that there would be independent truckers. That was the local agreement.

We would have had all gas and diesel yarders. There was very little cat logging going on here, the country was too rough. There was some odd shows. The yarders would have been mostly Skagits. This valley went high to Skagits. As long it was a Skagit winch, then they could have any kind of a motor on it they wanted. Then they came in with all these other things, Gearmatics, and everything else, later on you know. It revolutionized diesel or gas donkeys because they were so fast. Changed gears instantaneously on them.

Al Hendrickson

When I graduated in '54, I went to work for BC Forest Products at Caycuse and it was just at that time they started to take up the railroad. They were a combination of railroad, truck, steam, gas, diesel. During the early fifties the conversion was started, but they were still all wooden trees.

When they started using trucks they still had a lot of machines on sleds, like cold deckers. They used to move the machines on a low bed. It still had a sled on it. Take it up and unload it wherever a wooden tree was. But most of the slackline machines and skidders they mounted on steel wheels. There was a big tongue and they used to tow them with a Cat. They were heavier than hell with these big flat steel wheels. It was a tough job moving these things. I guess they didn't have low beds big enough.

Some of the slackline machines they left on sleds. All the smaller cold deck machines and smaller slackline machines were on sleds so they could move them either on logging trucks or low beds. They used to move Tyee slackline machines, as I remember at Homfray Creek, putting it on two trucks, one backing up and the other going ahead with the slackline machine on it, moving it up the hill. The smaller machines they left on sleds, the bigger ones they put on steel wheels.

That created a need for more compact, mobile yarders. After they left the railroad the loading units were always separate from the yarding units. Although Madill did build one steel spar that tried to incorporate a loading boom in the tower, it never got off the ground. It was at this time they were starting to develop the mobile spars. Some of the first mobile spars I saw were smaller machines mounted on a truck called a goat. They used to refer to the machine that went around with the bull gang to pre-rig wooden trees as a goat. On the railroad it was a goat. When they went to trucks they had a goat. A small machine with a steel tower on it that used to fold down and raise up. They called it the gunboat or the goat, and it was only used for rig-up, not yarding. It used to go and pre-rig the trees, stand the trees up and tighten up the guylines so the big machines could just move in. Then they went from there to the steel spar and it developed from there on.

Monty Mosher

Rick Doney (left) and driver Rusty Borrow with Norrie Brothers Logging's new Mack at Burgoyne Bay, 1948. (Frank Norrie collection)

The Hayes truck plant with two HDX logging trucks ready for delivery, mid-1960s. Left to right: Tom Everett, Hayes vice-president; Jack Curcio, president; Roy Saunders, a Nanaimo trucking contractor. (Ed Pryor photo)

The second Butler Brothers truck, built at the Madill plant, at MacMillan Bloedel's Menzies Bay division. (CRM 10713)

year. These were handmade, custom trucks, as opposed to the assembly-line models produced by major manufacturers such as Ford, GM and Chrysler. While the company built many models for use in various industries, the heart of its operation was in logging, a field it dominated until the 1970s. In 1972 Mack Trucks bought two-thirds of Hayes' shares and began to mount Mack cabs on Hayes frames. Then, in 1975, the trucking and logging industries were stunned to learn that Paccar, the old Pacific Car and Foundry Company of Seattle, which by then owned Peterbilt and Kenworth, had bought Hayes from Mack and was eliminating the company. A great deal of bitterness and anger over this decision can still be found almost twenty years later.

The demise of Hayes left one local manufacturer: Pacific Truck & Trailer, started in 1947 by a group of former Hayes employees. Beginning with a plant on Franklin Street in Vancouver, Pacific produced custom, handmade, heavy-duty trucks, primarily for the off-highway logging trade. A larger plant was eventually built in North Vancouver, employing more than two hundred workers at its peak. About 1970, the company was sold to International Harvester, which continued to produce the Pacific line of trucks. During its fifteen-year ownership, International built a few Pacific models with its own cabs. In the mid-eighties Pacific was acquired by Inchcape, a British-owned Singapore company, which closed the assembly plant in 1991. The company still manufactures parts for the more than 2,000 Pacific trucks built over the past forty-five years.

There were two other truck plants in BC, both branches of US manufacturers. Kenworth produced custom-built trucks at a Burnaby factory from 1956 to 1982, when truck manufacturing for the Canadian market was shifted to Montreal. In 1967, White opened a plant in Kelowna to build its Western Star trucks. It is the only major truck factory still building trucks in BC.

The demand on all these plants in the 1950s and 1960s was enormous, with new logging divisions opening and conversion from rail proceeding at a rapid pace. In one order alone, in the fall of 1965, MacMillan, Bloedel & Powell River Company bought forty-one off-highway trucks from Hayes, Kenworth and Pacific's Vancouver plants. This order brought to eighty the number of trucks bought by this one company in a one-year period.

The search for bigger and more powerful trucks led to the creation and use of at least two unusual machines on the BC coast. One was produced by Robert LeTourneau, the inventive genius who had developed several machines using big, wide rubber tires. The LeTourneau Log Transporter was a massive machine with six wide-faced tires, each seventy-three inches in diameter, mounted on independently driven wheels, each powered by an electric motor. When a wheel lost traction, all the power was transferred to the other wheels, giving the machine an unprecedented ability to traverse rough

A Hayes self-loading trailer that appeared in the early 1960s. A small winch behind the cab enabled the driver to load the trailer without leaving his seat. (Commercial Illustrators photo; Tom Everett collection)

and steep ground. It had no brake bands or drums, and was slowed or stopped with the electric motors. The Transporter's generator was powered by a 335-horsepower Cummins diesel.

These machines carried loads down thirty-five to forty-five percent grades and took loads of 12,000 to 15,000 feet up hills no truck ever considered tackling. With their clearance and traction, they required much less in the way of road construction. Herb Pohl and Tom Brown used the first Transporter on the coast at Crabapple Creek in Jervis Inlet in 1958. Former TLA president Bruce Russell drove this machine, which now sits at the Wajax Industries plant in Coquitlam. He said it could out-haul any truck ever built, but required too much maintenance and repair work. In the 1970s Courtenay contractor Bob Woods used Transporters to log tree-length second-growth hemlock for Crown Zellerbach on Sonora Island without building any roads. Limbs and tops were spread over soft areas and the machines travelled on top of them.

The other unique truck was built for Butler Brothers Logging at Sooke by Barney Oldfield in his Saanich welding shop. This machine consisted of a thirty-seven-foot heavy-duty chassis mounted on two Mack double-axle rear ends. A 435-horsepower V-12 GMC diesel engine sat beside a small, one-man cab at the extreme front end, well forward of the front set of wheels. The driveshaft ran to a transmission amidships, from which two shafts ran to each set of front and rear wheels, providing an eight-wheel, sixteen-tire drive machine. Turntables on both the front and rear sets of wheels allowed both ends to be steered, giving the truck a very small turning radius. Those who drove it said that it was a frightening experience, sitting in the cab ahead of the front wheels, to manoeuvre the truck around a corner with a steep dropoff and look straight down out the side window. The bunks were twelve feet wide and twenty feet apart, but because the cab top was below the bunk level, long logs could protrude over the front end. Its top speed was forty-six miles per hour in overdrive. It was an expensive piece of equipment, the first model costing $50,000 to build.

The Madill plant in Nanaimo built a second, larger model with an 880-horsepower, V-16 engine. It was designed to tow a pup trailer, with a combined load of 200 tons. Although both trucks were tried in several locations, they were not a great success. The first model eventually retired to the boneyard at Andy Byrne Trucking at Myrtle Point. The second truck, after working for a while at MacMillan Bloedel's Menzies Bay camp, was being used by Dougan Brothers Logging when it was wrecked. Coulson Forest Products salvaged the rear ends and used one set under a Madill steel spar.

Throughout the 1950s and 1960s, numerous improvements and modifications were made on logging trucks, many of them devised in the dozens of local garages and machine shops scattered along the coast. Several companies, including Hayes and Columbia, developed

In the late 1960s and early 1970s, Crown Zellerbach used a five-unit pup trailer train at its Courtenay and Nanaimo Lake operations. The truck is a Hayes HDX. This setup was not enormously successful, and the industry adopted the single-pup configuration for use on long hauls. (Ed Pryor photo)

increasingly sophisticated trailers. Right after the war, with trucks carrying bigger loads and smaller logs, cheeseblocks were abandoned in favour of stakes. And to the dismay of loaders who had to carefully lift the logs over them, the stakes kept getting higher and higher.

In the early 1960s, self-loading trucks—trucks equipped with log-loading machinery—were introduced. A variety of these devices reached the market. The first ones used tongs on a winch line and a heel-boom system. Venture Welding in Nanaimo sold the Venture Loader, and the Courtenay Machine shop marketed the EE-ZEE Loader in the mid-1960s. In 1966, some of the first grapple loaders appeared.

The availability of self-loading trucks had some far-reaching effects. Separate, expensive machines were no longer needed to load logs. A small operator had only to get his logs to roadside, where they were picked up. In a larger show, small patches of timber far from the main logging area could be logged without the expense of moving in a loader. Another consequence was the appearance of many trucking contractors providing loading and hauling services to widely scattered small operations.

Pre-loading systems were another innovation. Basically, pre-loaders provided a means of building a load at the landing which could be put onto a truck in a single pass. The objective was to reduce the time trucks spent waiting to be loaded. At a plant in Courtenay, Archie McKone designed and sold a pre-loader employing a set of four skids and ramps. The trailer was

A hydraulic self-loader mounted on an off-highway truck weighing in on the scales at Nanaimo Lakes, 1979. (FERIC)

backed under pre-loaded bunks resting on the skids and attached to the bunks. When the truck pulled ahead, the skids slid down the ramps, lowering the bunks onto the truck. It was a simple and efficient rig, but it required a large, level landing which was not always available.

Another type of pre-loading system was developed in 1959 at Crown Zellerbach's Bella Coola operation, by superintendent Bill Wainwright and Pacific Truck & Trailer. This method required two trailers per truck, each with a set of folding legs to support the front bunks when not attached to the truck. While the truck was hauling one load, the other trailer could be loaded. An advantage of this system was that no pre-loading equipment was left at the landing, which did not have to be as large as the McKone system required.

The development of road building techniques and equipment made possible the rapid growth of truck logging. For example, the creation of functional bulldozers for use on crawler tractors in the late 1930s made it possible to build roads economically and quickly. Another critical development was the air drill, which enormously speeded up the drilling of holes to blast rock. Before the air drill, blasting holes were hand drilled, a slow and laborious process. The first air drills were mounted on tractors, trucks and sleds. When combined with improved blasting techniques and materials, this equipment was capable not only of clearing rock from a right-of-way, but of pulverizing it into small pieces suitable for surfacing roads in locations where gravel was scarce. In the mid-1960s, self-propelled air track drills appeared, simplifying the process even further. One of the first of these machines was built in Vancouver in 1966 for MacMillan Bloedel to use at Sproat Lake.

The advent of laminated timber beams in the 1960s simplified bridge construction and allowed the erection of long-span bridges which were built quickly and economically and which could be re-used at different locations. Pressure treatment prolonged their lives substantially. As truck loggers moved farther back into rougher areas, such bridges became essential.

This elegant suspension bridge was built by BC Forest Products engineers thirteen miles up the San Juan River in the 1960s. (BCFM 4–2)

Below: A Tamrock hydraulic crawler rock drill working at Fletcher Challenge's Caycuse operation, 1988. (FERIC)

As the various components of full-scale truck logging came into use, the abandonment of railway logging accelerated. In 1945, the BC Railway Department reported there were 660 miles of logging railway on the coast, down from 850 miles at the high point in the late 1920s. Except for the Powell River Company's show at Cumshewa Inlet and O'Brien Logging at Stillwater, all the railway shows were on Vancouver Island. By 1955, the total coastal trackage had declined to 285 miles, and over the next five years another 100 miles was abandoned. Truck logging technology had evolved to the point where it could haul more logs at a lower rate than a comparable railway operation. A comparison of the equipment exchange involved can be seen in the conversion of Western Forest Industries' operation on Cowichan Lake. Five locomotives hauling 135 log cars over fourteen miles of mainline railway were replaced by thirteen Kenworth and four Pacific trucks, hauling over ten miles of road.

In most cases there was a two-stage conversion, with trucks hauling from the landing to transfer points on the railway. However, reloading logs from trucks to rail cars incurred an added expense, usually requiring a donkey and A-frame setup capable of lifting a full truckload at a time. This system also inhibited increased trailer size, as the reload process was most efficient when truck and rail loads were the same size.

By the time the conversion process got under way, almost all the railway operations were in the hands of four of the new companies that had emerged from the takeover and merger binge of

We fed 550 men in the cookhouse and we had eighty-two families at Franklin Camp. There was over 800 in the camp counting the children and women. I ran it from '45 to when I left there in '66. It was a good camp, big. When we switched to trucks it became larger, and for two years there we were producing anywhere from 20 to 25 million board feet a month. Our budget was around 200, 250 million board feet a year. We were dumping 100 to 140 truckloads a day.

Jack Bell

We developed China Creek, where the marina is now. We put a dump in there, and we logged all the China bottom with truck, all that where you got the Cameron Camp now. All that was truck logged, and we trucked it down and dumped it in China Creek. We had a truck show going there in 1946. Just that lower valley, we started a truck show, we were the first big truck show in the valley. We were one of the first to do tractor logging in the valley, with the old RD8 Caterpillars, with the old AC arch on behind. They came in in 1936 and we started tractor logging patches and pulling down to the rail. Isolated patches. And we'd have an old little steam pot sit at the track with a heel boom to load. Then eventually we got so we'd go out a mile, a mile and a half with the tractors, so it got a pretty good show going. And that was the first Cat show. The first trucks we had was the Mack trucks. And then when we bought new trucks we changed and bought Hayes. We still had the cold deckers, you see, so we used the cold deckers at roadside. There was quite a change in the yarding distance, too, because under the old railroad system, with the cold deckers, we used to go in 1000, 1200 feet. It cut down on the grading costs. Because with the old railroad, nine percent was the steepest we could go on, and that took one of the ninety-ton Shays and all he could take up was ten empties. So most of the grade anywhere was from five to seven percent.

Then we switched to rod engines for speed. As we got out farther, the old Shays took so long, so we switched to the mainline, or rod engines. They could take a maximum four percent grade, and twenty-four-degree curve. The first railroad that we had from Camp A to the first Camp B, we were using Shays then, see, because we built that railroad. But this railroad was built by the Canadian National, that was all laid out for rod engines. So that's why we switched to the rod engine mainliner because, the old Shays, the best they could go was maybe eight, ten miles an hour. We dropped down to 600 feet yarding distance when we went to trucks. We started a big truck show down on the Nitinat, too. See, when we were down there, what we did, we transferred. We had trucks logging the Nitinat, and bringing them down to the end of the railroad there, we had a what we called a transfer system. And you picked the whole load off the truck and swung and put it on rail cars. In '52. The trucks were a lot cheaper. What was killing us here, what made the big decision, was that we had twenty-seven bridges from our log dump here and these bridges here were all built starting in 1940, and they all started to deteriorate. They would all have to be replaced, and that was a tremendous cost to replace all those bridges. We had a bad winter, real heavy rains, and we lost a couple of bridges. So they said that's enough of that. A lot of those bridges were a hundred feet high, two or three hundred feet long.

Jack Bell

Combining trucks with railways required that logs be reloaded from trucks to rail cars, done here with a Hayrack boom at BC Forest Products' Harris Creek operation near Port San Juan in the early 1950s. (BCARS 82152)

With the advent of truck logging only two companies, Comox Logging at Ladysmith and Canadian Forest Products at Nimpkish, converted from steam to diesel locomotives. Canfor's #301, built in 1956, was a General Motors Diesel switcher engine modified for logging road use. (BCARS 81743)

the 1950s: MacMillan Bloedel, Crown Zellerbach, BC Forest Products and Canadian Forest Products. The first three began their conversion to trucks in the 1950s. MacMillan Bloedel and BC Forest Products went from steam railway to truck operations in one step, while Crown Zellerbach used diesel locomotives at its Ladysmith operation until 1984. Canadian Forest Products' Nimpkish operation was the only logging company in BC to retain its railway, which converted to diesel beginning in 1951. The company's first foray into the world of diesel-electric locomotives involved the experimental conversion of two steam locies, a Shay and a Climax. The first rebuild was done by Tyee Machinery at Granville Island; the company did the second at its Nimpkish shop. They were not an enormous success, and after that time Canfor used conventional production-line diesel locies.

The end of steam railway logging at MacMillan Bloedel did not come until December 1969, when the line between Nanaimo Lakes and Chemainus was closed. BC Forest Products had completed its conversion to trucks in 1957, when it closed the last steam rail lines at Nitinat on Cowichan Lake and the Bear Creek camp with its line to Port Renfrew. The last of the small independent railway logging operations, Hillcrest Lumber's line to the mills at Mesachie Lake and Honeymoon Bay, closed in the late 1950s, although the company kept its two Climax locies working in the mill yards until 1968. When the mills closed, Hillcrest's locies were given to the British Columbia Forest Museum, where they still run.

The rapid abandonment of steam-powered railway logging in BC threw onto an already dying market scores of locomotives and other rail equipment that no one wanted. Some of the newer and better maintained locies found homes in other industrial applications. Largely due to the efforts of former railway inspector Bob Swanson, retired logger Gerry Wellburn and others with a nostalgic fondness for an obsolete technology, a few were preserved in museums. But most were dismembered with cutting torches at the beach camps and log dumps to which they had once hauled logs, and sold for scrap.

The abandonment of the rail lines had a significant effect on other phases of logging. With the move out of the valley bottoms and into higher, rougher terrain, a demand arose for more mobile, more powerful yarding machines. With new improved diesel engines and vast amounts of surplus military machinery available after the war, many new-model yarding machines were produced during the 1950s. Between them, the three major BC yarder manufacturers—Hayes–Lawrence in Vancouver, Murdie in Victoria and Westminster in New Westminster—produced about twenty different models of yarding donkeys by 1950. Skagit, just across the border in Sedro Woolley, provided a dozen different models. All these machines were gas or diesel powered.

The late forties and early fifties saw a thriving business converting steam donkeys to diesel power. In 1956, for instance, Tyee Machinery repowered a mammoth Willamette Iron Works steam skidder with a 400-horsepower Cummins diesel. To increase mobility, it mounted the frame on fifteen large steel wheels so it could be towed along logging roads and used, as before, at a wooden spar. At forty-five feet long and 110 tons, the skidder had limited mobility, but Alaska Pine, which took the unit to Mahatta River, felt it was worth the cost of conversion because of lower operating costs and reduced fire hazard.

The change in power source facilitated the adoption of better controls. As one observer noted: "Gone are the unhappy days when the donkey operator could be seen standing on a front drum foot brake with one foot, doing a

After the end of the war we went down to Port Angeles and bought what they called a quad, which was four General Motors 671 diesel engines arranged around a gearbox. It was used as a power unit for a tank landing ship. Military surplus sold these things in the States by the dozens. We brought these four engines back and proceeded to build up machines using these engines. Not as a quad, but using the engines separately.

It got a little complicated. If you can picture four engines arranged around one gearbox. Each of these engines had a blower on the side of it. So you had two engines with a right-hand rotation. One has a blower on this side, the other one has a blower on the opposite side. The other two engines, at the back, have a left-hand rotation, and their blowers are opposite too. So none of those engines worked the same, they were all orphans. We had to take the left-hand rotation ones and put in new camshafts and lord knows what else to convert them to right-hand rotation. It turned out to be a lot more expensive than we ever expected. But it worked.

We bought an old steam donkey that was built in 1929. It was a good old machine and it was big. We had to build a sled for it. We got C.C. Murdie to build us an engine bed and a drive. We logged with that machine for quite a few years, until the mid-sixties. It was very successful. We eventually put it onto a steel spar and then we were really modern.

Frank Norrie

split to get the other foot on a clutch pedal while he hung over a revolving drum to yank at a gearshift lever and punched the throttle lever with an unused elbow. Today the paraphernalia of control are cleverly grouped where a man of any size can get at them without fatigue."

It was fairly simple to obtain machines with more power, but a different matter to build yarders that could be easily moved over ever-steeper logging roads and quickly rigged. The first stage was to mount the more compact yarding machines on trucks, trailers, sleds and surplus tank carriers. This made it easier to get the yarders to the spar trees. But it still took a day or more to raise and rig the spar. Since settings were getting smaller than they had been during the railway era, the hazardous business of raising a wooden tree had to be done more often. In the 1930s and 1940s, a typical railway setting might yield 20 million feet of timber; by the early 1960s settings of less than a million feet were being logged.

In October 1951, the first Skagit SJ8 arrived in BC and was taken to Elk River Timber for demonstration purposes. It was a mobile combination yarder-loader. Mounted on ten rubber tires, it could travel at twenty miles an hour. The machine could be set up and working in a few hours, and could yard logs at distances of 500 feet. It was powered by a 165-horsepower Waukesha gas engine. This was the beginning of a long line of Skagit yarding machines. Also appearing in 1951 was the Washington Track Loader. It, too, was advertised as a yarder-loader, but while it was not a great yarder, it became one of the most successful loading machines ever built.

Three years later, in December 1954, Chuck Madill and the Baikie brothers got together over a bottle of rye and roughed out the design of another revolutionary piece of yarding equipment. Within a few months, Madill delivered BC's first new-generation steel spar. The idea for such a spar was not new: they had been used before in BC on some of the early Lidgerwood skidders, and were mounted on rail cars. But this was quite a different machine. It consisted of a Skagit yarder mounted behind a lightweight ninety-foot steel spar that was raised hydraulically and had hydraulic-powered drums on its guylines. Apart from travel time, it took only two to three hours to lower, move and raise the spar. During the summer of 1955, Madill made a second steel spar for Vanwest Logging, with the use of fairleads instead of the banjo-style head used on the Baikie machine. It also had a Skagit yarder, powered with a Ford V-8 gas engine. Largely because of the fairleads, it was a much better machine than the first model.

Coastal loggers, particularly the small independents and contractors able to persuade the bigger companies to let them try it, beat a path to the Madill shop in Nanaimo where, over vast quantities of rum and whisky, more improvements and variations were devised. By 1958 Tom MacKenzie, a Royston contractor, had

A Caterpillar 953 track loader (background), equipped with log forks, at work in an unidentified location near Terrace, late 1960s. In the following decade these machines were replaced with Caterpillar 900 series wheeled loaders. (Northern Photo Service; Finning Tractor collection)

The old S-Js were pretty good yarders at the time. We had an S-J4 one time. It was something you had to sit right at the foot of the tree and, jeez, it was vulnerable if you're logging straight down the hill. Some turns would be coming down there and the guys would be running like hell. The leverman on the S-J would be running like hell trying to get away from the turn. We never had anybody hurt there, fortunately. Probably it wasn't good management but it was bullshit luck, that's all.

Al Hendrickson

three Madill spars. The first used a Ford-powered Lawrence 10-10 with an eighty-five-foot spar on a Hayes truck. The second model, mounted on a Kenworth, was ninety feet high and had a Clyde winch powered with a 200-horsepower Cummins. The third one, another ninety-footer, had a GM engine on a Tyee winch and was mounted on a sled so it could be used for cold-decking.

By the early 1960s, everyone was using or wanting steel spars. During 1963, the new Madill plant, a far cry from the blacksmith shop that Chuck's father, Sam, had opened to serve early horse loggers around Nanaimo, was turning out a fully equipped steel spar every three days. They were the world's biggest makers of steel spars. One indication of their popularity was that MacMillan Bloedel's Franklin River Division had sixteen Madill spars by 1969.

Naturally, such success induced others to develop similar machines. In 1956, the Berger Engineering Company in Seattle had a steel spar built at Vancouver Iron & Engineering Works that it sold Columbia Cellulose for use at Terrace. Called the Porta-Tower, it cost $18,000. It was a sled-mounted machine with a ninety-

About ten days before Christmas in 1954 I was up on the hill helping the rigger raise and rig a spar tree. It was one of those miserable days not fit for man or beast. I had been up the tree hanging blocks and threading lines in a howling snowstorm. It was very cold. When I finished I was as cold and miserable as the weather. I got in the pickup and headed for the office in Campbell River.

When I arrived there was a real Christmas party in full swing. Chuck Madill and his top salesman, Dave Russell, were on their annual trip to spread some good will, cheer and Christmas spirit around. There were Wallace, Jack, Pat Martin, Jim Luckhurst and anyone else who happened to poke their heads in the office. By the time I arrived they were all in a talkative mood. I had some catching up to do. That was not easy as I was wet, cold, hungry and suffering a bad batch of being cantankerous. Chuck Madill, in his usual joyful manner, undertook to change my foul mood by pouring me a very big shot of rye. He also proceeded to give me hell for being out in a snowstorm raising a tree in the first place. My answer was that neither he or any other wise machinery outfit had come up with anything better, such as a steel spar or some other contrivance which could be raised in an hour and used from one setting to another.

After a few more drinks we had the paper out and were drawing all sorts of mysterious designs. Everyone got in the act with different ideas. Finally we agreed on one design which we thought might fill the bill. Chuck seemed very interested and asked us to leave it with him for a spell.

He phoned two weeks later and said he had talked the matter over with his brother, Norman. They had decided to try and build something to suit our needs. They needed a set of drums or a donkey and an old truck. We had a big old six-wheel drive White which had been used as a tank retriever during the war and a good Skagit donkey with no sleigh. We loaded the donkey on the truck and sent them down to Nanaimo.

Three months later Chuck phoned to say they had a steel spar ready and wanted to bring it up. They arrived with this monstrosity and we gave it a try. It was not the answer but it was a hell of a good start in the right direction. As usual, we ran into resistance from the crew. We overcame that problem by putting an old engineer who did not mind where he was on the donkey, and I ended up hooking along with a couple of guys who did not realize what they were working on.

This steel spar had a big banjo-type head to hang the blocks on. This meant when you got around the quarter lead the lines would rub on the frames, and unless your guylines were all the same length and tightened to the same tension the banjo head would crack and warp. When this happened you had to have a welder there to mend all the cracks. With all its faults we persevered with the new spar tree all summer.

Around the first of November Chuck, who had been watching the performance all summer, came up and told us to send the whole outfit back to them. During the summer they had built two more steel spars that were a great improvement on their first effort. The new ones had a swivel top on the spar, so no matter which way you were yarding your lines were facing the right way. The guyline winches had hydraulic power both ways. The system of raising it was much better. There were numerous other changes. All in all it was a pretty good outfit, as were all Madill's spars for the next twenty years.

When we got the improved version back in the woods, everybody wanted to work on it. It put an end to rigging spar trees in a snowstorm, in fact it put an end to rigging spar trees period.

In a couple of years we had two more of these steel spars. A little bit later we got a TD25 International tractor. We mounted a triple-drum yarder and had Madills build a steel spar on it. This proved to be a very handy outfit. If there was a small patch of timber where a Cat road could be made, this machine could climb up there, be rigged and ready to go in about half an hour.

Harper Baikie

When the steel spars first came in everyone thought they were too short. They used to refer to them as tin stumps. They had all sorts of disparaging names for them. They eventually caught on as they became more mobile. They went from track machines to rubber mount. The first ones were mounted on trucks. Then a lot of them were on these tank carriers they used during the Second World War. Butler Brothers were quite into developing mobile spars. They had combined some trees with some of their machines and mounted them on big trucks, some of them tank carriers and some of them logging trucks. There was a lot of people had home-built machines.

Monty Mosher

foot tower that was towed to site by a crawler tractor equipped with a double-drum winch. With the sled in position, the tractor was backed onto it and bolted down, with the winch used for yarding. The tower was raised with heel blocks and the guylines were tightened with a winch-equipped McCulloch chain saw.

A similar machine was produced the same year by Burrard Dry Dock in North Vancouver. It used a RD8 Caterpillar with a double-drum Hyster winch that raised the ninety-foot tower, then performed the yarding tasks. This unit was on a frame resting on about fifty steel railway wheels.

The prolifically inventive Frank Lawrence built a number of steel spars, called Sparmatics, mounted on rubber wheels. They were three-drum slackline machines powered with 275 Cummins diesels. Their spars were telescoping, which provided a more mobile machine in tight quarters. Among others, George Percy, one of the major coastal contractors, was a great fan of Sparmatics and owned several of them. In the 1960s, Vancouver Iron & Engineering Works built Sparmatic telescoping spars in several lengths up to 120 feet. Tyee Machinery also produced a G-series steel spar in the 1950s which moved on a sled and sat on the ground when it was raised, separate from the yarding donkey.

During the 1950s, some of the logging companies that still had the large machine shops required during the railway era built their own steel spars. BC Forest Products built one at the Nitinat shop using a 165 Cummins on a Skagit winch, working off a ninety-two-foot tower. It was mounted on steel wheels and towed by a Cat. Master mechanic Ross Ray and engineer Frank Moran worked out the design on a scale model built for woods foremen and logging superintendents to use in planning sessions.

In 1957, Crown Zellerbach built three forty-five-foot truck-mounted steel spars to use for pre-logging small timber at Ladysmith so it would not be smashed when larger trees were felled on it. These machines used Ford V-8-powered Skagit yarders. By 1961, typical of the industry, CZ was logging only twenty-five percent of its timber with wooden spars.

In 1961, Skagit tried an experimental truss-type leaning spar to windrow logs at Sproat Lake. This was the precursor of the PT-4Y tower yarder. But the company also pursued a different type of machine. In 1964 Skagit came out with the SY-2, a mobile yarder with a single guyline mounted above the centre of a turntable which allowed the machine to swing logs into a convenient position for loading. A further development was the SJ4R-GT in 1965, a combination yarder-loader which did not require guylines while loading and stabilized itself for yarding by dropping the head of the loading boom to the ground. The Skagit machines, with towers so short they hardly deserved the name, anticipated the grapple yarder, a machine that would sweep the market in the 1970s.

There were some other interesting innovations in yarding equipment during the 1950s.

I think the first steel spar built was in one of our [BC Forest Products] shops. We had two camps in Cowichan Lake, Camp 3 and Camp 6. Camp 3 was at the west end of the lake. It was a railway show and had a big shop where they did all the repair work. They built a steel spar up there. Madill built after that. They were certainly the first ones to market them. The company had a design engineer come over from Vancouver to help design this, his name was Frank Moran. He died just a few years after that.

We didn't get a Madill for a long time because we wanted a slackline system. Because it was steep country both at Caycuse and Renfrew you wanted to reach a long way from the main roads to save building roads up in the mountains. Madill only had a high-lead machine that could reach 600 feet or so. Then an outfit in Vancouver, Vancouver Iron and Engineering Works, came out with a slackline machine. It was called a Viewspar. We got a couple of those. Madill sort of lost out on slackline operations. I think they built a few slackline spars but they never did get into it very heavily.

Because of the steep ground we stuck to these slackline operations. At other places where there was flatter ground, like Northwest Bay, the Madill was ideal. They could build roads everywhere, yard 600 feet.

Ken Hallberg

One was the Wyssen skyline crane, a device that originated in Switzerland where it was used extensively in steep mountain logging. It consisted of a tight skyline up to a mile and a half long, down which a carriage lowered logs with the aid of a diesel engine on a single-drum winch. When the carriage was lowered to a predetermined position on the skyline, it released the logs, and was pulled back up for another turn. They did not yard uphill, but with a great deal of effort, or the assistance of a chain saw winch, the mainline could be hauled out 150 feet or more on each side of the skyline to hook logs.

When the Wyssens were first introduced, loggers discovered they were too light for heavy coastal timber. Vic Walters tried them out at two locations on Salmon Inlet and Sonora Island, but without great success. Heavier models were designed and tried in several places. Widgeon Creek Logging installed two at Pitt River in 1959, and in 1961 MacMillan Bloedel used one at Sproat Lake. Canadian Colliers used a Wyssen at the head of Indian Arm to lower logs 2700 feet over a one-mile horizontal distance. The five-man logging crew wore mountaineering boots to work on the steep bluffs and were trained by a crew of Swiss loggers. They lived in a small, portable camp which moved along the mountainside as logging progressed. But although it has been used occasionally ever since, the Wyssen system has never become an important part of coastal logging.

In about 1950, an Oregon logger, Phil Grabinski, developed a machine called the Skyhook, designed to log rough country. The machine was used briefly in BC. It had a large

---

They wanted a steel engineer at OB Logging just up here in Jervis, one of the last places I worked. I went up there and I was up there for four or five years I guess. I probably run all the steel spars but mostly what I run were Skagits. Telescoping. I liked them. When I was at Woss they still had two wooden trees going, and two steel spars. I was on the wooden tree but I did go fill in for the odd day. I was head rigger and maybe the hooker was off steel spar and I went in. These were high-lead machines. They never had a steel spar skidder. They tried it a little bit over here. They were pretty well all high-lead but they had slackline too. BC Forest Products had two Skagit slacklines and one that Vancouver Iron and Engineering Works built. They had one of them. There were quite a few of them around.

When I went over to Tahsis, they had a Skagit tower and I went on there. You see, I hooked on them a little bit, when you were raising and lowering, you watch everything and you tell the engineer what you want him to do. So I had experience on the Madills. I had no experience on Skagits, they were different, they were telescoping spars. You just stood the first half up and then you strung your guylines. The spar went up with heel blocks on the bottom end of it. That stood it straight up and then your lines—five, ten, three-quarter lines were holding it there. And then you could string your guylines when it was standing up. It was a lot easier out there than putting up a Madill.

And then the Skagit drums were freewheeling. They had a light brake on them, and you left that on and then you skinned them off the drums with the strawline as fast as you wanted to. Whereas the Madill, you had to wait for that old slow hydraulic thing all the time. Skagits were a lot faster to rig up when the crew got used to them. When I went over there, I told them I'd never run one of these. They sent someone to help me with the first two or three moves. There wasn't that much difference.

I never worked on the (steel spar) slacklines. I worked on slacklines, but with the wood trees. Apparently they were the same thing, basically. Just the rig-up time, that's all they saved. They lost a little bit too, in length. You see, an average wooden tree was about 150 feet to the bull block. The longest of these Skagits was 110, I think. Maybe 120.

The VIEW one, the Sparmatic was a little more cumbersome than the others. You had to put buckle guys out. You put the first half up and then you had to put buckle guys. You had three seven-eighths guylines to hold the first half and then you pushed her up. Or you could string your guylines and then shove her up. It was big and heavy and cumbersome. I only stood one up twice. The Tahsis Company had one when I worked for them. They were mounted on rubber. Four big high wheels. I never worked on it long enough to really get used to it. Seems to me compared to the Skagit they were . . . the Skagit was so good. You only had a hundred foot, so your tower was only fifty feet long. With a Madill, you got that big long end sticking out. How do you go around the corner when there is a big bank there or something? You got to put the tower halfway up and then teeter it around the corner with the tower halfway up to get around it. With the Skagit you never had any trouble.

I worked around the towers from 1966 until I retired in 1978. I was sixty-six years old. I retired but then in the summertime I made four trips up to the Queen Charlottes for Beban and two up Jervis Inlet here.

Cliff Hallberg

A unique mobile unloading crane, built for BC Forest Products by Hayes, working at the Port Renfrew dryland sort, late 1960s. (BCFM 4–2)

carriage, hung from twin skylines by four sets of twin blocks and powered by a small engine. The operator rode on the carriage, several hundred feet in the air. After he raised the logs, the carriage travelled along the skylines to the dump, where he lowered and released them. The carriage was also fitted out with rubber tires so it was mobile on roads. Empire Mills used one of these machines at Squamish, and Kelsey–Johnson Logging used one at Ruby Creek, near Harrison.

A more successful device was the Skycar, which travelled on a two-inch skyline but, mercifully, was radio controlled from the ground. This machine was built by Skagit, and in 1960 Crown Zellerbach used one on the Cruickshank River near Courtenay. The Skycar was powered by a 100-horsepower GM diesel which raised the logs off the ground and held them while the carriage was lowered down the skyline by a more powerful snubbing engine.

Phil Grabinski's Skyhook, used in several locations in BC in the 1950s, as it appeared on the cover of the Pacific Logging Congress Annual Report 1983. (Viv Williams collection)

Up to now you had the wooden tree with the hayracks and the Maclean booms, straight duplexes and all the rest of that stuff from the steam era. But now the wooden tree is gone so you have to have some kind of a loading machine. This is when they started taking shovels, the Model 6 or the Americans, and putting on these booms. The Bohemian boom was developed by Bohemia Lumber in Oregon, it was a gooseneck boom. Then they ran the drums out through blocks and had tongs on the end of them. They were pretty jerky and hard to work with, but they swung and with the Bohemian boom you could heel the logs and get them onto the trucks. Then they developed an air tong that went on the end of the boom of a road building shovel. There was a tong on the end of the stick instead of a bucket, a tong actuated by air and then you heeled the log to put it on the truck.

The trucks weren't all that big then. You're talking about ten and twelve-foot bunks, smaller Kenworths, Hayes trucks, so you didn't have to reach quite as high. That evolved into bigger trucks and raised cabs. Washington track loaders came out with a high cab on them, with fast-moving tracks. They still had tongs, but they put in extra drums so you could pull the tongs out in the woods and do a little yarding. If your logs were downhill you could take the strawline out and tie it to a tree and slide the tong down and cherry-pick 100 or 150 feet off the road. That wasn't a new concept either, because they used to drag the tongs out in the woods with the steam cranes.

They had the big steam cherry pickers that swung on the railroad. The standard steam cherry picker used on the railroads during the thirties was a steam machine with an A-frame built on a bull car with a couple of tongs hanging from it. They didn't go out and pre-log the settings like we refer to cherry picking now. They used to go up and down the railroad with that thing picking up the logs that came off the rail cars. They never had binders or high stakes, just cheeseblocks and the loader had to build the loads so the logs would stay on. But of course they didn't so they had these machines that every once in a while started at the back end and loaded all the spilled logs onto skeleton cars and shipped them off to the beach. It was a natural offshoot to go to the diesel shovels and put booms on them with straight tongs.

When Washington came out with a track loader that could spin really quickly, they used to get those things and a good operator could get that tong swinging and throw it 100 or 150 feet downhill. We'd be waiting downhill. You'd stand behind a tree or something and he'd fire that tong down to you. It'd clatter and bang down to you and you'd go and pick it up, put it on the log and he'd drag it up, heel it and put it on the truck.

They could really fire those tongs out. One guy who used to be really good on it was Gordy Margetish. He worked out of Camp 6 and he and I were sent over with another fellow to Camp 3 and we took the track loader to work with. You'd stand down there and wave at him to show him where you wanted the tong, then run and hide. He'd fire that tong down and you didn't want to be in the way when it got there. But it was a lot better than running all the way up the hill, grabbing and running all the way back down again.

Those Mondays when the loader operator had too much to drink on the weekend was when we young guys got a chance to learn to run the machine. Loader operators were like that. I used to work with a guy called Tuesday Charlie Ross. He was a duplex operator on the railroad at Copper Canyon and eventually became a dump operator at Honeymoon Bay. It was pretty hard working with him on Mondays so we'd leave him sit and somebody else would run levers. It was a good thing; we got to learn how to run these machines too. I set a family record once, doing just that. Tuesday Charlie Ross was not feeling too good so he told me it was time I learned how to run levers. This was on the Mosher side at Copper Canyon. My Uncle Luke was running the yarder. My cousin Bill was chasing and my brother and I were second loading, all in one landing. So Charlie said I should learn to run the levers so I got up there and did. Those duplexes are really touchy, and really powerful. Those logs get swinging and you sort of lose control of them. They not only get swinging sideways, but swinging endo. I got this one swinging endo and was going to drop it, but instead of dropping it I pulled the lever the wrong way and all I did was speed it up. I hit the peaker on the loaded car, and that peaker went down and hit the peaker on the next car and knocked it off. My uncle, who was driving the yarder, shut everything down and came up and patted me on the shoulder and said: "Young fellow"—he always called us young fellows—"I do believe you've just set the Mosher record. To my knowledge no Mosher has ever, either in Canada or the US of A, has ever got the peaker two cars down."

Monty Mosher

A grapple loader with a Hayes truck at CIPA Industries' Rennell Sound operation on the Queen Charlotte Islands. (Truck Logger)

The advent of smaller settings and faster-paced yarding produced a need for independent loading systems to replace the guyline and boom loaders used with wooden spars. The development of steel spars spelled the end of these earlier loading devices, although Madill briefly tried one steel spar in 1956 with a thirty six foot loading boom. One of the earliest solutions to the loading problem was to modify the tracked shovels used for road building. Heel booms and tongs were added in various configurations and, although slow and somewhat cumbersome, they did the job. As turntables were added to shovels, operators discovered they were able to swing the booms and throw the tongs.

In the late 1950s, log grapples were designed that did not damage the valuable clear outer wood on logs nearly as much as tongs did. Pioneer grapple users such as Cattermole Trethewey found they were useful in taking logs out of cold decks or loading out of the water. In 1959, that company and several others used grapples to great advantage loading logs out of Buttle Lake, which was being flooded by a hydroelectric dam.

In 1965, Skagit came out with the Model 555 loader equipped with a drum to operate a tag line, which was attached to a tree or stump some distance from the loader. The grapple unit was lowered down the line and dropped onto a log. This was a rubber tire-mounted machine, whose compact design provided good mobility. A single loader of this type could be used to load logs from two yarders working some distance apart. Like yarding machines being developed around the same time, this loader was moving Skagit closer to the development of the grapple yarder. Along with most other loading machines of this period, the Skagit loader featured a raised cab, which enabled the operator to build higher loads on the logging trucks.

Viv Williams's LeTourneau electric arch skidding spruce in the Queen Charlotte Islands, 1957. (Viv Williams collection)

By 1956 we were getting farther back on the claim all the time and we were just beating the Cats to death because it was all big rock. There's so much overburden in that country that when you went up a road, even to get it built you had to mud it out. You'd be down from anything from two to ten feet. There was so much rock, a set of tracks would last only about three months. It was costing more than we got out of Northern Pulp for logging.

A fella told me about this rubber-tired machine (a LeTourneau electric arch) working up Indian Arm. He told me, "Viv, if you could keep that thing working it would do the job." Ronnie Krug was a young fella then working for me, and we went down to Vancouver. Russ Guest, a salesman for Logan Mayhew, took us up Indian Arm and fired this thing up and showed us what it would do. My God, I'd never seen anything like it. So I ordered one. Had it shipped to the Charlottes. That just about upset the whole logging industry, that a contractor, of all people, would buy a machine like that and ship on a freighter to the Charlottes.

I told Ronnie, if you can put up with it we'll make it run. I knew if anything happened to it we had to live with it and fix it. We got it up there and within three or four days, on a trip back to the woods, it just stopped, just like that.

We were going up a creek bed and there it was. We hadn't had a chance to get to know the machine yet. We had a big parts bible, I always insisted on that. We got the book out and went through it page by page, trying to figure out where the trouble was. It was diesel electric and we weren't too familiar with all this electric equipment. We got into the book a little ways, and had to laugh. In one place it said if you have loud reports and black smoke you have trouble in that area. We didn't have any black smoke and there hadn't been any loud reports. We phoned the serviceman and he told us what the problem was. That was the only time we had to call in a serviceman.

The weather didn't affect them. These were the LeTourneaus that were built after LeTourneau sold out to Westinghouse. He couldn't build dirt-moving equipment after he sold the first outfit. So he went on to build logging equipment, using larger and bigger motors.

Because of the height of the fairlead we could take more than double the amount of logs as a Cat. With the rubber you could travel so fast. It done the work of three Cats, for a mile or a mile and a quarter back. You couldn't possibly use Cats for that. It would do anywhere from nine to fourteen miles an hour, a Cat would only do three or four. It made me nothing but money. It was the ninth one in Canada.

Viv Williams

War surplus trucks like this six-wheel-drive model, possibly at work in the Interior, were precursors of the rubber-tired skidder that became popular in the late 1950s. (BCFM, Box 2)

During the 1950s, one other major piece of logging equipment was developed that had a lasting effect on the industry. Before World War Two, Cat logging had evolved as the most efficient means of skidding logs over limited distances on suitable ground. However, crawler tractors were slow and did not stand up to continued use on rocky ground. The earliest attempt to mount a tractor on rubber tires was made by the ubiquitous equipment designer, Robert LeTourneau. The unsuitability of diesel electric equipment in the Pacific Northwest, combined with loggers' resistance to this kind of power source, restricted the use of the LeTourneau skidder. Only a few operators with mechanical ability and intellectual curiosity actually liked the LeTourneau machines. However, they were fast, lasted well on rough ground, and provided a lot more clearance than tracked machines. A similar machine, called a Tournaskidder, was sold by Westinghouse after it bought the LeTourneau company.

After the war, with diesel engines and good tires available, there were several attempts to build a rubber-tired machine for skidding logs over rough, rocky ground at a faster speed than a crawler tractor. One of the initial attempts in BC was made by Bill Long in Vancouver. Long added winches and arches to the deck of four- and six-wheel-drive army surplus trucks. They worked fine on a decent road, but were useless off the road and in tight quarters.

Caterpillar was a global supplier of heavy equipment in the early 1950s when it tried adapting some of its other tractors to the task of log skidding, with little success. MacMillan Bloedel tried out a DW20 with an added integral logging arch, but it did not perform well. By 1951, Paul Westfall, a Portland designer, had spent five years working with scale models to come up with a skidder that had a long wheelbase for traction, an integral arch to locate some of the load over the driving wheels, and a power and drivetrain combining high power and speed. His Performer was driven by a 225-horsepower Cummins diesel. It used big, wide tires and a Carco logging winch. His objective was to build a machine of components familiar to loggers so they could maintain and repair it themselves.

Dumping cordwood at the Powell River Company log pond. It was cut in an experimental commercial thinning operation near Powell River in 1949. (PRM 3243A)

In 1954 another Portland manufacturer, Wagner Tractor, sold a skidder called the Loggermobile which carried most of the features of the Performer, but also had an articulated frame that swivelled in the middle for steering. Although this was a key feature of successful skidder design, it took a while to catch on. The major difficulty with the Loggermobile was to keep it running. Several were used on the coast, but they were plagued with breakdowns.

The new skidders were not really suited to skidding big logs, but this did not stop loggers from using them. They made it possible to move logs quickly a mile and a half from a landing to the dump, and to return for another turn even more quickly, with little or no road building. And the skidder was a reasonably priced machine that could handle forty percent slopes.

The biggest problem with the early models was braking. Skidders had the capacity to pick up three or four tree-length logs and move them downhill to the beach. But once a skid trail got

Freil Lake Logging's Kenworth 300 skidder pulled this 22,000-board-foot spruce log to the beach in 1961. It was used as a stiff-leg for an A-frame in Ououkinsh Inlet. (Rudy Deering collection)

worn in, and especially with a bit of rain to lubricate it, these machines picked up speed and would soon be hurtling toward the salt chuck at fifteen or twenty miles an hour. With their big tires they would be bouncing and leaping about, the turn of logs thrashing along behind. Available braking technology was simply not capable of controlling the size of turns the machines could skid. When the situation got too dicey, the driver had to bail out, a tricky proposition with three or four logs flying and bouncing behind. The best bet was to wait for a hole to show up alongside the skid road and jump into it, letting the logs fly overhead. The skidder was then retrieved and repaired, usually without a great deal of difficulty.

Repeated trips down the same route created cuts ten or twenty feet deep in the hillside, churned out by the tires and swept down by the logs and rain. Once a driver was headed down one of these chutes, there was no stopping or jumping. He had to ride it out to the bottom, then make his escape. Rudy Deering of Freil Lake Logging, a West Coast skidder pioneer, had one site in Ououkinsh Inlet where a runaway skidder came through a long cut at about twenty miles an hour, across the flat and into the bullpen, where it sank out of sight. The winch line and chokers were still attached to the logs, the ends of which would float around like deadheads. They used another skidder to fish out the logs, with the drowned skidder on the end.

Deering and his uncles Emil, Ed and Herman, who owned Grandview Logging at Port Renfrew, were the first on the coast to log seriously with skidders. The Deerings came to BC in 1941, after logging and milling in Saskatchewan and Manitoba, where they used a six-by-six truck converted to a skidder. Grandview bought the first Tournaskidder in BC and skidded more than a million feet a month over a one-mile haul. When Rudy's first attempt at retirement ended

---

I did the first bundling. It was a unique situation. The reason for it was that just above Skidegate village is Miller Creek. Northern Pulp owned all the timber north of the reserve. There is no protected water here, it looks straight at Sandspit into the southeast winds. So when Northern Pulp was going to start logging I was interested in the falling contract. I asked them for the contract so they said sure. I took in three sets of fallers, and they were going to log it themselves. They had logged a lot around the inlet with Cats where they just yarded the logs out on the beach at low tide, dropped them and ran a set of boomsticks around them. At high water they towed them away to the bay. Well, they tried that on the east coast but with the extreme tides, the current going into Alliford Bay, and of course there's always a swell. Well, in two and a half months I guess they only got about 150,000 feet into Alliford Bay. Every time the swells would drive the logs back up on the beach.

Hugh Hodgkins was the timber man. He came along one day, up to me and George Fife, and naturally they weren't getting any logs out, so he says "Viv, how about taking the Cats and taking the whole show on?" And I said, "Do you think I'm crazy?" He knew I'd spent all my life on Cats. But I was happy doing the falling. He said, "Well, that's the situation, if we can't get the logs out we don't need a falling contractor." So, with that they talked me into it.

They supplied the equipment and I took on the contract. My idea of how to get the logs off was to take the logs down on the beach and make bundles. We laid cables down and with two Cats, one on each side, pushed the logs together into a bundle. We got logs off that way and found out the bundles would stay together. Then I got the idea of a skidway. We got a bunch of long, long spruce timbers and built a skidway up the beach to dry land. We could build the bundles there. We had a rig with stakes we could fill with logs, put a cable around them and trip the stakes. Then we put a Cat on each side and shoved them out on the skidway at low tide. Then we strung a master tow line over the bundles and used a choker to attach each one. We built a special shallow draft tug for use on the beach, fairly powerful, and the skipper would come in with the boat and throw his anchor overboard. At low tide our boys would take the tow line out and put a bobber on it. When the tide came in he'd pick it up and start to pull. For ten years and one week I ran that claim and we never put the tow in at the same time twice. It was all according to the tides. You'd get up at two o'clock in the morning, high tide, low tide, you worked it.

We used to get 100- to 125-mile-an-hour winds there. The spruce skid logs would float and the waves danced them up and down 'til they were chewed off. I had to do something else. I knew the system was good but I had to do something else. I had been in a junkyard in Vancouver and saw a bunch of paravane booms off sub chasers from the war, pipes. My bright idea was to buy up all Active had in their yard, get them on a freighter going north and lay them out on the beach. They worked fine. We used it from then on.

We used Cats, had three of them. You'd always have more than one Cat working on the beach, in case one got stuck.

Viv Williams

Above: An off-highway Kenworth at an unidentified dryland sort, early 1970s. After the logs are unloaded, the pup trailer is loaded onto the bunks for the back haul, below. (Ed Pryor photo)

in 1961 and he found himself flat broke, he got a contract from the Tahsis Company and cobbled together an operation based around a 10-10 Lawrence yarder and a rubber-tired skidder. After trying and giving up on just about every make then available, he bought a Kenworth, built by the truck company in Seattle. Liking it, he began to work with the Kenworth people to improve it for West Coast conditions. The breakthrough came with the use of the Allison brake, which worked off the transmission, turning it and the engine into a part of the braking system. He got the first Model 300 off the assembly line in Seattle. Now it was possible to control the turns of big spruce Friell Lake was logging.

Kenworth eventually began building skidders at its Vancouver truck plant, but by then West Coast skidder logging was in decline. There was a limited amount of timber available within the

skidding limit of one to one and a half miles. When it was gone, it was necessary to bring in trucks. Although skidders were still used on the coast, their future lay elsewhere, where the trees were smaller and the conditions easier.

Timberland Machines, an Ontario company founded in 1947, began to adapt western-designed machines for loggers in the East. In 1956, it introduced the Timberskidder, advertising it as a replacement for horses. It was not until 1961 that the company developed an articulated skidder called the Timberjack, a machine that soon set the world standard for rubber-tired skidders.

The old established equipment company John Deere developed an experimental skidder in 1958 which, although it was never sold, paved the way for its Model 440 skidder in 1965, propelling the company into logging equipment in a big way. There were several other makes to appear during the 1950s and 1960s on the BC coast. Canadian Car built the Tree Farmer at Fort William and it was sold in BC by Finning. Clark and the BC-made Timber Toter were sold in the 1960s.

These skidders were another boon to small independent loggers like Deering. With one piece of equipment, he could log from the stump right to the dump, without having to build much in the way of roads. Alternatively, skidders were used to skid logs from a spar landing to the dump. In Deering's case, he used skidders to get on his feet so that when the time came he was able to acquire trucks and all the other equipment—road building machinery, loaders, dumps—he required.

A typical merging of old and new technologies in the late 1960s was Pedneault Brothers' use of a Kenworth skidder to haul logs from one of the last wooden spars used on the coast to the dump in Thurston Bay, about a mile away. Through the sixties and seventies, skidders were essential to scores of small loggers running similar operations.

By 1970, coastal logging technology was transformed. There was no steam machinery left, trains were all but gone, wooden spars were a thing of the past, and the surviving hand saws were stored in basements and attics for occasional use in bucking competitions at loggers' sports days. The postwar years had been a period of immense and rapid change. There was even more to come.

An early, camp-made boom boat in action at O'Brien Logging's Stillwater dump, 1954. (PRM 1665B)

Wide, high-flotation tires, as used on this Timberjack, were developed for use on skidders in the 1980s to reduce site damage and allow them to operate in soft ground conditions. (FERIC)

IX Modern Logging

Previous page: The BC Forest Products log dump at Wakeman Sound, 1978. (Harry Foster photo)

After 1970, the pace of change in the coastal forest industry slowed. With the exception of some minor dips in the market, the seventies were a decade of steady, sustained growth and consolidation. With a few exceptions, there were no major technical alterations in the machinery used for logging. Instead, there was a steady refinement and improvement of the equipment introduced during the twenty-five years of rapid technical change following the war.

Likewise, the economic and political setting in which the logging industry functioned continued to change, but at a slower pace. In 1972, the first change of provincial government in more than twenty years led to a fourth royal commission on forests. Among the many observations in his report, delivered to yet another new government in 1976, Peter Pearse directly addressed the issue of corporate concentration and control of cutting rights: "In my opinion the continuing consolidation of the industry, and especially the rights to Crown timber, into a handful of large corporations is a matter of urgent public concern." Pearse, a forest economist at the University of British Columbia, went on to recommend a series of policy changes aimed at restoring some balance and diversity among various sized firms in the industry.

For the most part these warnings and recommendations were ignored, and the new Forest Act passed in 1978 contained numerous changes consolidating the position of the large integrated companies. Although the government introduced a Small Business Enterprise program aimed at independent logging and forest management operators, its intended objectives were far from realized. In the late 1980s, the government announced its intention to implement a recommendation of the Council of Forest Industries and transfer almost two-thirds of the commercial timber land into TFLs, to be held by the large companies. Public response to this announcement forced the appointment of a Forest Resources Commission to examine and report on a series of policy issues. This commission issued a report in 1991 which showed that, since Peter Pearse had issued his warning, the five largest corporations had increased their control of cutting rights from forty-one to fifty percent, while the ten largest companies had increased their stake from fifty-nine to sixty-nine percent. A mere twenty companies controlled eighty-six percent of the provincial timber harvest in 1990. The recommendations of the FRC for major changes in the distribution of cutting rights appear to have been ignored by both the Social Credit government in power when the report was released, and the New Democratic Party government which replaced it a few months later.

The most obvious trend in the 1970–1990 period was the transformation of big companies into even bigger ones. Crown Zellerbach and BC Forest Products were merged into Fletcher Challenge, whose owners were new corporate players from New Zealand. Control of MacMillan Bloedel was acquired by Noranda Forest, which in turn is owned by an elusive series of corporate entities, ultimately ending up with Montreal's Bronfman family. The Tahsis Company became part of the Canadian Pacific corporate group, and along with Pacific Logging was folded into Canadian Pacific Forest Products. Columbia Cellulose was bought by the government and, after reappearing under a series of identities, was being carved up and sold to several smaller operators. Canadian Forest Products remained intact, with a member of the founding Bentley family still in charge, but much of the ownership and control is now in Japanese hands.

Two large companies emerged during this period which bucked that aspect of the trend toward absentee ownership. Doman Industries, a Duncan-based family company, acquired full control of what had been Western Forest Products, including two pulp mills. Whonnock Industries, controlled by Vancouver's Sauder family, took over several small to medium-sized

A Kockums feller-skidder working at MacMillan Bloedel's Menzies Bay Division, 1989. (FERIC)

sawmill and logging operations and in 1989 transformed itself into International Forest Products (Interfor). In 1991, Interfor bought a large portion of Fletcher Challenge's cutting rights along with some sawmills, making it the second largest timber holding company on the coast next to MacMillan Bloedel.

Changes occurred in the independent logging sector as well. In 1990, there were about 1100 coastal firms registered in the Small Business Enterprise Program. They bid on more than 400 small timber sales, which together accounted for about twenty percent of the timber logged on public land that fiscal year. The major coastal companies that logged most of the coast's seventeen TFLs and seventy-eight Forest Licences, cut the other eighty percent.

But independent operators made some significant gains in the 1970s and 1980s, particularly those in the contracting sector. The 1978 legislative changes guaranteed them the right to log fifty percent of the timber cut on TFLs and Forest Licences. As well, after ten years of futile negotiation with the major companies, and almost as many lobbying the provincial government, they won the right to written logging contracts containing an arbitration provision. For some contract loggers, this ended decades of working under verbal agreements that were often violated by the constantly changing management personnel of the big corporations.

Bucking the trend, a few small loggers demonstrated that it was still possible, if difficult, to obtain cutting rights of their own and build an independent operation. Going into the 1990s, Pretty Timber was still logging its own timber at a number of locations in the upper end of the Fraser Valley. Its founder, Charles Pretty, had been in the timber business for more than eighty years and was still president at age 100. Cliff Coulson, the Port Alberni logger who started as a contractor and acquired his own quota, had a substantial operation logging a Forest Licence on Barkley Sound. With his son Wayne and daughter Darlene running the company, Coulson Forest Products became one of the coast's major helicopter logging companies and even began construction of a sawmill at the same time as the area's dominant corporation, MacMillan Bloedel, was in the process of systematically closing its mills. Husby Forest Products, another independent company, also owns part of a helicopter logging operation. And over the past few years many contractors have also secured small amounts of timber through the Small Business Enterprise Program. In what may be the reversal of a decades-long trend, Lineham Logging acquired cutting rights from the major company for which it was contracting, Canadian Pacific Forest Products, in 1992.

These concerns, along with many others that had preoccupied loggers in previous decades, were overshadowed in the late 1970s and through the eighties by widespread public opposition to logging. Coastal logging practices that had evolved during the railway era and continued after the conversion to truck logging—essentially, clear-cut logging—came under intense

When it was built, the Giesbrecht's 771 feller-buncher was the biggest in the world. Here it is shown working in second-growth timber near Port Alberni. (FERIC)

fire from a widespread environmental movement. The movement had its first major impact on the logging industry in the southern Queen Charlotte Islands where Beban Logging, under third-generation logger Frank Beban, was contracting for Western Forest Products. After an intense international campaign, logging opponents succeeded in having most of the area turned into a national park, with all logging excluded. This decision precipitated dozens of similar protests against logging at other locations on the coast, with some individuals and organizations calling for an end to all old-growth coastal logging.

By the late 1980s, when the Forest Resources Commission was receiving 1700 submissions, public concern about logging was widespread, if not always well informed. By this time the focus had shifted slightly; there was growing alarm at the declining number of jobs in the woods and the mills. The FRC made several recommendations aimed at resolving the growing forest resource conflicts, most of which are still awaiting action.

The result of all the corporate, social, economic and political changes that have occurred since the war is a great deal of uncertainty about the future of the industry. Which forests will be logged? By whom and how? These are unresolved questions. About the only certainty is that logging on the BC coast in the future will be much different than it has been in the past.

In some respects, the forces of change are just now beginning to affect logging technology. For most of the period since 1970, logging equipment and systems continued to evolve on the basis of the industry's internal needs and opportunities. The one overwhelming feature of the logging equipment manufacturing sector during this era was a kind of mirror image of the corporate concentration that had occurred in the woods.

In the falling and bucking phase, few changes in procedures were introduced between 1970 and 1990. The biggest changes were in the saws available to fallers, which steadily became lighter, more powerful and more reliable. By the late 1980s, a saw capable of handling the biggest coastal timber weighed less than twenty-five pounds. Its sealed electronic components were impervious to coastal weather conditions, its noise was substantially reduced, and the vibration had largely disappeared. Most of the North American manufacturers were gone, and while the chain saw market was still growing, it was dominated by two European manufacturers, Stihl and Husqvarna.

During the latter half of the 1970s, a falling technology widely used in other parts of the world was introduced: feller-bunchers. This machine consists of a set of hydraulic clamps that hold a tree while a cutting head saws or shears it at the stump, then lays it on the ground. Initially, coastal timber was too large for the falling machine to handle. Then, in 1974, Crown Zellerbach began logging second-

Skagit's SJ-6 grapple yarder bringing in a turn at MacMillan Bloedel's Shawnigan Division, 1969. This was the first all-grapple operation on the coast. (UBC 1930/321)

growth timber on some of Hastings Sawmill's first sites along Johnstone Strait. Being of uniform age, the trees on some of these stands were also of fairly uniform size, and small enough to be handled by feller-bunchers. Some sites were flat enough for the machines to be able to travel and work. In 1981, CZ's Courtenay Division utilized one of the first feller-bunchers on the coast at Turn Harbour on East Thurlow Island. The following year, Dave Balinski contracted to CZ with a Drott feller-buncher on Quadra Island, while Ted LeRoy contracted with a Timbco on East Cracroft Island. LeRoy has since become the major user of mechanical harvesting machinery on the coast, contracting for CZ's successor company, Fletcher Challenge. A major lesson learned from these operations was that the biggest cost-cutting gains from feller-bunchers came from bunching tree-length logs, and from grapple yarding the bunches.

CZ also was one of the first coastal companies to use a processing machine, a Hahn, to limb and buck the full-length trees logged at Turn Harbour. Since 1981 several other companies have used a variety of processors, most of them in second-growth stands with uniform timber sizes.

One of the bottlenecks in the early development of a coastal feller-buncher was the lack of a suitable machine on which to mount the cutting head. The evolution of feller-bunchers paralleled the development of tracked excavators, but those machines were generally too light and could not work on the steep slopes typical of coastal logging sites. One solution to these problems was designed in the mid-1980s by the Giesbrecht family of Port Alberni, owners of Antler Creek Logging, a contractor at MacMillan Bloedel's Cameron Division. Alberni Shipyard built the Giesbrecht feller-buncher, a modified Timbco capable of cutting through a thirty-inch tree and handling 150-foot timber.

As with many of the new, sophisticated and expensive types of logging equipment, the desire to adopt the feller-buncher grew out of mixed motives. On one hand, there is a tendency toward a more capital-intensive industry in which higher-priced machines are substituted for labour, thereby eliminating jobs. But safety is an additional concern. Feller-bunchers and log processors are much safer for the operator than chain saws, just as grapples are safer than chokers for the chokermen and chasers whose jobs are displaced.

After 1970, there was a general movement away from cable yarding systems using chokers to grapple yarding. The appeal of grapple yarding was simple: lower costs. The machines were

A Caterpillar with a modified arch used as a tail hold for the Skagit SJ-6 at Shawnigan. (UBC 1930/321)

expensive, but savings were achieved because grapple yarders used a smaller crew, took less time to set up than a steel spar, and speeded up loading at roadside. A grapple yarder's main advantage lay in its mobility. Whereas a spar hauls all the logs in a setting to a single landing, including those lying beside the roads to and from the landing, a grapple yarder moves along the road, yarding logs the shortest distance to the road. Because they require fewer men and their crews work clear of the rigging, they are safer than conventional cable systems.

In basic form, a grapple yarder consists of a yarder with interlocking mainline and haulback drums, as well as a drum to operate the grapple. The grapple itself, of which there are several designs and manufacturers, is hung from a carriage suspended on a skyline. The operator hauls the grapple out to the setting with the haulback, keeping the skyline tight. When guided into position, often with the assistance of a radio-equipped spotter, the skyline is lowered with the open grapple over the log. The grapple is closed by hauling in on the grapple line, the skyline is raised and the mainline is hauled in.

Initially there were two types of grapple yarding machines in use. One used a modified high-lead tower, such as a 90-foot Madill steel spar, with a two-drum yarder. The mainline drum was replaced with two smaller drums, one of which controlled the grapple. The second type was the yarding crane, or swing yarder, made by Washington, Berger, Skagit, Caterpillar and others. They were mounted on excavator bodies and utilized either hydraulic or mechanical winches. Their advantage was their better mobility and their ability to swing their boom, either to facilitate yarding around obstacles or to yard logs into a better position for loading at roadside.

The development of grapple yarders in BC probably began at CZ's Nitinat division in the mid-1960s under manager Bill Wainwright. In 1963, he tried using a grapple on a steel spar and found that the guylines slowed down yarding

Rigging a Cypress 7280-B grapple yarder at the Cypress plant in Delta, early 1980s. (Harry Foster photo)

and moving. Wainwright took his ideas to Skagit, which modified an SJ-7 loader with a 40-foot boom. This machine took about ten minutes to rig and yarded from a tail-hold stump 450 feet from the landing. Compared to the six men a steel spar required, this grapple yarder needed a four-man crew. Ed Lewis operated the machine, and Dave White acted as a spotter with a walkie-talkie. Two others rigged the tail holds. Following the lead of MacMillan Bloedel's Shawnigan Division, many camps replaced all of their steel spars with grapple yarders over the next decade.

By 1969 the Nitinat division was yarding at night and using a mobile tail hold, usually a Cat or a shovel. Night logging started at many operations in the late 1960s, and by 1969 MacMillan Bloedel was doing it on a regular basis at half of its twenty logging divisions. Equipment, especially the new grapple yarders, was becoming more expensive, and there was a drive to keep the machinery working as constantly as possible. Banks of mercury vapour lamps powered by diesel generators were mounted on mobile trailers and used to illuminate the logging site. Production was slower at first—sixty percent of daytime output at Nitinat—but with improved equipment and experience it often approached regular daytime rates.

Grapple yarders were used much more extensively in BC than in Washington and Oregon, reflecting some basic differences in forest policy between the two countries. As a general rule, the new type of yarder required more roads than tower logging systems. In the US, private timber owners like Weyerhaeuser were less inclined to spend money on additional roads or sacrifice productive forest land to roads and landings. In publicly owned US forests, there was a reluctance to build roads into mid-slope or other

A BC Forest Products' Cypress 7280-B grapple yarder working at Narrows Inlet, early 1980s. The company purchased two identical yarders and equipped them both with $150,000 lighting systems for night logging. The other machine, shown on p. 285, worked at Pitt Lake. (Harry Foster photo)

environmentally sensitive areas. The BC industry did not operate under the same constraints. Also, during the period of grapple yarder development, some road costs were deducted from stumpage payments, which tended to pass some of the grapple yarders' costs onto the government, the owner of the land. As a consequence, there was a more widespread use of grapple yarders in BC.

Grapple yarders have been the dominant yarding system used on the coast until the present time. However, a revival of skyline logging systems is now under way. The resurgence of this technology stems from the expense of building roads into the steep terrain where much of the old-growth timber supply is now found, the environmental sensitivity of these sites, and the fact that getting to some of this timber would entail building roads through second-growth stands. Using a skyline that can yard from distances of up to 6000 feet saves on access costs, reduces the impact of logging on the environment and, because turns are not dragged through the slash, increases log values.

The first of the new generation of skyline systems was engineered by John Chittick at Canfor's Englewood division in the late 1980s, using a Washington four-drum slackline yarder. A couple of years later, Fletcher Challenge's Sandspit division, under manager Doug Mosher, put together an experimental skyline system using a Madill yarder on an old 120-foot Vancouver Iron & Engineering Works steel spar. By the fall of 1991, there were at least seventeen skyline operations up and running. MacMillan Bloedel had seven skyline sides, and Canfor had added a second one. Several contractors had also got into the game, using a variety of machines. Hull & Sons were using a Thunderbird near Terrace, Mars Industries was using a Skagit on MacMillan Bloedel timber at Kildonan, and Kwatna Timber, McKay Trucking and Timfor were using Madills. Hans Lee, the only logger on the coast to stick with the Wyssen skyline machines introduced in the late 1960s, was operating two of them, one for the Greater Vancouver Water Board and the other at Port Neville. His rigs were using single-drum

Skyhook Enterprises' mainline Washington Aero Yarder at Phillips Arm, 1992. The haulback yarder is a separate, track-mounted unit. Using two self-propelled machines permits more flexibility in layouts and reduces the length of the haulback line. (John Doyle photo; Truck Logger)

yarders to lower turns down a fixed skyline suspended from intermediate supports.

The basic skyline system consists of a four-drum yarder, a steel spar and a set of wire ropes rigged in one of the various configurations devised during the heyday of steam yarding before World War Two. The major modern addition, apart from the more powerful engines and mobile spars, is a carriage that hangs off the skyline and yards logs laterally to the yarding road before pulling them into the landing. Carriages are equipped with their own diesel motors and are radio controlled. The most popular machines in use at this stage are Madills.

The irony of the present situation is that with the enthusiastic adoption of grapple yarding twenty years earlier, a large number of skyline systems owned by BC logging companies were sold to US loggers. The recessions of the 1980s devastated the logging equipment manufacturing sector, putting many builders out of the skyline business—if not out of business altogether. Now that there is a renewed need for skyline machines in BC, loggers are forced to make do with what is still available locally, or try to buy back equipment from the US.

During the 1980s, the search for equipment to log inaccessible stands of high-value timber focussed attention on unconventional aerial-logging systems—balloons and helicopters. For many years the logger's dream of a big hook descending out of the sky, plucking logs out of the slash and carrying them over the roadless land below, was just that—a dream. The first attempt at balloon logging was in 1965 at MacMillan Bloedel's Sproat Lake division where two well-used 75,000-cubic-foot surplus ballons provided lift on a cable yarding system. They were attached to a carriage, along with the chokers and a mainline and haulback line. This system was moderately successful at yarding a 300-foot swath, but was abandoned, largely because the balloons were in poor condition.

A more advanced balloon system was launched the following year by Chester Matheson to bring logs off a ridge between Seymour and Lynn creeks in North Vancouver for the Greater Vancouver Water Board. Matheson was a forest engineer who also developed the MacMillan Bloedel system. He had a new, 100,000-cubic-foot balloon designed to provide lift for a rigging setup similar to the one used at Sproat Lake. He also had a hydraulic winch with interlocking drums, a Skylift built by Frank Lawrence, which overcame many of the problems encountered in using conventional yarding winches—basically by adding line spoolers to the drums. While Matheson's system was much improved, his company, Balloon Transport, went broke, and no further attempts were made to revive this method for more than twenty years.

Both these balloon systems were actually high-lead systems with balloons providing deflection—what could be termed balloon-assisted logging. This type of setup was developed further in Oregon by Faye Stewart of Bohemia Lumber, with the involvement of Frank

International Forest Products' Vertol coming in to pick up a turn at Broughton Island, c. 1987. (Harry Foster photo)

Mosher, yet another member of the ubiquitous coastal logging family. Stewart formed Flying Scotsman Enterprises, which developed an improved balloon and a new yarder, the Aero-Yarder, built for him by Washington Iron Works. It had a 525-horsepower Detroit diesel engine on a twin-drum winch designed to hold down the 25,000-pound lift balloon. In principle this system is the same as the one Matheson built in the 1960s, except the lifting capability of the balloon is enough to carry the logs off the ground.

In 1990, Flying Scotsman and Dave Balinski formed Skyhook Enterprises and used the system to log in Seymour Inlet. Two years later Belinski sold his interest to Ted LeRoy, Don Williams of Lineham Logging and Beban Logging. They were operating two Skyhook systems, one at Seymour Inlet and the other on a contract of LeRoy's with MacMillan Bloedel at Phillips Arm.

For many years, through the 1970s and early eighties, attempts were made to develop a free-flying heavy-lift balloon that could be used for logging. The most serious of these ventures was undertaken by Aero Lift Inc., with an aircraft called the Cyclo-Crane. The company was put together by a syndicate including BC Forest Products, MacMillan Bloedel, Tahsis Company, Pacific Logging and Silver Grizzly Helicopters, a California helicopter logging company owned by Jack Erickson. The Cyclo-Crane was a positively buoyant powered balloon that generated additional lift by rotating. A prototype model was built, but was destroyed in a windstorm in 1982. Attempts to rebuild failed, and in 1992 the aircraft's inventors announced the launching of a different type of heavy-lift machine called the Buoyant-Copter. To date, no free-flying balloon has been successful at logging.

Helicopter logging has been a more successful aerial system. As early as the mid-1950s, loggers and helicopter people began discussing the possibility of using the machines for hauling logs. In 1957, ten years after helicopters first went into commercial use, the first North American use of helicopters in logging took place in Quebec when Consolidated Paper flew bundles of pulpwood with a Sikorsky. That same year Alistair Smiley of Okanagan Helicopters told the fourteenth annual Truck Loggers' convention that there were two major obstacles to logging with helicopters: limited lifting power and high costs. When extensive heli-logging trials were undertaken in Russia in the late 1950s, they revealed two things: the pickup system was critical, and it was too dangerous to lift from the stump without pre-falling.

Through the sixties and into the seventies, little progress was made in heli-logging in BC. In California, however, veteran lumberman Jack Erickson began serious helicopter logging in the late sixties. In 1975, BC's first helicopter logging operation was launched on Redonda Island with a Bell 214 by a group made up of logging contractor Ian K. Mahood, forest engineer T.M. Thomson and Okanagan Helicopters. Although the Bell lifted only seven thousand pounds, and clearly was not the right machine for logging, this operation opened the door to heli-logging on the BC coast.

Two years later Erickson came to BC and set up Silver Grizzly Logging, the province's first large-scale operation, in the Khutzeymateen Valley north of Prince Rupert. T.M. Thomson did the engineering on this show, which provided a whole range of BC loggers—engineers, fallers, hookers, foresters—with Erickson's extensive US experience.

Ralph Torney, one of the Thomson people who had worked at Redonda and Silver Grizzly, joined with Husby Forest Products in 1986 to form Air Log Canada using a Sikorsky Skycrane with a lift of 20,000 pounds. Although the Skycrane costs about $100 a minute to operate, it can move as much timber as six steel spars. Since it started, Air Log has worked at numerous coastal locations, logging Husby's timber as well as contracting to others.

A Bucyrus–Erie line loader decking logs that have been yarded with a ninety-foot Madill Model S, at Upper Valley Logging's Stave Lake show, c. 1980. (Harry Foster photo)

The most experienced player in the heli-logging business is Helifor, which started with a leased Boeing Vertol 107 in 1978 under the Whonnock Industries name. Since then the heli division has logged more than 40 million feet a year for its parent company, Interfor, and on contracts for other coastal companies.

A second logging company to enter the heli-logging field was Coulson Forest Products in 1988. Using a Sikorsky S-61 capable of lifting 10,000 pounds, Wayne Coulson, along with chief pilot Bob Hawthorne and manager Garry Colinge, developed the first successful log grapple used on a helicopter. In some situations, logging with the grapple is much faster because there is no ground crew with whose safety the pilots need be concerned. Coulson Aircrane logs the company's own timber, as well as contracting from Alaska to California.

Another logger-owned heli-logging operation, Vancouver Island Helicopter's VIH Logging, grew out of the established helicopter charter company bought by truck logger Frank Norrie when he retired in 1984. In 1991, after acquiring some heli-logging experience with a Sikorsky S-61, VIH leased a Russian-made Kamov KA-32 with a unique double rotor design, which lifts a bit more than an S-61 and is more compact and manoeuvrable. After a summer flying logs at Cowichan Lake alongside Russian crew members, VIH people were enthusiastic about the Kamov's future in logging.

One of the most versatile pieces of ground-based equipment applied to logging after 1965 was the swing machine, originally developed as an excavator. Various companies, including European and Japanese manufacturers, were actively developing and marketing a version of this machine. In North America, Caterpillar played a major role after introducing its Model 225 hydraulic excavator in 1972. The basic swing machine consists of an engine, cab and turntable, which can be mounted on tracks, wheels, the back of a truck and various other bases. It was developed primarily to be part of the excavator built to replace track shovels, a machine that found instant popularity among loggers because of its utility in building roads. Soon a variety of other applications evolved, one of the first being to use it as a heel-boom loader. As log sizes decreased and utilization standards got tighter, cable loading machines became less practical and hydraulic machines gained popularity. By the late eighties, they accounted for almost ninety percent of loading equipment sales.

Along the way, loggers found a variety of other uses for these machines. They were used extensively as carriers for feller-buncher attachments, yarders and log processors. As fast as the various manufacturers began producing modified machines, loggers found more uses for them. One of the latest uses is called "hoe chucking." By 1990, it was widely practised on the

An American line loader working at Fletcher Challenge's operation at Camp 2 Bay near Bella Coola, 1985. (Harry Foster photo)

coast. It involves taking the loader out into the felled and bucked timber on suitable ground, and literally throwing the logs toward the road in a series of swaths that eventually has them decked at roadside. It is used efficiently at distances as far as 500 feet from the road.

Another widely practised yarding technique to evolve from loading in the 1980s is called "super snorkelling," in which a cable loader equipped with a wooden boom or snorkel casts a grapple up to 150 feet from roadside.

Apart from minor design improvements, there were few significant changes in logging truck manufacturing after 1970. One trend was toward the increased production of custom-made trucks, with practically all builders providing choices in cabs, engines, transmissions and other components. After the deliberate destruction of the popular Hayes truck company in 1975, there were about six makes remaining in the logging sector, half of them from the US. Kenworth's Model 850 with a 600-horsepower Cummins was still a heavily used brand in 1990. International's Navistar and Peterbilt's 357 were two other imports. In BC, the Kelowna-made Western Star was still selling well. Another White-related truck that played a minor role in the logging industry—the Freightliner, built at a plant in Burnaby—ceased production in June 1992. In North Vancouver, Pacific Truck and Trailer produced its P-16, one of the favourite off-highway workhorses on the coast, until October 1991 when the assembly plant closed.

The biggest truck news for coastal loggers was the launching of a new make of truck in 1986, the Challenger. It was developed by John Casanave, a former Coulson Forest Products driver and owner of Lois Lake Logging. Casanave picked up where Hayes left off, producing in a small Port Alberni shop what is essentially a tougher version of the Hayes HDX,

A Madill super snorkel working at Port McNeill, 1991. (FERIC)

Caterpillar D8N with dozer and rippers building road at MacMillan Bloedel's Eve River Division, 1992. (Finning Tractor collection)

An American line loader loading from a ninety-foot Madill for Norm Bowen's G&F Logging in the Skagit Valley, c. 1982. The truck is a Peterbilt. (Harry Foster photo)

A Kenworth at Whonnock Industries' Drury Inlet camp, 1987. (Harry Foster photo)

considered by many loggers to be the best logging truck ever built. The Challenger is a truck built specifically for logging and is designed to appeal to drivers and mechanics. Everything on it is simpler, tougher and easier to service than any other truck available. Every unit is custom built, down to the choice of tape deck. Standard engines, axles and transmissions are used, with Casanave-designed frames, cabs and suspension. Eight units had been built by 1992. It is a curious coincidence that the most refined and sophisticated logging truck produced today in the Pacific Northwest is built on the very same site where BC's first loggers used oxen to skid logs to the Anderson mill in 1860.

Challenger Manufacturing's fifth off-highway truck, custom-built for MacMillan Bloedel in 1989, driven by Don Watts at Franklin River. Founded by John Casanaves to rebuild other trucks, Challenger designed an entirely new truck that was stronger, and easier to drive and to service. (Marc Stenberg photo)

In the early 1980s, a relatively new type of logging began on the coast, the selective harvesting of immature second-growth forests, usually referred to as commercial thinning. Long a standard procedure elsewhere in the world, it was slow to start on the BC coast. There were several selective logging operations during the late 1920s and early 1930s, most of them on Vancouver Island, which were established in response to objections from the public and some loggers to railway-era clear-cut logging. In 1934, Buck and Turner selectively logged on Mount Benson near Nanaimo with crawler tractors and log pans. Most of this kind of logging at that time was, in reality, a form of high-grading. The best trees were logged and the residual stand consisted of commercially less valuable species, damaged trees and genetically inferior trees, which reproduced to create inferior stands.

In 1958, Henry Nygard's Garibaldi Timber used tractors to log selectively at the UBC research forest near Haney. In the mid-1960s, MacMillan Bloedel engaged in probably the most serious commercial thinning operation to that time at its Sproat Lake, Northwest Bay and Stillwater divisions. Most of this work was done by contractors using the recently developed Timberjack rubber-tired skidders.

Apart from a small amount of thinning done on private woodlots, very little of this sort of harvesting occurred on public land until the late 1970s, when the BC Ministry of Forests announced it intended to do 1000 acres a year in the Sayward forest. A variety of logging systems were tried before the Forest Service curtailed the operation for bureaucratic reasons. A revival of horse logging, using essentially the same methods developed early in the century, accounted for part of the operation, including a show of Howie Griessel's Frontier Logging.

Bob Woods and Jim Lambrick ran one of the most successful thinning operations at John Hart Lake near Campbell River using the European-made Mini-Alp skyline system. Lambrick was later involved in the development of a similar small skyline machine built by Skylead Logging Equipment in Enderby, and in 1992 was thinning with a Timbco feller-buncher at Cowichan Lake. An enormous volume of second-growth timber is available on the coast through commercial thinning, but by the early 1990s the Ministry of Forests still had not overcome the bureaucratic obstacles preventing its widespread application.

From the vantage point of 1992, the future of logging on the coast of BC is hard to predict, although an assessment of the timber available indicates what might lie ahead. There is still a substantial volume of old-growth timber—somewhere in the neighbourhood of half the original forest area, depending on the definitions used. Some of this timber can be logged with existing logging technology, primarily grapple yarders or high-lead systems. Much of it, though, is on hard-to-reach sites, such as the steep mid-slopes of mountains that have been logged from the valleys below and the ridges above. There are hanging valleys that cannot be reached easily by trucks, remote pockets of high-value timber, and patches of timber in sensitive areas along rivers and estuaries and in remote valleys along the North Coast. Clearly, considering the technology available, much of this timber can only be logged using helicopters, skyline systems or some type of balloon or balloon-assisted systems.

An increasing share of the timber supply is going to be provided by second-growth wood off land that has been previously logged. Two general methods will be used to harvest this timber. One will employ a capital-intensive, mechanized system using feller-bunchers, grapple yarders, grapple skidders, forwarders and other sophisticated, high-production machines. This approach will involve clear-cut logging and plantation forestry on low rotations designed to produce high volumes of wood fibre, primarily for pulp mills. The second approach will entail more labour-intensive methods, tending toward selective cutting, and the creation of intensively managed, uneven-aged stands that will be harvested with the kind of small-scale, relatively low-cost harvesting machinery now used in many European forests. The objective in these forests will be to produce high-value logs to supply timber to a new generation of coastal sawmills, plywood plants and other processing facilities, which will also make better use of the available old-growth timber.

The balance of the various approaches to logging used in the future involves questions of public policy. There will be environmental constraints, economic development objectives, short-term political considerations, non-timber uses of the forests, and any number of other factors brought to bear on the issue.

Decision-making is clouded and confused by a tendency to adopt a short-sighted solution to the perceived problems of logging the temperate rain forest. Over the past decade or so, governments, often under public pressure from individuals and organizations concerned about various aspects of the environment, have chosen to ban

Unloading cedar with a Caterpillar 966 at Whonnock Industries' Kingcome Inlet dryland sort, 1985. (Harry Foster photo)

logging in some areas. Some interests have chosen to portray loggers as pillagers of a pristine natural environment and demand that logging be curtailed even further. For many, the choices are clearly defined, black and white issues—logging is good, logging is bad; log, don't log.

While decisions to ban logging certainly eliminate any problems it may cause in the forests, they also create serious economic and social problems for the entire region. Jobs are lost; families and communities are threatened. The economy of the entire Northwest is jeopardized.

Instead of slowly eliminating the logging industry by banning it from more and more areas, a more sensible option is to redefine and redirect it. Such a redirection is something BC loggers themselves have been trying to achieve for the past fifty years. In many respects, the BC coastal logging industry has come to its present circumstances for reasons that have nothing to do with the environment in which it works. The situation facing loggers today is largely a consequence of decades of monopolistic corporate greed, bureaucratic ineptitude and political opportunism. The people who actually go to work in the woods every day are not the people who created this situation. They, as much or more than anyone, would be happy to see it change. But they also know, quite apart from their own personal stake in the matter, that the way to make the change is not to chase loggers out of the forest.

The entire coastal forest industry in BC is at a critical stage in its history. It is a monolithic industry under the ultimate control of a single entity, the provincial government. No matter which political party is in power, the government is composed of elected members who know very little about forests and logging, but who are faced with the impossible task of devising policies that make sense on every acre of forest land in the province. Then, these policies must be put into practice by a large bureaucracy which exercises its authority through a massive body of regulations defining how those who actually work in the woods are supposed to do their jobs. Standing between this bureaucracy and the loggers is another layer of large corporations, many of which are controlled and directed from out-of-province centres by people with no real interest in or concern about the forests or the people who work in them. Beyond all these layers of politicians, bureaucrats and corporate managers are the people who actually

Wild Duck Trucking evolved from this single Kenworth gravel truck to a full-phase contracting operation during the late 1980s. The company is owned and operated by the Swanson family, headed by coastal logging veteran Don and six logger sons. Art Swanson at Knight Inlet in 1991. (Harry Foster photo; Truck Logger)

put on their rain gear and caulk boots, and go to work in the forests. They are fallers, chokermen, loaders, foresters, engineers, owners of small companies, foremen and superintendents of big companies, machine operators, tree planters, spacers and scores of others with direct, firsthand knowledge and experience of both the forest and the industrial apparatus upon which the provincial economy is built. It would be a big mistake to solve perceived environmental problems by chasing these people out of the woods, as has been done in Washington and Oregon. It makes much more sense to involve them intimately in the redefinition of the industry and to give them the opportunity to rebuild it. Obtaining such a redefinition will not be done by turning the task over to those who, in H.R. MacMillan's words, have never had rain in their lunch buckets. In the end, when it comes to making decisions about real trees in specific forests, the only people who can make realistic, environmentally sound decisions are those who know a particular forest and how to work in it without destroying it.

The creative ability required to figure out how to move fifty-ton logs off remote mountainsides, and the willingness to do it, day after day, rain or shine, attracted a rare breed of people from all over the world. For more than a century they have adapted and evolved, creating the unique cultural tradition of the coastal logger. It is a tradition that places great value on inventiveness, innovation, creativity, curiosity, and adaptability. It is an entrepreneurial tradition which rewards both individual effort and the ability to function in an interdependent community of equipment makers, processing mills and a global marketplace. And, more than most occupations, it is a tradition that has grown out of the unique physical environment of the Pacific Northwest.

The history of coastal logging is, in the end, the story of how a regional culture came to be. Ignoring the history and rejecting that culture would be a grave and serious mistake.

Sources

In addition to the books I have listed under each chapter below, I have consulted a number of more general works containing aspects of BC forest history. Chief among these is G.W. Taylor, *Timber* (Vancouver: J.J. Douglas, 1975). Also useful were the chapter on coastal logging in Donald MacKay, *The Lumberjacks* (Toronto: McGraw–Hill Ryerson, 1978); two books by Joe Garner: *Never Chop Your Rope* (Nanaimo: Cinnabar, 1988) and *Never Under the Table* (Nanaimo: Cinnabar, 1991); *Raincoast Chronicles* No. 3 (Harbour Publishing); Ralph C. Bryant, *Logging* (New York: John Wiley & Sons, 1923); Nelson C. Brown, *Logging* (New York: John Wiley & Sons, 1949); and various issues of the *British Columbia Lumber Trade Directory*, published by Progress Publishing in Vancouver since 1916. The reference used for spelling and definition of logging terms was Walter F. McCulloch, *Woods Words* (Corvalis: Oregon Historical Society, 1977). Also of great assistance were the local histories of various coastal communities.

Chapter One

A good description of the Pacific Northwest's coastal rain forests is found in Elliott A. Norse, *Ancient Forests of the Pacific Northwest* (Covelo, CA: 1990). In Cameron Young, *The Forests of British Columbia* (North Vancouver: Whitecap Books, 1985) there is a detailed description of British Columbia's coastal forests and tree species. A thorough description of BC forests as they were at the beginning of this century can be found in H.N. Whitford and R.D. Craig, *Forests of British Columbia* (Ottawa: Commission of Conservation, 1918).

Chapter Two

The acknowledged authority on early Vancouver Island forest history is W.K. Lamb, *Early Lumbering on Vancouver Island: 1846–66* (BC Historical Quarterly, April 1938). A good account of the first coastal mainland mills and early logging in the Fraser Valley is found in Arnold McCombs and W.W. Chittenden, *The Fraser Valley Challenge* (Harrison Hot Springs: Treeline Publishing, 1990). An early account of Northwest lumber markets is Thomas R. Cox, *Mills and Markets* (Seattle: University of Washington Press, 1974). Also see James E. Lawrence, *Markets and Capital* (unpublished thesis, University of British Columbia, 1957); James E. Flynn, *Early Lumbering on Burrard Inlet: 1862–1891* (unpublished thesis, University of British Columbia, 1942); Frank Gunderson, *The Rise and Fall of the Forest Industry on the North Shore of Burrard Inlet Between the Capilano and Seymour Rivers* (unpublished thesis, University of British Columbia, 1980); David C.S. Martin, *Maritime Aspects of the Early Logging History of British Columbia* (unpublished thesis, University of British Columbia, 1978).

A good account of traction engines, including the type used for logging by Jeremiah Rogers, is in W.J. Hughes, *A Century of Traction Engines* (London: Percival Marshall & Co., 1959). River driving in BC is described in Donna Watson, Edith Stephens and Dick Isenor, *Land of Plenty* (Campbell River: Ptarmigan Press, 1987); John Saywell, *Kaatza* (Duncan, 1967); and Arnold McCombs and W.W. Chittenden, *The Harrison–Chehalis Challenge* (Harrison Lake: Treeline Publishing, 1988). The classic fictional account of river driving is Stewart Edward White, *The Riverman* (New York: Grosset & Dunlap, 1908). Much of the early history of the Rock Bay area is from notes by Susan Smith of the Link and Pin Museum at Roberts Lake. Details of early logging in Vancouver have been obtained from *Early Vancouver*, an unpublished collection of documents in the Vancouver City Archives.

Chapter Three

See James Stevens, *Paul Bunyan* (New York: Alfred A. Knopf, 1925) for a full account of this legendary logger. Information on early steam logging in the Campbell River area is in notes and tapes collected during the Campbell River & District Museum's Industrial History Project, 1983. Descriptions of the first logging railways are found in "A Brief Sketch of Small Logging Railroads on Vancouver Island's East Coast," an unpublished manuscript held by the Courtenay Museum; Ken Bradley, *Historic Railways of the Powell River Area* (Victoria: BC Railway Historical Association, 1982); David M. Rees-Thomas, *Timber Down the Capilano* (Victoria: BC Railway Historical Association, 1979); various issues of *Callboard*, newsletter of the BC Railway Historical Association, whose members' passion for railway history has, incidentally, preserved much early logging history; and Robert D. Turner, *Logging by Rail* (Victoria: Sono Nis, 1990), without doubt the most authoritative account ever published. The transformation of a handlogger into a mechanized A-frame logger is told dramatically in Martin Allerdale Grainger, *Woodsmen of the West* (London: Musson, 1908).

Chapter Four

The best social and economic history of the forest industry during the first third of this century is found in a series of papers by Richard Rajala: "Bill and the Boss," *Journal of Forest History* (October 1989); "Managerial Crisis: The emergence and role of the West Coast logging engineer, 1900–1930," *Canadian Papers in Business History* (Vol. 1, 1989); and "Lumbering on the Lake," unpublished, Kaatza Museum, 1987. The best source on the technical aspects of yarding equipment is *The Timberman* magazine, which covered the logging industry throughout the Pacific Northwest from 1889 to 1962. An excellent personal history is Harper Baikie, *A Boy and His Axe* (private publisher, 1991). A good account of the first use of steam yarding equipment in the Chemainus area is in Donald MacKay, *Empire of Wood* (Vancouver: Douglas & McIntyre, 1982). Union history was obtained, in part, from Jerry Lembcke and W.M. Tattam, *One Union in Wood* (Harbour Publishing, 1984) and Myrtle Bergren, *Tough Timber* (Toronto: Progress, 1967). The best logging novel ever published, Roderick Haig-Brown, *Timber* (Toronto: Collins, 1942), is set in a big railway camp probably based on one of the Menzies Bay camps of that era.

Chapter Five

In addition to Turner's *Logging by Rail*, material in this chapter comes from Kramer Adams, *Logging Railroads of the West* (Seattle: Superior Publishing, 1961); and John Labbe and Vernon Goe, *Railroads in the Woods* (Berkeley CA: Howell–North, 1961). Additional material was obtained from *The Timberman*. Information on logging in the Cowichan Valley during the 1920s was found in a series of three articles by Darryl Muralt in *Whistle Punk* magazine, Vol. 1, Nos. 2–4, 1984–86. Also see Robert Swanson, *A History of Railroad Logging* (Victoria: Queen's Printer, 1960).

Chapter Six

Valuable sources on truck logging include the two books by McCombs and Chittenden; James Young and Jerry Budy, *Endless Tracks in the Woods* (Sarasota: Crestline, 1989); Tad Burness, *American Truck & Bus Spotter's Guide 1920–1985* (Osceola FL: Motorbooks International, 1985). Also useful were *The Timberman* and *Western Lumberman* magazines.

Chapter Seven

Major sources for the section on tractors were Young & Budy's *Endless Tracks in the Woods*, *The Caterpillar Story* (Peoria IL: Caterpillar Inc., 1990), and back issues of *The Timberman* and *Western Lumberman*. For the power saw section, articles from *Chain Saw Age* and *Truck Logger* magazines.

Chapters Eight and Nine

The major work on modern logging techniques is Steve Conway, *Logging Practices* (San Francisco: Miller Freeman, 1976). The greatest source of information is in back issues of *Truck Logger* magazine, with additional material from *BC Lumberman* magazine.

Oral History Sources

Ed Aldridge Interviewed in Squamish, November 1991, by the author.

A.P. Allison From an article in *Western Lumberman*, December 1921.

Dewey Anderson Interviewed at Vancouver, February 1957, by C.D. Orchard. UBC Special Collections.

P.B. Anderson From an unpublished memoir, "Logging in British Columbia," UBC Special Collections.

Harper Baikie From his book, *A Boy and His Axe* (Campbell River, 1991).

Wallace Baikie From his book, *Rolling With the Times* (Campbell River, 1985).

Jack Bell Interviewed at Qualicum Beach, December 1991, by the author.

Jack Warren Bell Interviewed by Major J.S. Matthews, 1946–50. CVA.

Art Bellerby Interviewed at Campbell River, 1989, by Marshall Beck. CRM.

Art Bendickson Interview in *Truck Logger*, July 1959.

Cliff Brackett Interviewed at White Rock, October 1991, by the author.

Alvin Brown Interviewed at Port Alberni, December 1991, by the author.

Robert Brown Quoted in *Kaatza*, by John F.T. Saywell, Cowichan Lake, 1967.

John Burke Interviewed at Vancouver, January 1957, by C.D. Orchard. UBC Special Collections.

Dave Challenger Interviewed at Burnaby, September 1991, by the author.

W.W. (Curly) Chittenden Interviewed at West Vancouver, September 1991, by the author.

Joe Cliffe Interviewed at Courtenay, July 1980, by Lynn Henderson. CDM.

E.K. (Ned) DeBeck Unpublished memoirs, BCARS.

Albert Drinkwater Interviewed at Vancouver, March 1964, by Imbert Orchard. BCARS.

Glen Duncan Interviewed at Roberts Lake, November 1991, by the author.

Cecil Harlow Edmond From an undated, unpublished manuscript, "Tan Bark Lease #432." CRM.

Olaf Fedje Interviewed at Nanaimo, October 1991, by the author.

Richard Fiddick Interviewed at Nanaimo, April 1961, by C.D. Orchard. UBC Special Collections.

Robert J. (Bob) Filberg Interviewed at Comox, June 1960, by C.D. Orchard. UBC Special Collections. And, text of a speech to Pacific Logging Congress, 1932, in *The Timberman*, November 1932.

Aird Flavelle Interviewed at Victoria, April 1957, by C.D. Orchard. UBC Special Collections.

Martin Fossum Interviewed at Campbell River, 1989, by Marshall Beck. CRM.

Joe Garner Interviewed at Nanaimo, September 1991, by the author.

Bob Grant Interviewed at Grantham, November 1965, by Tom Menzies and Ben Hughes. CDM.

Charles H. Grant Interviewed at Royston, February 1961, by C.D. Orchard. UBC Special Collections.

Tom Hall Interviewed at Gowland Harbour, 1989, by Marshall Beck. CRM.

Cliff Hallberg Interviewed at Madeira Park, November 1991, by the author.

Ken Hallberg Interviewed at Nanaimo, September 1991, by the author.

Skate Hames Interviewed at Campbell River, May 1989, by Marshall Beck. CRM.

Dan Hanuse Interviewed at Vancouver, September 1991, by the author.

Al Hendrickson Interviewed at Squamish, November 1991, by the author.

Mose Ireland From a memoir recorded in 1905, in *Western Lumberman*, July 1926.

Archibald Kerr From an unpublished manuscript, "Lumberjack," in UBC Special Collections.

Bazil Kier Quoted in *Kaatza*, by John F.T. Saywell, Cowichan Lake, 1967.

George Lutz Interviewed at Woss, July 1974, by Derek Reimer. BCARS.

Hiram McCormack Interviewed by Eric Flesher at Phillips Arm, 1922. *Truck Logger*, June 1969.

Tom MacInnes From a memoir in *Western Lumberman*, August 1926.

H.R. MacMillan Unpublished manuscript in Lawrence Collection. UBC Special Collections. And memoir in *Water Over the Wheel*, W.H. Olsen, Chemainus Valley Historical Society, 1963.

Monty Mosher Interviewed at Vancouver, September 1991, by the author.

Frank Norrie Interviewed at Cobble Hill, September 1991, by the author.

E.J. Palmer Unpublished manuscript. Monty Mosher private collection.

Mel Parker Interviewed at Campbell River, 1989, by Marshall Beck. CRM.

Al Parkin Interviewed by Howie Smith. BCARS.

Mauno Pelto Interviewed at Nanaimo, September 1991, by the author.

Ben Ployat Interviewed at Vancouver by Imbert Orchard. BCARS.

Lloyd Rodgers Interviewed at Victoria, December 1956, by C.D. Orchard. UBC Special Collections.

Woodrow Runnells Interviewed at Stillwater, November 1991, by the author.

Mark Smaby From an unpublished memoir in Orchard Collection. UBC Special Collections.

Eustace Smith Unpublished manuscript, British Columbia Forests, February 1944. Vancouver City Archives. And, unpublished manuscript, "Sign of S," B.C. McKelvie. BCARS.

Sidney G. Smith Interviewed at Vancouver, December 1960, by C.D. Orchard. UBC Special Collections.

R.V. Stewart Interviewed January 1960 by C.D. Orchard. UBC Special Collections.

Robert Swanson Interviewed at Vancouver, September 1991, by the author.

Sam Telosky Interviewed at Campbell River, November 1991, by the author.

Al West Interviewed at Sproat Lake, December 1991, by the author.

Viv Williams Interviewed at Vancouver, September 1991, by the author.

Appendix

BC Coastal Log Production
(In millions of board feet)

Year	Production
1912	1075
1913	1075
1914	607
1915	830
1916	1012
1917	1335
1918	1430
1919	1410
1920	1596
1921	1407
1922	1555
1923	2099
1924	2066
1925	2160
1926	2242
1927	2411
1928	2723
1929	2823
1930	2243
1931	1660
1932	1441
1933	1711
1934	1983
1935	2369
1936	2705
1937	2850
1938	2416
1939	3041
1940	3323
1941	3226
1942	2711
1943	2525
1944	2519
1945	2482
1946	2520
1947	3286
1948	3266
1949	3137
1950	3476
1951	3332
1952	3335
1953	3702
1955	3939
1960	4215
1965	5147
1970	6100
1975	4522
1980	6505
1985	5871
1990	5339

Index

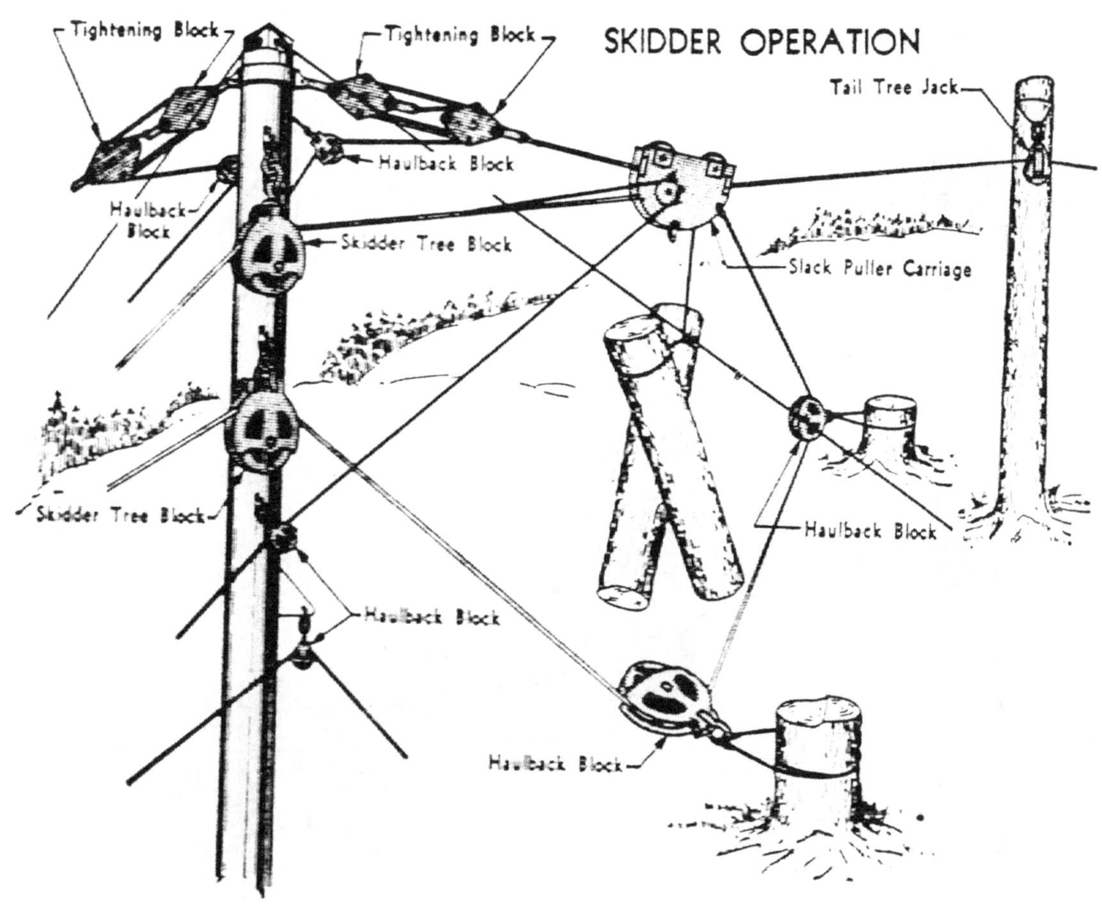
SKIDDER OPERATION
Tightening Block
Tightening Block
Tail Tree Jack
Haulback Block
Haulback Block
Skidder Tree Block
Slack Puller Carriage
Skidder Tree Block
Haulback Block
Haulback Block
Haulback Block

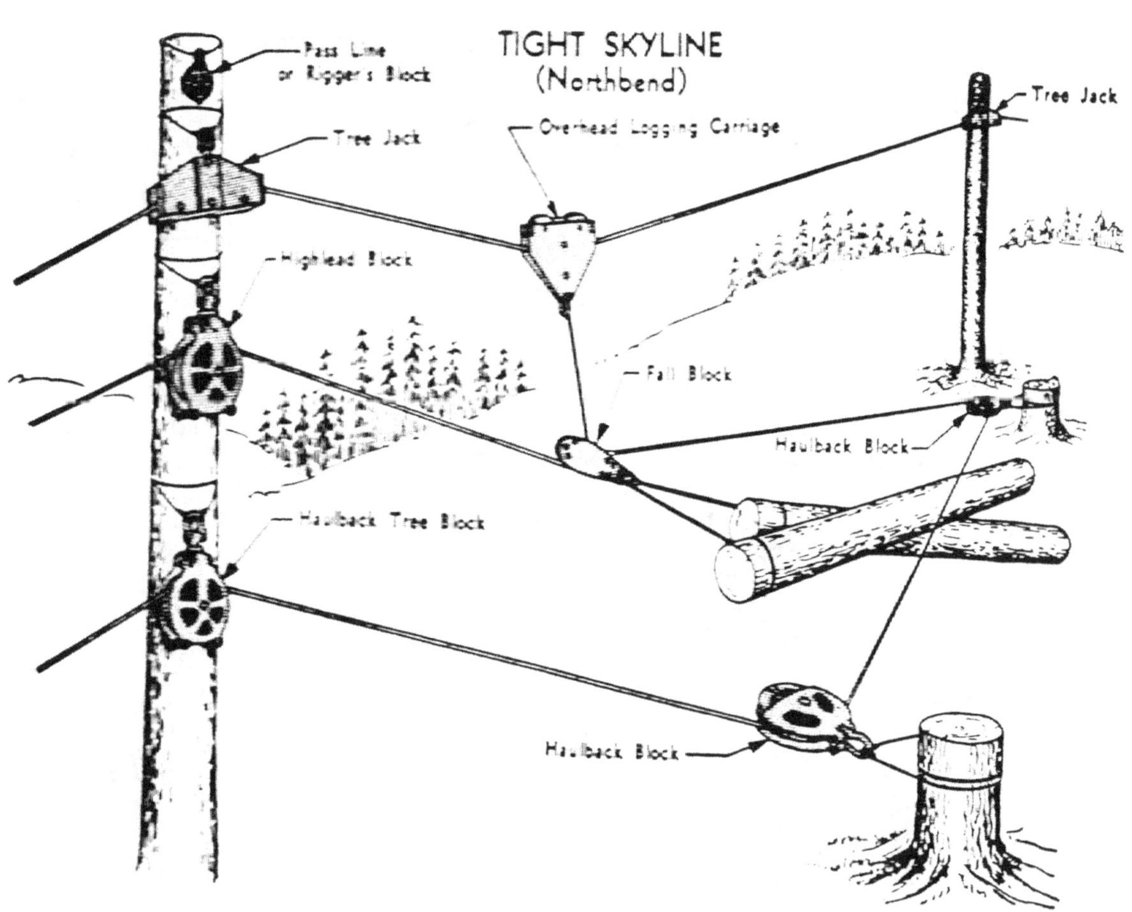
TIGHT SKYLINE
(Northbend)
Pass Line or Rigger's Block
Tree Jack
Tree Jack
Overhead Logging Carriage
Highlead Block
Fall Block
Haulback Block
Haulback Tree Block
Haulback Block

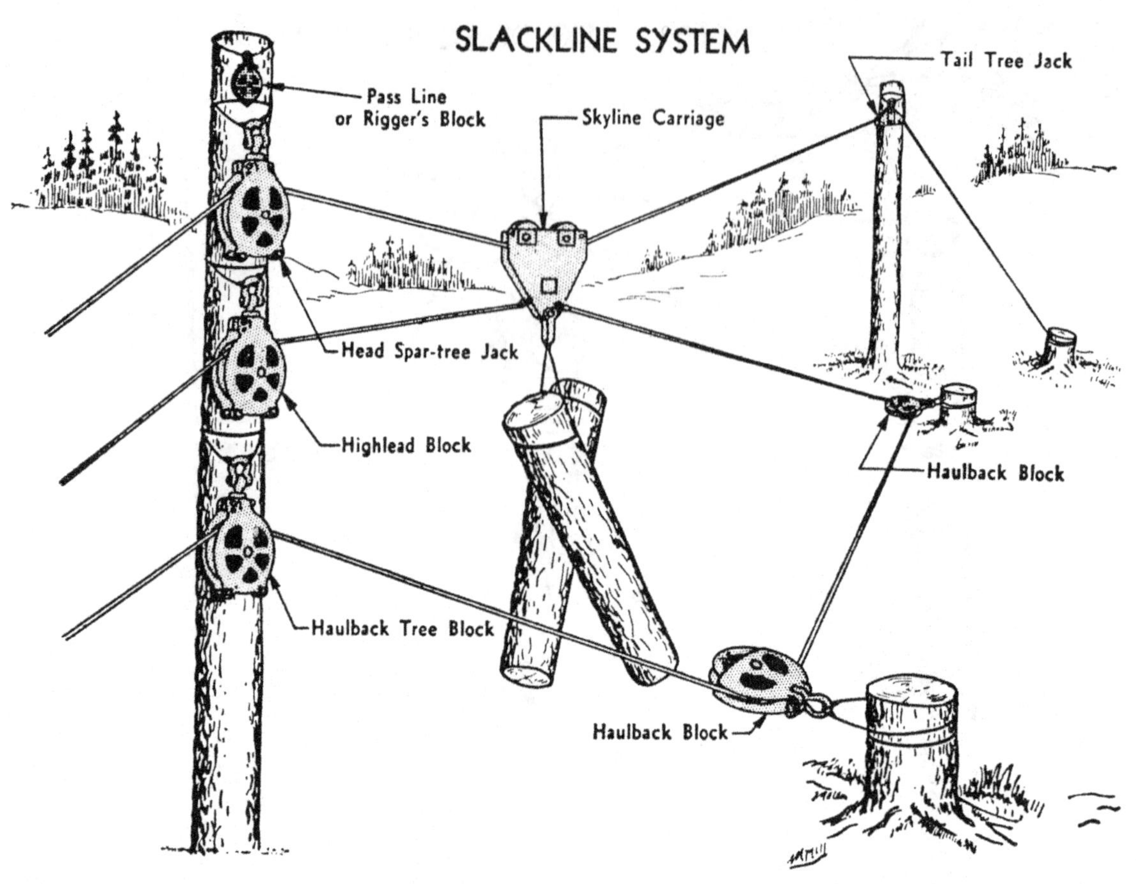
SLACKLINE SYSTEM
Pass Line
or Rigger's Block
Skyline Carriage
Tail Tree Jack
Head Spar-tree Jack
Highlead Block
Haulback Block
Haulback Tree Block
Haulback Block

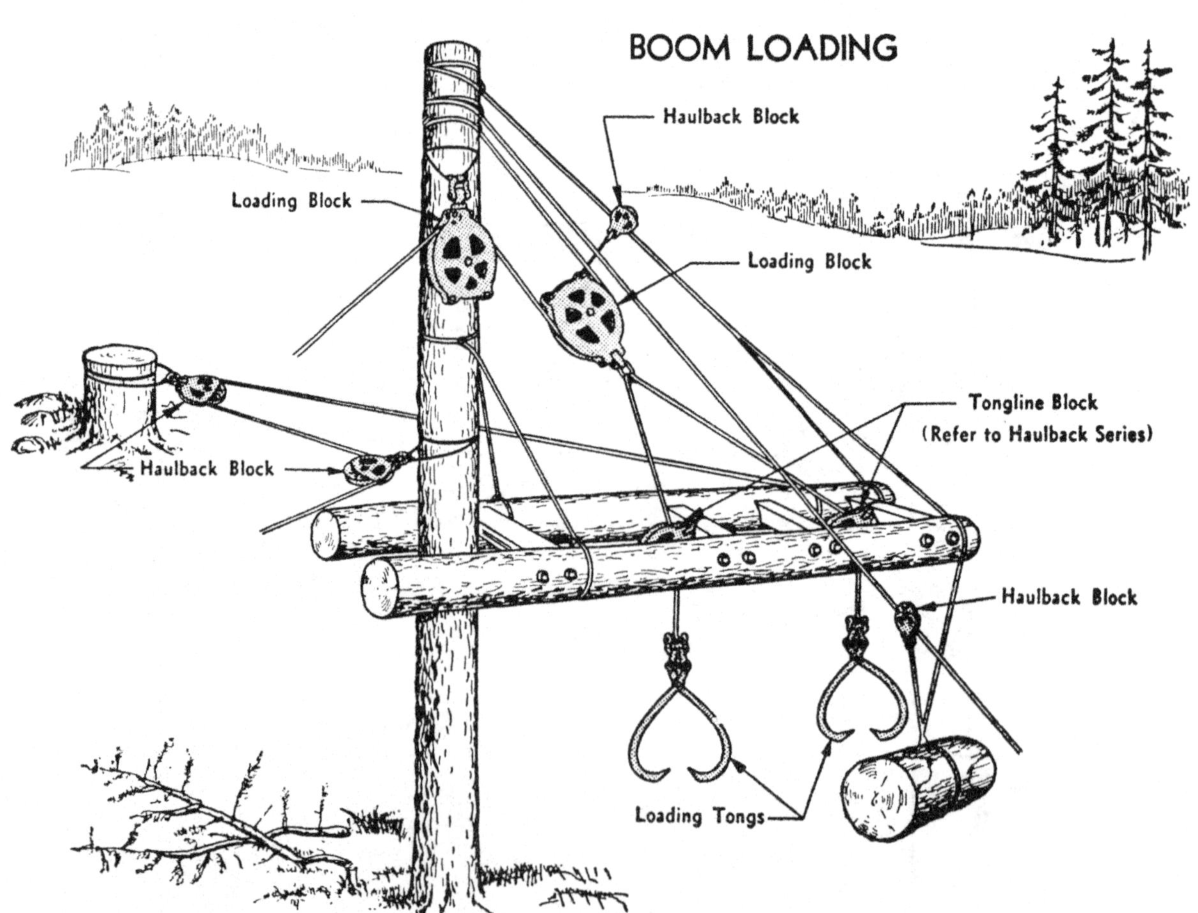
BOOM LOADING
Haulback Block
Loading Block
Loading Block
Tongline Block
(Refer to Haulback Series)
Haulback Block
Haulback Block
Loading Tongs

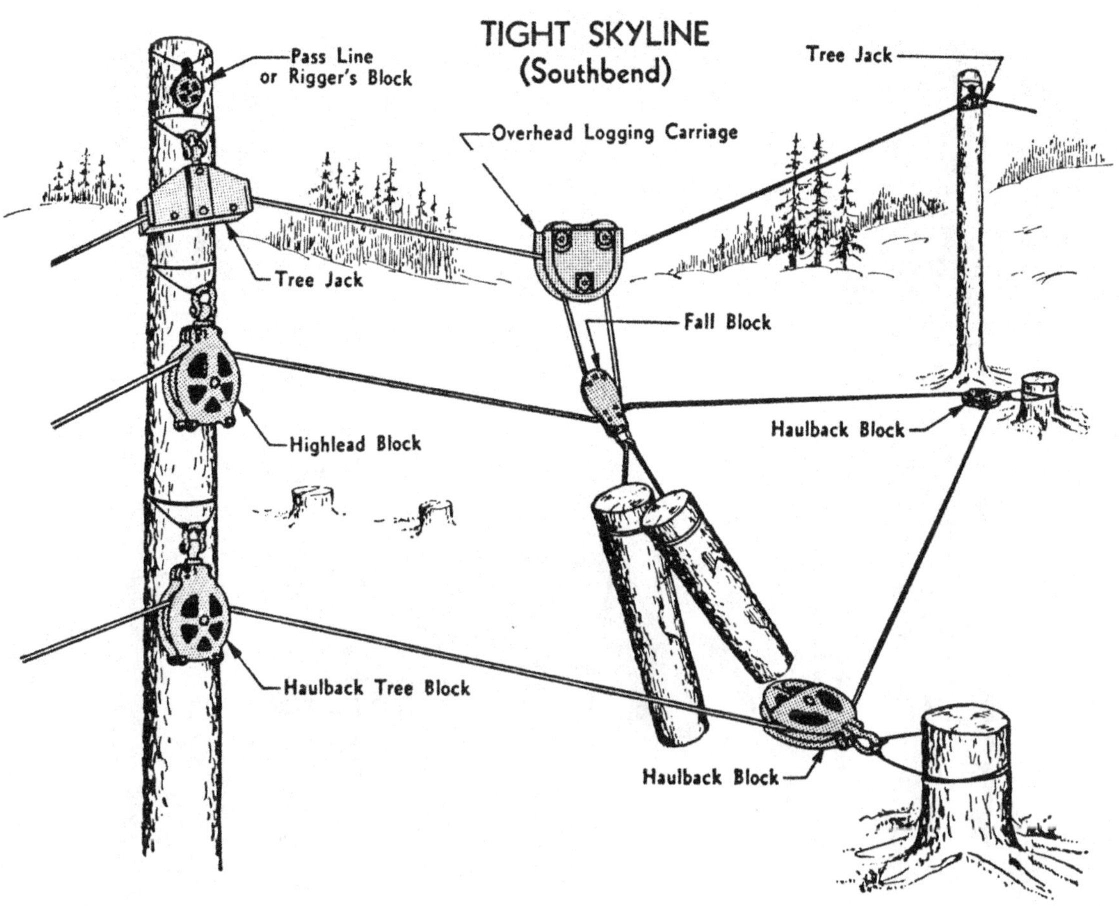
TIGHT SKYLINE
(Southbend)
Pass Line
or Rigger's Block
Tree Jack
Overhead Logging Carriage
Tree Jack
Fall Block
Highlead Block
Haulback Block
Haulback Tree Block
Haulback Block

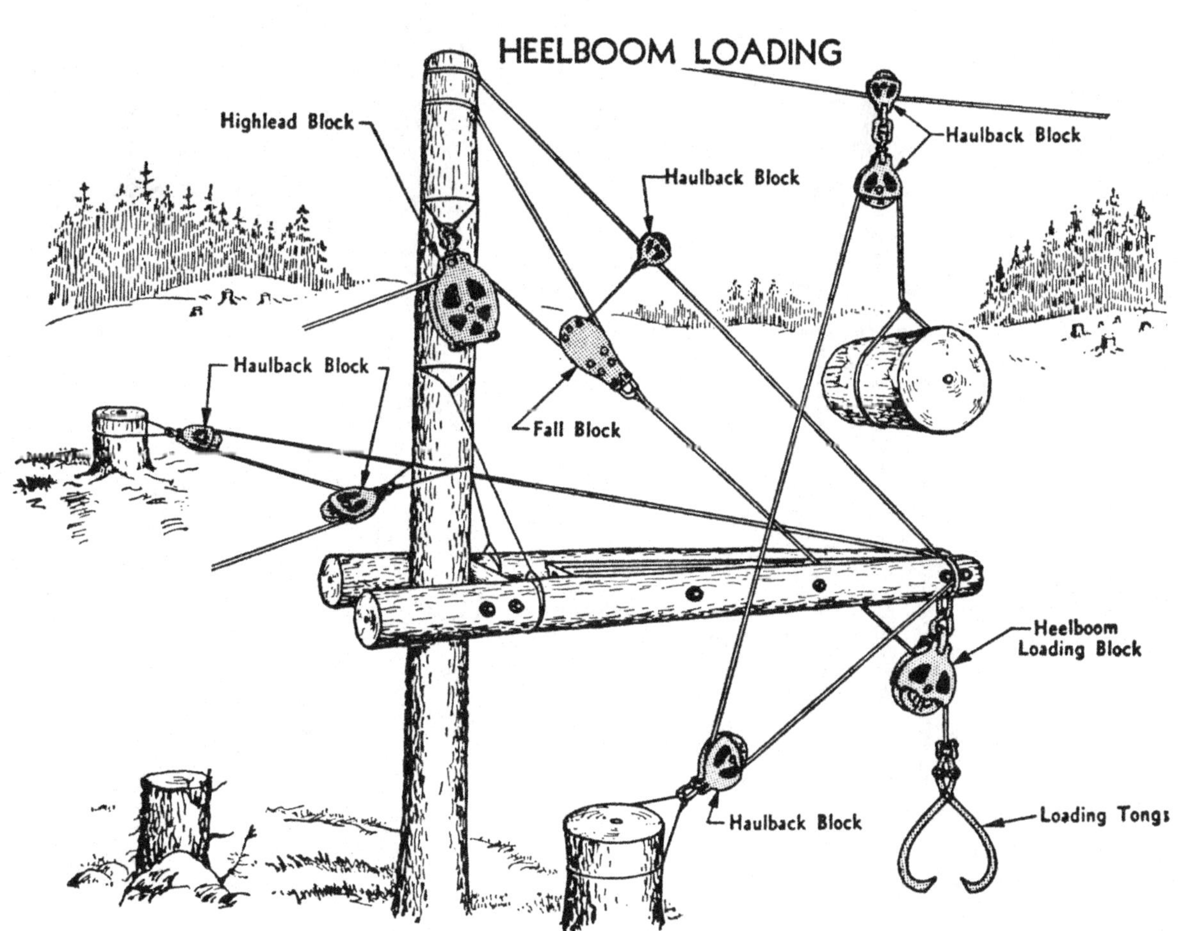
HEELBOOM LOADING
Highlead Block
Haulback Block
Haulback Block
Haulback Block
Fall Block
Heelboom
Loading Block
Haulback Block
Loading Tongs

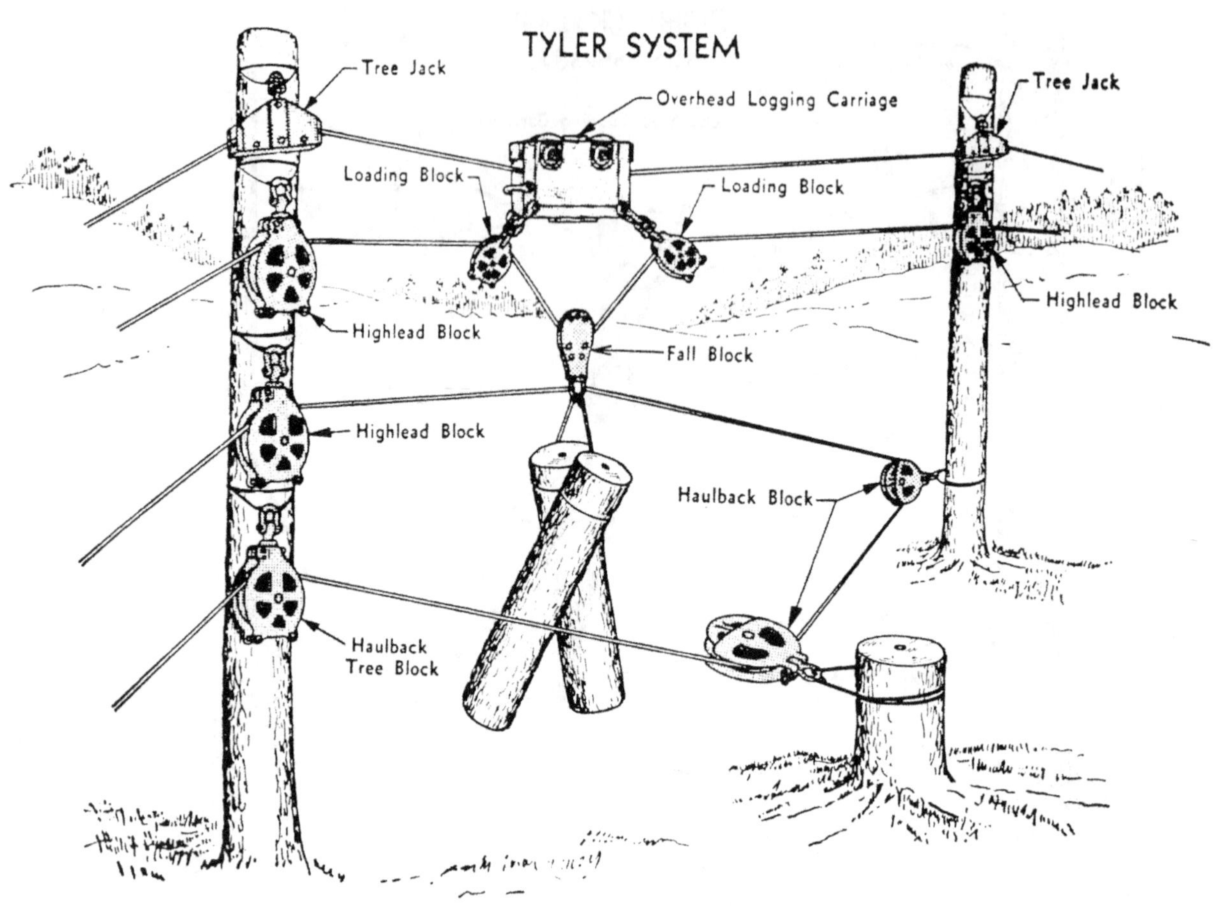
TYLER SYSTEM
Tree Jack
Overhead Logging Carriage
Tree Jack
Loading Block
Loading Block
Highlead Block
Highlead Block
Fall Block
Highlead Block
Haulback Block
Haulback
Tree Block

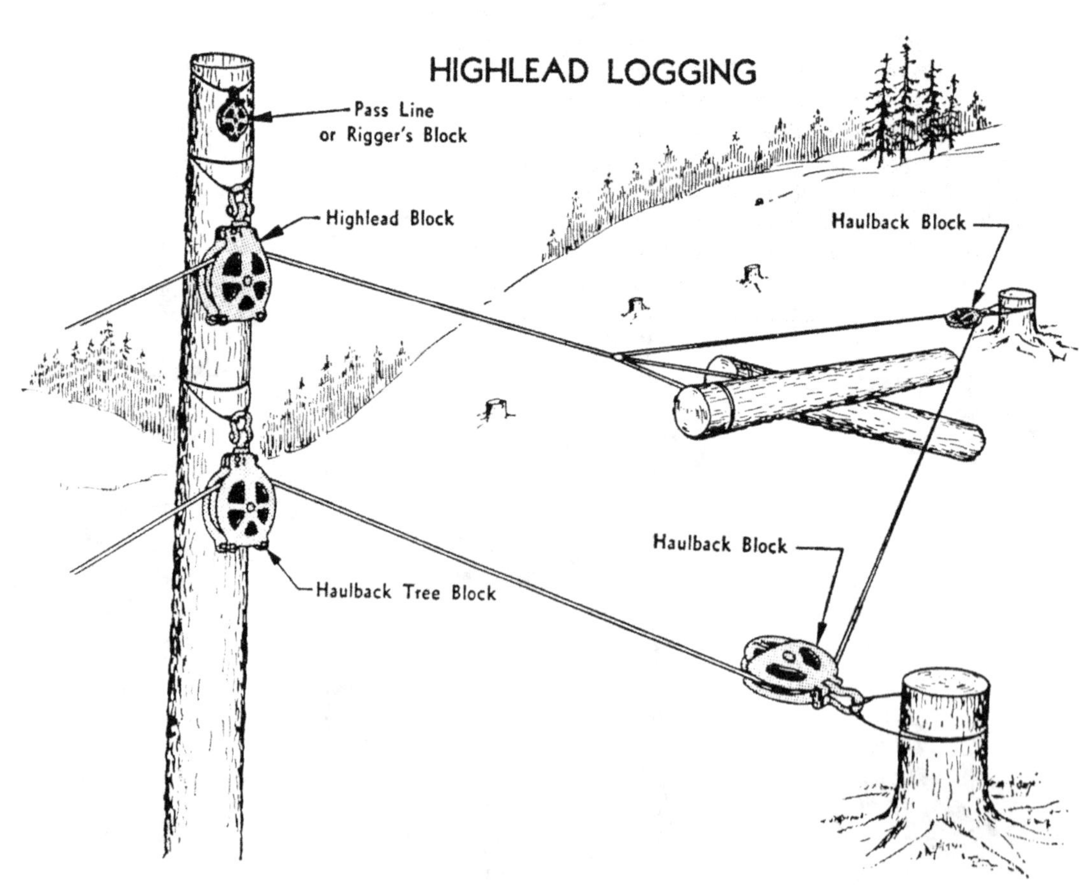
HIGHLEAD LOGGING
Pass Line
or Rigger's Block
Highlead Block
Haulback Block
Haulback Block
Haulback Tree Block